THE PRINCIPLES OF SCIENTIFIC THINKING

The
PRINCIPLES
of
SCIENTIFIC THINKING

Rom Harré

THE UNIVERSITY OF CHICAGO PRESS

501
H23p
76945
Dec. 1971

THE UNIVERSITY OF CHICAGO PRESS, *Chicago* 60637

MACMILLAN AND CO LTD, *London and Basingstoke*

Published 1970

International Standard Book Number: 0-226-31708-0
Library of Congress Catalog Card Number: 78-126074

Printed in the United States of America

Cygnus Melancoryphus, the Black and White Swan, habitat South America

An awful warning to those who suppose that 'All swans are white' and its confirmation or falsification by instances exhausts the logic of the laws of nature.

Preface

In this study I have attempted to set out a reasonably complete and systematic exposition of the realist point of view in the philosophy of science. The text derives from graduate seminars given in Oxford, and several universities abroad, during the last ten years. The book is meant for those whose interest in science is mainly philosophical. I hope, however, that the manner of exposition of the theory makes the major themes sufficiently clear to be of interest not only to scientists concerned about the nature and assumptions of their work, but also to historians of science.

An extensive bibliography has been included, both to support the argument, and to provide a guide to the reading necessary for a graduate seminar in the philosophy of science. I have tried to select those papers and books which present the best arguments for views which are in opposition to the line of thought I develop in this book, as well as those supporting it. Since the items in the bibliography are intended to form a continuous argument, supporting the necessarily restricted arguments of the text, a number of items referred to incidentally in the text, do not appear in the bibliography. Those that do are cited by their serial number in the bibliography.

The realist point of view in the philosophy of science eschews simplifications and tries to present a theory of science with some resemblance to scientific theory and practice. The inevitable result of this is a rather complex argument. It has, however, been very carefully organized, both in structure and order of exposition. The argument, for all its length, is a whole, and complete only on the very last page. Each chapter begins with an anticipatory summary, so that the structure of every step in the exposition of the theory shall be entirely clear. The summaries may be read successively as a shortened presentation of the themes of the book.

This book can also be looked upon as an essay on conceptual change. Not only is there a theoretical apparatus for understanding the dynamics of theory construction, but the limits within which theories can be constructed are delineated. In Chapters Two, Three and Four the limits imposed by the current state of scientific knowledge are explored. The theory of models explains how the possibilities of theoretical innovation are constrained by what is already known. In Chapters Nine, Ten and Eleven the more general constraints imposed by certain transcendental considerations are investigated. For instance neither discontinuous time with continuous space, nor continuous time with discontinuous space are possible bases for any kinematics of material things. By developing the ideas of Boscovich it is shown that a fundamental theory cannot employ entities modelled upon the ordinary concept of a material thing. Considerations such as these set general limits to conceptual possibilities.

Chapters Six and Seven, on Probability and Confirmation, and Description and Truth deal with topics each of which deserves a book to itself. The exigencies of space have required a somewhat schematic treatment of these important subjects. Enough has been said, I hope, for it to be clear what a full scale realist treatment of these topics would be like.

A great many people have contributed to the book by their criticisms of the ideas developed in it, during the years of their genesis. I am particularly grateful to Wilfrid Sellars, Roger Buck, Angus Ross, Richard Campbell and Ann-Marie Jacobson for very specific criticisms the import of which I hope the text will show I have understood.

I must also thank the editors of *Mind, Ratio,* and the *British Journal for the Philosophy of Science,* for their permission to use material which appeared originally in their journals and to the editors of the *Proc.IV. Int.Con. on Electron Microscopy,* and Professor M. Drechsler, for their permission to reproduce the photograph facing page 63.

Oxford
November 1969

Contents

Preface   vii

Chapter **1**   The Mythology of Deductivism   1

Chapter **2**   Models in Theories   33

Chapter **3**   Existential Hypotheses   63

Chapter **4**   Laws of Nature   92

Chapter **5**   Protolaws   130

Chapter **6**   Probability and Confirmation   157

Chapter **7**   Description and Truth   178

Chapter **8**   Principles   203

Chapter **9**   Principles of Indifference   234

Chapter **10**   The Ultimate Structure of the World   257

Chapter **11**   The Constitution of the Ultimate   295

Index   316

1. The Mythology of Deductivism

The Argument

1. Three main traditions in Philosophy of Science: Positivism, Conventionalism, Realism; these are distinguished by their views concerning the status of hypothetical entities, and hence their views as to the meaning of theoretical terms.

2. Common to Positivism and Conventionalism is a theory about theories, which is to be called 'Deductivism'. It has generated a string of 'problems' in the endeavour to incorporate our intuitions of rationality into too narrow a system.

3. Deductivism, particularly in its positivist form, that is combined with positivistic metaphysics and epistemology, is internally consistent, so that it can be attacked only by 'confrontation', that is by showing that the problems identified in (2) are a *reductio ad absurdum* of the theory if it is offered as an account of science. Realism, which accepts neither deductivism nor positivism nor conventionalism, resolves the problems by preserving the intuitions and developing a richer system of rational principles than is in deductive logic.

4. *The Great Myths*

   A. *The Myth of Events as the Prime Objects of Knowledge*
   - (i) It is derived from Locke's simple ideas through Hume's impressions, and appears sometimes as phenomenalism.
   - (ii) It leads to a characteristic idea of the form of laws, e.g. 'If X *occurs* then Y will *happen*'.
   - (iii) This generates the problem of induction because it is easy to show that no amount of event–pair occurrences is adequate as evidence for the truth of a generalization.
   - (iv) It leads to the problem of causation, namely how to distinguish a causal regularity from an accidental sequence, because all sequences are logically identical, and open.

   B. *The Myth of the Vehicles of Thought.* The only vehicles of rational thought are sentences, from which deductivism is a natural consequence.

   C. *The Myth of Deductive Systems.* This myth has its origins in Descartes.
   - (i) logical order matches natural order.
   - (ii) scientific method is reduced to analysis, and subsequent synthesis by logical conjunction.
   - (iii) mathematics, which is capable of expression in a deductive system or systems, is adopted as an ideal of knowledge.

5. *Counter Principles*

   (*a*) The world is a collection of semi-permanent structures, and there is knowledge both of how these structures behave (conditional law statements), and of what they are (categorical descriptions of structure, etc.). These two fields of knowledge are linked together by the idea of powers, i.e. behaviour is seen as the exercise of capacities things have in virtue of their natures.

(*b*) The vehicles for thought are not wholly propositional but 'pictorial' as well, so that considerations affecting the judgement of likeness and unlikeness become important as principles of rationality.

(*c*) A theory is a complex of
  (i) a representation or description of a permanent structure which is responsible for the phenomena explained by the theory;
  (ii) a set of conditional statements describing how that structure reacts, in particular circumstances.
  (iii) if the structure responsible for the pattern of phenomena is unknown then an iconic model of it must be constructed.

The Copernican Revolution in the Philosophy of Science is to see the traditional view that a deductive system of laws is the heart of a theory, and an associated picture of the mechanisms and permanent objects are but a heuristic device, turned upside down. It is to see the model as essential and the achievement of a deductive system among the laws as a desirable heuristic device.

6. *The Critique of Deductivism*

(*a*) Its principle of explanation is unrealistic: if the propositions, from which what is to be explained is deduced, are not subject to any restriction as to number of alternatives, then a demand for empirical content for the explainers is fatal: either
  (i) the nouns and verbs in the explainers do not refer to given objects and processes, but to putative new objects and processes, (but then analogy is required to give them meaning): or
  (ii) they are reduced to logical functions of given terms, but then the theory is not an explanation but a redescription.

(*b*) The Prediction = Explanation symmetry is objectionable.
  (i) Explanation with no prediction, e.g. where prediction involves a tight logical circle. In this case explanation is achieved by describing suitable conditions for the circle to obtain.
  (ii) Prediction with no explanation: e.g. Babylonian astronomy *v.* Greek astronomy. Brodie's chemistry *v.* atomic chemistry, etc.

(*c*) Introducing theoretical terms in such a way that they reduce to logical functions of observables, destroys any putative causal relations from states of unobservables (things and processes) to observables.

The meaning of theoretical terms must be independent of the means used to describe phenomena, because a theoretical term must describe or refer to a possible cause.

N.B. GASSENDI'S PRINCIPLE. The logical form of statements introducing theoretical terms is 'if T then O', where T is the theoretical term, and O the term for some effect that might be observed. This is proper *only if* we affirm the consequent, that is treat T as a possible cause of O.

(*d*) A destructive dilemma can be constructed between the Problem of Induction and Principle of Falsifiability if we adopt Nicod's Criterion as a complete account of the relation of evidence to law.

(*e*) Goodman's Paradox is a paradox only under the condition that we accept Nicod's Criterion. It is better to accept the paradox and deny the criterion.

7. *The Structure of a Theory*

    Triple Structure

    A. Laws describing phenomena.

    B. 'Laws' describing model of unknown mechanism modelled on.

    C. Laws describing phenomena observed with a certain set of things.

A, B and C may or may not be deductively organized, they may be organized only around their subjects. A to B to C will not in general be relations capable of deductive expression, typically they will be analogy, i.e. no set of correspondence rules between A and B, or B and C, can be known to be complete.

The study of Philosophy concerns the way we think and reason about different subject matters. It is concerned with the organization, nature and modes of generation of the products of intellectual activity: that is with thoughts, theories, judgements, assessments and so on. The final objects of this study exist in the individual consciousness of individual men, since it is individuals who perceive, have thoughts, formulate theories, prove theorems and so on. It is the fact of communication that makes possible a rational, critical discipline of philosophy – since to communicate thoughts and feelings they must be expressed in some public way. So, for example, statements and diagrams, signs and models constitute the actual objects of philosophical study. The study of intellectual methodology, as one might say, comes in the end to be studied in the operations on and manipulations of vehicles for the communication of thought.

Recent philosophy, though in many ways an improvement on the philosophy of the past, has been particularly prone to myths, to charismatically proclaimed standpoints from which the extremely complex facts of human intellectual activity have seemed suddenly to be capable of definitive, explicative analysis in relatively simple terms. Contemporary philosophy has been dominated by a particular style of logic, exemplified in a group of systems forged in the study of the foundations of mathematics. The main analytical tool in the philosophy of science has been the propositional and predicate logic developed from Frege's and Russell's studies of mathematical reasoning. Intimately associated with the use of mathematical logic has been a predilection for a metaphysical system derived from Mach, and ultimately from Hume, formed in the mould of classical positivism. Not all of the adherents of the propositional and predicate logic have embraced a full-blown positivism. Some, for instance, have combined the view that Russell's material implication is the proper analysis of conditionality, with various and different fragments of positivist metaphysics. For instance the view that causality is but regularity of sequence has been combined with a belief in the ultimacy of material things as well as with a belief in the ultimacy of sense data. But there has been a sufficiently close connection between logical and metaphysical style, to suggest that the connection is not wholly accidental. In the course of this chapter I hope to show how natural it is to combine extensionalism in logic with positivistic empiricism and phenomenalism in metaphysics and epistemology. This complex of doctrines I shall call 'deductivism'.

The rise of linguistic philosophy has had little direct effect upon the main stream of philosophy of science. Its main influence in this field has been indirect but has, I believe, been to contribute some additional myths of its own. Among those that seem to have reinforced deductivism are the idea that language using is an activity governed by rules and that the limits of thought and the limits of the possibilities of language are co-extensive. This has had the consequence that many philosophers have taken for granted that rationality can be maintained only through the use of the true/false logic of statements. Connected with these ideas is the view that propositions are the only vehicles for thought. The upshot of this for the philosophy of science has been the widespread assumption that scientific theories are essentially deductive systems of propositions, or even, it has been said, of sentences.

This whole loosely ordered structure is compounded of myths. At least in the philosophy of science adherence to some part of these principles creates an illusion of clarity and understanding while leading inexorably to a certain characteristic kind of problem. The kind of problem I have in mind arises in this way: we have certain intuitions of what ways of thinking are rational, for example what features distinguish laws of nature from other general statements. In accordance with such intuitions we have been conducting our intellectual lives very successfully. The adoption of a principle of the kind I have described as a myth, leads us to say that these intuitions are astray – that we had no right to follow them. But the adherent of deductivism has at first no wish to abandon large slices of important facts and procedures. He tries in vain to find a way of preserving some of the intuitions in terms of his principles. Finally he abandons the intuitions altogether, and in the last stage his principles become dogmas, dominating and absorbing him completely. For instance, the insight that it is sometimes a sufficient condition for a theory to be said to explain a fact, that the statements expressing the fact be deducible from the statements expressing the theory, together with statements of conditions, has been transformed into the dogma that explanation consists in nothing but the setting up of a deductive link from what is to be explained to what explains it. We watch with consternation the consequent contortions of the dogmatist as he endeavours to stretch his dogma to accommodate the cases which do not fit; cases for instance when the explanation is not given in deductive form at all, but say in a form which essentially involves an analogy.

My method of dealing with such myths as express philosophical dogmas will be 'confrontation'. I shall show how adherence to the myth generates problems insoluble in its terms. I shall then give an account of the true facts of the matter, and show how characteristically philosophical problems melt away, and show, too, in what way many important intuitions can be preserved under newer, wider conditions of rationality. Gradually, one hopes, a reader brought up within the myth, will come to abandon the dogma. In this way we shall come to acquire a more just, though more complex, understanding of our intellectual activities. Among the myths characteristic of our age, and perhaps the most deeply entrenched, is the belief that propositions, expressed by sentences, are the only proper vehicles for rational thought. Then there is the myth that scientific theories are like logical systems, particularly that they are like proofs in mathematics. This is the oldest element in our contemporary mythology. We shall glance at its origins in the seventeenth century. The third great myth, whose origin in the eighteenth century we shall investigate, is atomistic epistemology. It is connected with the myth that the ultimate elements are events, and that our knowledge of nature consists in the knowledge of regularities amongst kinds of events, and nothing else. We shall come to see that the three great myths are connected, and indeed form a systematic metaphysics and theory of knowledge which is tacitly subscribed to by most contemporary philosophers and scientists, and from which our most characteristic contemporary philosophical problems stem. So we are, or have recently been, under the spell of a myth about the objects of knowledge, a myth about the means of knowledge, and a myth about the nature of knowledge.

The Myth of Events

Locke thought that our knowledge was acquired by the organization of simple ideas. Simple ideas were particular impressions of sense: a colour, a shape, a taste, and so on. In a man's consciousness there would be a succession of such simple ideas, and if simple ideas exhausted his empirical experience, then knowledge of their succession and the patterns of that succession was all that he could have. And since such knowledge can be based only on previously acquired simple ideas it was a short step for Hume to argue that our knowledge does consist only of atomic impressions, *and* that we can have no surety that the pattern of those impressions with which we have become familiar will continue to be experienced. Impressions, or the acquisitions of simple ideas, are events in the consciousness, and what they are impressions of, if we allow that epistemological leap, are events in the world. If we do not allow the leap, then the impressions themselves constitute the world of events. On either view, an object as we can know it can really only be an organized group of events. If events are all we can know about, the laws of nature will be such as to describe the patterns of their succession. They will be characteristically of the form 'If such and such events have already been observed or experienced, then so and so others may be expected to occur'. Our knowledge will be propositional, for all we can entertain are hypotheses about successions. Even to talk of the 'pattern of events' is a very dangerous metaphor, on this view, for by the contemporaneous presentation of the thought of non-contemporaneous events, it encourages the misleading conception of the simultaneous subsistence of non-contemporaneous events.

If statements describing the succession of events with which we have some acquaintance, exhaust our knowledge of nature, then no general statement including in its subject matter so far unobserved events of similar kinds can be known to be true. There will always be insufficient evidence for it. Since to predict, we need to generalize our past knowledge to new as yet unobserved cases, we arrive very simply and swiftly at the *Problem of Induction*. How do we justify logically our entirely practically justified confidence in the continuance of certain, carefully distinguished, event 'patterns'? Events are, it is alleged, entirely independent of one another. One event must have ceased to exist before another in the sequence can be. There can be no carry over. If an event of a certain kind has happened it seems that an event of any other kind whatever *could* succeed it, no matter what had come before. Yet we know that this is not so. We do distinguish an accidental succession of events from a causal succession. This leads to the *Problem of Causation*. The universe as we know it is *characterized* by the striking phenomenon of causation. It is not the case that an event of any logically possible kind follows one of a given kind. Provided we ensure, or are satisfied of the stability of the conditions in which the events occur, just one kind of event follows one of a certain kind, in most situations. Even when the outcome is not fully known, a determinate range of probabilities for the likely outcomes can be found. This is so striking as to lead one to the idea that events are generative, that one happening generates another, and also that things have quite determinate and characteristic powers to produce definite kinds of events. But this could not be if it is believed both that events are all that there is, and that events are independent one of another. *Of course*, events are not all that there is, nor is the content of our experience exhausted by our own sensations. Just because events are changes in and among things, they are not always independent. When events are independent the exact pattern of their succession is accidental, and could have been different in any number of ways.

This will need careful qualification later because even in accidental sequences the member events are parts of causal sequences.

The myth of the independence of events directs our attention in wrong directions. It makes us assume that what we should expect by nature, as it were, is an incoherent sequence of events of unordered kinds, and that our discovery that this is not so should call for explanation. It seems as if we should be searching for an explanation of why any regularities of sequence among kinds of events persist. And we can find no general explanations. Rather it is characteristic of science to explain only why certain regularities do not persist. Mere persistence never needs explanation and later we shall see why. Only changes, whether of things or of processes, need to be explained, though we must add to this those pseudo-persistences which are created by equilibria of tendencies to change. This is one of the most profound principles of natural science. Events are themselves changes or sequences of changes in things, in their properties or arrangements, so each event needs an explanation, and it gets it from the events which generate it under appropriate conditions and the mechanism by which this generation occurs. But the regular pattern of events is not itself a change, and provided the pattern remains stable, that it persists needs no explanation. Only a change in pattern needs to be explained.

The Myth of the Vehicles of Thought

Linguistic philosophy arose from the insight that the way a statement was expressed, could, by unnoticed analogy with other statements, expressing quite different propositions, but in a similar form, mislead one as to the actual content of the original statement, and so lead to pointless puzzles. For instance many nouns are used for things and substances. What can be said by means of an adjective can also be said by means of an abstract noun. Compare 'This is red' with 'Redness is here'. It might then be asked, 'What is that thing or substance "Redness" and where might it be found?'; as one might say 'What is "Uranium" and where can it be found?' or more mysteriously 'What is "Soltamentrul" and where can it be found?' But the question is misplaced, because, it is argued, at best there is only a grammatical analogy between 'Redness is here' and 'Uranium is here'. The propositions expressed differ in such a way that while we can go on to ask 'What is Uranium and where can it be found?' it would be a mistake to ask 'What is Redness and where can it be found' unless we can be satisfied with the answer 'Redness is a hue and it is found on the surfaces of things'. By further careful classification of meaning we can distinguish that answer from 'Uranium is a metal and it is found in Labrador' in such a way as to make us dissatisfied with the very question which prompted that answer. That hues and substances are categorially different is easy to see, and so by carrying on this progress the problem of the status of Redness is slowly resolved. This discovery of the power of verbal analogy to mislead, led to a concentration of attention upon language as a vehicle for thought. In the manner of the generation of myths in philosophy, it soon came to be assumed that language was *the* vehicle for thought. 'Thinking is talking to myself' is a principle of some antiquity and the cycle of this myth had been gone through before. This was bad enough, but worse was to come. The critique of language produced a kind of linguistic puritanism. The only words which were to be used, in a safe language, had either to be defined ostensively, by pointing to samples (another myth – that such a procedure could provide a definition), or had to be the particles of logical syntax, like 'and' and 'if', whose grammar was shown in the way they were used. Finally only those sentences were admitted which could be used to make statements the question of whose truth or falsity was in principle decidable by an empirical investigation of a very simple kind. If grammar is confined to the logic of true and false statements, it is evident that the only way sentences of the admitted kind can be arranged is in deductive systems; and the only relations that can hold between statements are those that derive from the need for statements to be consistent with one another, and the prohibition of inconsistency. The idea of material implication led to the identification of the relation of statements, in which the statements are deducible one from another with the prohibition of inconsistency. This gave encouragement to another myth.

The Myth of Deductive Systems

The most pervasive and seriously misleading myth of contemporary philosophy is the belief that the ideal form for knowledge, and particularly for scientific knowledge, is the deductive system. This error goes back at least as far as the seventeenth century, and the distasteful consequences of it in the form of insoluble problems permeates the twentieth. It appears at its most explicit in the works of Descartes. Here the ideal form of deductive system is connected with three characteristic items with which it has continued to be associated.

(A) The belief that logical order somehow reflects the natural order, that logical power of having many consequences is somehow related to the natural power of a process, the most profound having the greater number of actual consequences. There is, too, the belief, usually tacit, that order in time of cause and consequence is reflected in logic in the order of premise and consequent.

(B) The belief that the method of science is that of analysis and synthesis – that we seek for the component strands of complex processes. The description of the complex process is supposed to consist of the logical product (conjunction) of the description of its components, just as it is supposed that the sum of the components constitutes the complex. The account of a process would then be given completely in a conjunction of descriptions of component events. This view of natural knowledge has been criticized frequently by philosophers of the idealist persuasion, though they have misunderstood the true import of their uneasiness. They too believed that scientific knowledge was analytically derived. They understood that simple synthesis by conjunction would not reconstitute the object of knowledge, so they maintained that scientific knowledge was worthless. Since scientific knowledge is not worthless their conclusion creates a problem. The solution is simple enough. Scientific knowledge does not consist in the logical products of statements describing the components of natural processes, though it includes them.

(C) The belief that mathematics, which is deductively systematizable, constitutes the ideal of knowledge, *towards which all other knowledge should develop*. Numerous mistakes are involved in this belief. There is the mistake that mathematical proof constitutes a superior way of arriving at true propositions; indeed it soon comes to seem that it is a way of arriving at conclusions of superior truth! There is the mistake that mathematical propositions, in so far as they can be proved, are about the world. There is the mistake that expressing natural laws and descriptions of objects in mathematical style, converts them into mathematical propositions. Descartes, though not perhaps originally responsible for these absurdities, propagated them with enthusiasm. Indeed in giving an example of scientific method he cites the method of finding the geometrical mean!

The extravagances of mathematicism are not new. The 'geometrical method' reached its apotheosis in the late seventeenth and early eighteenth centuries. Every kind of serious work was expressed in the vocabulary of geometry and set out as propositions, theorems, corollaries, lemmas etc., in the attempt to attain the deductive rigour of geometry, as a guarantee of correctness. For some projects the language did reflect the structure. Newton's *Principia* is, in part, a deductive system; Spinoza's *Ethics* is too. But in Newton's *Opticks* the language does not reflect the structure. In the *Opticks* the order is actually exploratory. The experiments move from more obvious phenomena to the more recondite, and the account always passes *from* effects *to* their causes. This is precisely the opposite direction from the geometrical and deductive, which must pass from causes to effects to avoid the fallacy of affirmation of the consequent. Nevertheless the vocabulary of the *Opticks* is that of propositions, theorems, corrollaries etc., as for the deductive order. At the beginning of the seventeenth century the geometrical style had not become universal. Gilbert's *De Magnete* which describes an investigation of exactly the same kind as Newton's *Opticks*, is explicitly in the exploratory order, the structure of which Gilbert very beautifully describes in Chapter 5. 'First we have to describe in popular language the potent and familiar properties of the stone; afterward, very many subtle properties, as yet recondite and unknown, being involved in obscurities, are to be unfolded; and the causes of all these (nature's secrets being unlocked) are in their place to be demonstrated in fitting words, and with the aid of apparatus.'

In fact, in actual science, deductive systems are quite rare: fragments of such systems can be found in physics, but mostly scientists come up with descriptions of structures, attributions of powers and laws of change, related by having a common object, not by being then and there deducible from a common set of axioms.

Modern philosophy of science consists largely of attempts to overcome, within the genre of the three great myths, problems created by the uncritical adoption of the doctrines constitutive of the myths. In the second part of this chapter I shall turn to a detailed exposition of contemporary philosophy of science and the problems with which it needlessly concerns itself. Now I shall set out briefly another, and as we shall come to see, a better account of the matters to which the great myths have provided partial and unsatisfactory answers.

Of what Nature is the world Scientists attempt to Describe and Understand?

The actual world is neither a Parmenidean block of unchanging structure, nor a Heraclitean flux of independent events. Here I use the term 'event' with its ordinary signification of both instants and processes. The world consists of numerous, fairly permanent structures, some compact enough to be called things, which are organized in various ways, that is there are numerous fairly permanent units, having internal structures and being parts of larger structures which persist through certain kinds of change, but not through other kinds. The constitutions of such units are what Locke called 'real essences', and their differentia constitute the natural kinds. The chemical analysis of a material, the genes inherited from a parent, the structure of a crystal, the nuclear and electronic constitution of an atom, are 'real essences' and are the constitution of things, and in them reside the powers of generation and production, through the operations of which the flux of events is generated. These structures are, however, metaphysically Aristotelian, in that while their persistence in the absence of external influences or internal instability is not itself a fact that needs to be explained, any changes which take place in or among them, and their generation and decay does require explanation. Similarly, not all events require explanation in terms of other events. A process can be regarded as having a structure, and, while it proceeds in the same way, no explanation of its sameness, over and above the causal explanation of the process itself is called for. In the absence of interference, causes and causal mechanisms operate in their usual way. This fact does not call for explanation. If some change in the normal sequence of a process occurs, some particular external influence or internal instability explains it. The permanent structures of the world need not be physical objects which are a metaphysician's abstraction from the material things that are perceived. They may be of many different kinds and the relation of kinds of objects both to each other and to the things we perceive is one of the branches of metaphysics. We shall return to this later.

This device, of explaining differences in kinds of things and materials by referring the differences to differences in structure of organized systems of other kinds of things and materials, is quite pervasive in science. The differences between the one-hundred-odd chemical elements are referred to structural differences of a systematic kind between one-hundred-odd differently organized systems of the three new kinds of things: protons, neutrons and electrons. The differences between these three new kinds of things can be left unresolved without upsetting their utility in the development of structural accounts of the variety of the chemical elements. Differences in organization of the one-hundred-odd different kinds of chemical atoms in molecular combination serve to account for differences of the order of a millionfold in the properties and behaviour of materials. Using the traditional terminology one might say that while the real essence of sodium is its electronic structure, its nominal essence is its physical characteristics and chemical powers, as a soft, alkali-producing metal. In a somewhat similar way the differences between organisms are expressed in differences in structure of the organized systems of cells of which they are constituted, leading to anatomy as the study of structure and histology as the study of the kinds of cell which are the units of the anatomical structure. Following these leads it would hardly be surprising if the differences between protons, neutrons and electrons might some day be accounted for by finding that each had a characteristic structure and was an organized system of some other entities which need not, of course, be thing-like. An end of a kind to scientific discovery can be foreseen when all the differences there are in things and substances can be referred to structural differences between organized systems whose elementary components are absolutely identical entities, all of the same kind, that is all having identical powers of interaction.

What are the Vehicles for Thought?

Of course there are linguistic vehicles for thought, and of course sentences used to make statements are external and public manifestations of thought. (They are *not* representations, because to manipulate sentences in certain ways *is* to think.) But the importance of this vehicle must not be allowed to conceal the existence of other vehicles for thought. The most important kind to describe for my purposes is that group of entities such as pictures, models and diagrams. Of what are these the outward and public manifestations? Certainly to draw a diagram, to make a model, is often to think. Here great caution must be exercised. It would be tempting, but wrong, to suppose that images are the inward manifestations of that of which pictures and models are the outward manifestations, so that images would be to models and pictures what propositions are to sentences. But this is wrong. Imagined speech is to sentences what images are to pictures. There is a difference between thinking of a structure (*imagining* a structure) and *imaging* that structure; just as there is between entertaining a proposition and saying over an appropriate sentence to oneself. Saying it over to oneself or imagining a picture are of the same kind as saying it out loud and drawing a picture. It is a fact of the mental life that many people who are perfectly capable of thinking structurally and pictorially are quite incapable of picturing a structure. I believe that we actually operate with a complex of vehicles for thought, one typical outward manifestation of which I shall call a statement-picture complex typically consisting of a working drawing and a statement of how the structure depicted will react to appropriate stimuli. We cannot fully understand the way sentences function in statement-making, I believe, if they are considered in isolation from the statement-picture complex to which they belong.

Sometimes the actual picture can be dispensed with in favour of a description of the structure, and then the function of the description is to inform us of the nature of the structure, to enable us to form a conception of it. Sometimes the conditional statements describing stimulus and response, in short describing the capacity for behaviour of the mechanism, can be dispensed with in favour of a flow chart depicting the possible responses of the structure laid out in time. A Markov diagram is such a chart; as will be seen in detail in Chapter 2, it is essentially propositional in nature. So though it is the case both that pictures can be described and the content of conditional statements

expressed diagrammatically, these are, I believe mostly notational accidents which leave the essential difference in meaning and rôle between
pictures and statements untouched.

Perhaps it is best to regard talk about the entertaining of pictures and
models and the entertaining of propositions as ultimately metaphorical.
It is not my purpose here to go deeply into the psycho-epistemological
issues raised by the problem of the vehicles for thought. But to give the
quasi-substantial status to propositions and pictures and models which
this way of talking suggests is surely wrong. I am sure it is better to speak
of imagining structures, and formulating propositions, of Picturing and
Sentencing. Thinking is surely an activity, but is not the acting upon,
or with, permanent units of thought. I shall frequently use the metaphors
associated with the manipulating conception of thinking, indeed the
very phrase 'vehicles for thought' is drawn from that system of metaphors. And the 'statement-picture complex' belongs in that system too,
unless it is being used to refer to one of the products of thinking, namely
a diagram, the structure as constitution of a thing, accompanied by an
account of how it behaves. I used these metaphors to draw attention to
certain facts about thinking, which seem to me indisputable. In this
book I am offering no theories about these facts.

To put this another way: scientific knowledge consists of knowledge of
the structures of which the world consists, and knowledge of how these
structures can change, that is of how they can behave. How they do
behave depends upon which particular stimuli they suffer. So our
knowledge of actual behaviour depends upon knowledge of actual
initial conditions. Since this is particular, traditionally it is not considered to be part of scientific knowledge proper. Characteristically,
structure is presented diagrammatically (by pictures and models), and
the possibilities of change sententially, as conditional statements. By
considering only conditional statements, and forgetting the structural
'picture' from which they have been abstracted, one gets the characteristic 'event' view of the world since it is just the successive states
of things, but not the things themselves, that conditions describe. But
the idea that event-series are the only true objects of knowledge can
be maintained only by considering the statement part of knowledge.
The picture represents the structures, the things, within whose relative
permanence the events are changes, or short, internally organized,
sequences of changes.

It seems clear, that with respect to what is the subject matter of
scientific thought, there are two kinds of symbol conventions, the two
aspects of the statement-picture complex considered separately. There
are the arbitrary conventions by which words possess their referential
and descriptive power, and there are what I shall call the representative

conventions by which pictures, models and diagrams possess their representational power. These kinds of conventions will be discussed in much greater detail in Chapter 2. It suffices now, that their existence and the necessity of our subscription to both, be noticed. Rationality will consist for me, not only in subscribing to the arbitrary conventions of a language and to the necessary laws of logic, but also to the representative conventions of the age, and the principles of propriety for the using of models, diagrams and pictures to which I shall later devote much attention.

What are the True Components of a Theory?

A theory consists of a representation of the structure of the enduring system in which those events occur which as phenomena are its subject matter, and by which they are generated. To this must be added the open set of conditional statements which express the possibilities of change in that structure, in so far as we know them, and so know the powers of such structures to generate or produce changes upon the obtaining of certain conditions. The conditional propositions may or may not be deductively related. They may be related only by being each an account of a different possibility of change in the structure. Of course a structure may be brought to mind by being described, but conditionality cannot be pictured. It is the possibility of coming somewhere near being able to put everything in words, without the possibility of being able to put everything into pictures, that has perhaps led to the mistake of supposing that only the conditional need be conveyed to others as scientific knowledge, as if the descriptive part merely stated the initial conditions for the conditional to be applied. But this would leave out the essential core of a theory, without which there could be no certainty that an event sequence could occur in anything but an accidental way.

In short, a theory usually consists of a set of conditional propositions, which are sometimes such as to form a deductive system, and a picture or model of the structure of the inner constitution and external relations of natural things which will be described in a set of categorical propositions, together with more conditional propositions which describe its mode of behaviour. The model may exclude, ultimately, things of the kind with which perception had made us familiar. Recent tradition has treated the set of conditional propositions organized as a deductive structure as the essential feature of a theory. The picture or model of the inner mechanism and the place of the whole in the structure of the world has come to be treated as if it were merely heuristic, as if it were nothing but a device which merely aids one in grasping the deductive system of conditional propositions. In keeping with this view of theories would be the view that the subject matter of science is sequences of events, and that the vehicles for scientific thought are propositions. The Copernican Revolution in the philosophy of science, which this book describes, consists simply in coming to see that the picture of the inner structure and constitution of things, and of the structure of the world, is what is essential to a theory, and that the existence of a deductive system among the conditional propositions which describe the possibilities of change for that structure is not essential, *but that its achievement is of great heuristic value*. It enables the conditional part of a theory to be expressed very succinctly by means of the axioms of the deductive system. It can lead to the deduction of new conditionalities. But these advantages are merely pragmatic. In keeping with this view of theories would be the view that scientists are at least as much concerned with the discovery of structures and constitutions as they are with the discovery of regularities in the course of events and with trying to discover how things behave. On this view the vehicles for scientific thought are not only propositions, but pictures, models and diagrams as well, because these latter control, to a large extent, what propositions appear in the heart of a theory.

The now traditional deductivist view of scientific theory has several characteristic theses or principles, with which it expresses itself concerning the main components of scientific thought. Perhaps the most damning refutation of the view can be obtained by drawing out the main principles and their consequences, consequences which have indeed been drawn by the practitioners of this view, despite their absurdity. Dogmatic adherence to the principles has led the advocates of the view to treat what are essentially *reductiones ad absurdum* of their theories as problems to be solved. We shall see them as refutations, which should entail the abandonment of the view from whence they are derived. The attempts to solve the 'problems' within the very set of principles which has created them has led to an enormous scholastic literature of an essentially trivial nature. It is to the principles of the deductivist view and their absurd consequences that I now turn.

The Deductivist Principle of Explanation

According to the deductivist, to explain is to produce some sentence or sentences from which a sentence, which could be used to make a statement of what was to be explained, could be deduced. The explaining sentence or sentences must be at least as general as the explained sentence, and usually the explainer will be more general than the explained sentence. The most typical case, according to this view, will be that of a natural law and a particular phenomenon. The phenomenon would be explained if a sentence which would be used to describe it, were able to be deduced from a law and some other sentence. Typically, again, the law would be a conditional sentence and the other sentence or sentences would be particular sentences which would be used to assert the obtaining of the conditions described in the antecedent of the conditional. If the phenomenon to be explained is say the red colour of a flame in which strontium salts have been introduced, then the colour of the flame is explained by the joint assertion that strontium salts have been introduced, and that if strontium salts are introduced into a suitable flame it will become red in colour. This would not, of course, except in the eyes of the deductivists, constitute a scientific explanation *at all.* In a scientific explanation the internal constitution of flame, salts and colour producer would have to be introduced, However, cheerfully ignoring the realities of scientific explanation, the authors of a typical modern version of this remarkable Cartesian travesty, give their 'conditions of adequacy' for an explanation!* First they give what they call 'logical' conditions. These are:

1. 'The explanadum must be a logical consequence of the explanans.'
2. 'The explanans must contain general laws, and these must actually be required in the derivation of the explanandum.'
3. 'The explanans must have empirical content.'

Each of these conditions is both ambiguous and inadequate.

One is tempted to ask immediately of the first condition, 'To what is logic confined?' Is this the logic appropriate to and evolved for mathematics, or are some other systems of the rational connection of propositions to be admitted? It is clear from the context that the original authors of this doctrine had the methods of reasoning used in mathematics in mind. But a large and important class of explanations in science depends upon relations of similarity and difference, or resemblance and lack of resemblance, between the subject matter of the explanans and explanandum. One such relation is expressed, albeit crudely, in the logic of analogy. If 'logic' is understood as the 'logic of mathematical reasoning' then a large class of explanations is dogmatically denied to be explanatory. The kind of explanation I have in mind can be illustrated with the explanation of X-ray diffraction as being due to a crystal 'lattice'. A lattice consists of intersecting lines, forming a grid of points. A crystal 'lattice' consists only of the points or vertices. It is like a lattice in that it consists of regularly arranged molecules or atoms, as the points of intersection of a lattice are regularly arranged. It is unlike a lattice in that the vertices of the crystal lattice are not the points of intersection of linear strips of anything. In a lattice the strips

* C. G. Hempel (*see* 21), Ch. 10.

are real and the vertices mathematical; in a crystal the vertices are real and the strips mathematical. To explain X-ray diffraction as being due to a lattice is to use an analogy.

Again, much the same objection can be brought against the second condition. If the 'derivation of the explanandum' is confined to the logical deduction of the explanandum, then the same objections apply, since in many cases the explanans explains the explanandum not because sentences describing the latter can be deduced from the sentences describing the former, but because what is described by the latter is something like what is described by the former. If, say, as in Harry F. Harlow's primate studies, we explain a piece of behaviour by saying that the monkey was angry, we are effectively offering this on the basis of certain similarities in the behaviour of monkeys and men, only to the latter of whom does 'being angry' unequivocally apply, because only among men can behavioural observations be supplemented by questioning as to the accompanying state of mind.

The trouble with the third condition is more complicated. To demand that the explanans have empirical content may be the innocuous demand that the explanans describes a logically possible state of nature. We must not put the demand so high as to demand a physically possible state of nature. Physical possibility is defined with respect to consistency with the known laws of nature and known ascertained facts about the nature of the world and what sorts of things there are in it. To demand physical possibility would force hypothetical guesses as to the nature of things and processes to be conservatively restricted to the laws of nature already known to us.

However it is more in keeping with the spirit of the philosophy within which this theory of explanation has, in its more recent appearances, to be formulated, to put a much higher demand on empirical terms. It is one of the subsidiary myths of this philosophy of science. It is the myth of ostensive definition. I shall discuss this in greater detail in a later chapter. Roughly, the idea is that empirical terms are introduced by saying the word and indicating an example of what the word is to be used to describe. Thus all empirical terms would be observational terms on this view, since nothing but what can be observed could be pointed to. Some controversy occurred as to what it was that was indicated in definition – sensations or objects. But this was merely another sterile controversy generated within this system of thought, of which the theory of ostensive definition was the element responsible for the controversy. So, on this theory, there could be no empirical terms whose content concerned structures, mechanisms, properties or processes, either not yet observed or indeed perhaps, with our senses, unobservable, or perhaps observable only in their effects. Thus the terms appearing in a theory could only be either observables of a kind already known, or merely logical devices. I shall return to this undesirable consequence in discussing a further phase of this philosophy: its method of introducing theoretical terms.

In the particular modern formulation of the Cartesian theory of explanation which I have chosen to discuss there was a fourth condition. It was that the explanans be true. Descartes himself had seen that this condition was not necessary, for, he pointed out, God could have created the world according to any number of different principles, which could have had as a consequence an accurate description of the world we now observe. Indeed an explanation may, in certain circumstances, still be acceptable though known to be false, for instance when the explanation has certain desirable pragmatic and heuristic characteristics, such as being capable of being expressed as a deductive system of sentences!

However this condition was soon withdrawn. The strongest case against it is perhaps not the fact that it is seldom, if ever, achieved, but that the demand for generality in the logical conditions introduces a kind of proposition whose truth, at least within this system of thought, cannot be established. General propositions of the kind required for this theory of explanation are notoriously unverifiable by any finite amount of evidence, however favourable. It would really have been better had the condition of truth been preserved and the logical conditions dropped to preserve consistency, since taken by themselves they lead to the most objectionable consequences of all. This consequence was noticed by Hempel and Oppenheim themselves in the following passage: 'the same formal analysis, including the four necessary conditions, applies to scientific prediction as well as to explanation. The difference between the two is of a pragmatic character . . . an explanation is not fully adequate, unless its explanans, if taken account of in time, could have served as the basis for predicting the phenomena under consideration.' Instead of realizing that this consequence was a *reductio ad absurdum* of the logical conditions, which should then have been dropped, and despite the storm of protest from within and without the genre of philosophy of which this theory was a part, the authors instead insisted, in effect, that explanation be redefined to fit these desiderata!

The absurdity of the alleged explanation/prediction symmetry is, or should be, immediately obvious. Since predictions, except in the trivial sense of expecting what has happened once to happen again under exactly the same conditions, is very rare in science, and explanation very common, a great deal of what counts as explanatory theory, indeed explanatory sciences such as chemistry*, biology, geology and so on have

* Mendeleev's predictions of new elements from his periodic table are the only cases I know of.

to be abandoned. The melancholy fact that 'predictions' of oil deposits from geological data are but rarely successful does not prevent the explanation by geological facts of why the oil deposits are where they are from being an explanation. The point is very simple. Even if the geological conditions are right, whether there is or is not oil there is a matter of contingent fact that has to be ascertained by boring, and nothing but the discovery of the presence of the oil is sufficient to predict its being discovered by some *logical* means.

In the extreme case of evolutionary biology I daresay no prediction has ever been made, or indeed is possible, since what the theory does not contain, is a detailed enough cosmology to predict just exactly what the environmental conditions will be like at any given second at any point in the universe. But the theory will give a fairly adequate explanation of why, under the conditions that did obtain, the species that *did* come to be were evolved. The circularity, often noticed, of the deduction of the geological record from the distribution of organic types, and the explanation of the order of the types from the 'facts' of the geological record, is just a reflection of the assymetry of explanation and prediction. As an explanation it is quite adequate, as a prediction schema it is indefensible. At the heart of the attempt to reduce explanation to deduction lies the dangerous myth that somehow the course of nature parallels a logical progression from premises to conclusions, a myth we have already noticed.

To see just what has gone wrong in cases like this the steps of reasoning must be looked at more closely. The theory of organic evolution describes the conditions under which change in the character of a population occurs, and consists largely of hypothesis as to the mechanisms by which such changes occur. To apply the theory in such a way as to make a prediction that a certain species will be evolved, the existing population and the existing circumstances must be specified in detail, so that the initial conditions of the process which issues in the presence in the environment of a differently characterized population, after so many generations, can be put in as the minor premise of the logical deduction of the prediction, whose major premise is the theory of organic evolution. But the choice of facts to settle on as the initial conditions depends upon our already knowing what was actually evolved. Between the initial conditions and the prediction a neat little logical circle obtains. This circle must always exist when the problem of deciding which, from the myriad circumstances that surround a change, are the important or operative circumstances, is solved by reference to the outcome. And the rôle of theory in this enterprise is not that of major premise, rather the theory lays down the general environment in which a search for this or that kind of operative conditions circle is fruitfully pursued.* Only when you have found the oil do you know that you are in the presence of oil bearing shale, so to predict from the discovery of oil bearing shale that oil will be found is fatuous.

However, it might be argued that though as a matter of fact prediction cannot be achieved in such theories prior to the discovery of what actually occurs, nevertheless, from a logical point of view the explanation *once achieved* has the logical form of a prediction that can be expressed deductively. Schematically: suppose a represents the necessary conditions for b, and T is the theory, then though as a matter of fact we may not be able to know a fully until b happens, rather say, than c; after b happens and we know that a obtained, we can express the explanation at T & $a \therefore b$.

But a much stronger argument is to hand. Let us concede that at least in some cases, when an explanation has been achieved it can be cast into deductive form, and hence have something of the logical air of an *ex post facto* prediction. If there is a symmetry between explanation and prediction then it ought to be the case if a prediction can be made an explanation must, a fortiori, be available. If I know that sodium salts turn a flame yellow then I can predict that a flame will turn yellow when sodium salts are introduced into it. This has the form

> All sodium affected flames are yellow
> This is a sodium affected flame
> It will be yellow.

Do we get an explanation by simply changing the tense of the conclusion? It would read like this

> All sodium affected flames are yellow
> This is a sodium affected flame
> This is yellow.

But now consider the explanatory regress of queries:

$Q1$ Why is the flame yellow?
$E1$ This is a sodium affected flame.
$Q2$ But why are sodium affected flames yellow? (Surely a legitimate and indeed typical scientific question.) To this query the deductivist must answer.
$E2$ All sodium affected flames are yellow.

And if we try to query that we get back to $Q2$ again, and our curiousity prompts only reiteration. But $E2$ is hardly an *explanation* of the characteristic colour of the sodium flame. For that we must try to bring in such topics as ionization, molecular orbital theory, quantum theory and so on. We must, in short, have some idea as to the nature of the pheno-

* Cf. the discussion by A. R. Manser, 'The Concept of Evolution', *Philosophy*, **40**, 18 ff.

mena we are witnessing. But in the deductivist theory, taken strictly, any further step beyond $E2$ must be gratuitous since the deduction of the phenomena from a law has, by then, already been achieved. Explanation is logically identical with prediction only if the deductivist theory is correct, so it is quite otiose for their alleged logical identity to be cited as evidence for the deductivist theory.

A further example of the same point can be found in the prognosis of the course of a disease from the symptoms. Very accurate prediction of the course of an infection is possible from the symptoms alone, but it would be quite absurd to say that the symptoms explained the subsequent syndrome.

Principle of the Introduction of Theoretical Terms

Nowhere is the scholasticism of this philosophy of science more clearly evident than in the treatment of theoretical terms. The incorrectness of the treatment is first clearly evidenced by the appearance of one of those famous 'problems'. It has been expressed as follows:* 'The use of theoretical terms in science gives rise to a perplexing problem: Why should science resort to the assumption of hypothetical entities when it is interested in establishing predictive and explanatory correlations among observables?' Actually there is no problem, since science is certainly not interested in establishing *only* predictive and explanatory correlations among observables. This is the assumption upon which the existence of a 'problem' depends. Science is actually interested in discovering the structure and inner constitution of natural things and their relations in the cosmos, in virtue of which phenomena display the regularities and irregularities they do. The use of theoretical terms is precisely the best way of achieving science's real aim, for they are just what lead to existential hypotheses about the unobserved. This belief in the exclusive interest of science in the correlation of observables is another aspect of the concealed metaphysical predilection for events, which is so characteristic of deductivists. This stems from the positivist strand in modern philosophy of science and though the connection of deductivism, in general, and positivism, had become intimate, it is clearly not necessary. For any time-dependent law those observables which naturally lend themselves to statistical treatment and correlation are events, particularly the atomistic instants of Humean epistemology. It leads too, as we shall see, to another famous 'problem', that concerning the distinction of laws from merely accidental generalization, where laws are understood as stating the fact that certain regularities or patterns have been observed in the course of a series of independent events.

* C. G. Hempel (*see* 21), p. 179.

The characteristic method by which new predicates, some to function as theoretical terms, are introduced into the language of science according to deductivists is by what have been called 'reduction sentences'. These sentences, so typical of the deductivist point of view, operate in such a way as to make theoretical terms not just simple correlations of observables. They are, nevertheless, such that their meaning is exhaustively determined, so far as its empirical possibilities are concerned, by a set of *already known* predicates used to predicate physical, observable properties. Let these be, as in the classical exposition of this view,* $Q1$, $Q2$, $Q4$ and $Q5$. To introduce a new predicate $Q3$ we stipulate that if whatever property $Q1$ stands for is observed to occur, then we infer the conditional that if $Q2$ occurs then $Q3$ does. However, in the conditions that $Q4$ occurs, we stipulate that if $Q5$ occurs then $Q3$ does not. In symbols the reduction pair is

$$Q1 \rightarrow (Q2 \rightarrow Q3)$$
$$Q4 \rightarrow (Q5 \rightarrow \sim Q3)$$

Another, rather simpler case logically, is the case where we infer on the occurrence of $Q1$ that $Q3$ occurs if and only if $Q2$ does.

The striking thing about this account is that it expresses precisely the way that new predicates are *not* introduced into science, or new properties hypothesized. The reasons for this are:

(i) the confinement of meaning of $Q3$ to some function of $Q1 - Q4$ is precisely contrary to scientific method. A scientific language does not grow by addition of predicates which are related in this exhaustive way to observables, in the first instance. For example, the predicate 'magnetic' is used to ascribe a *definite and independent* property or power to substances, and furthermore a property which, because of the accidental constitution of our sense organs, is never among the observable properties of things. We know something is magnetic because of certain movements and patterns among other things, but its magnetism is not those movements and patterns, it is the property or power responsible for them. What that property or power is we may come to grasp only by what may be a quite complicated system of analogies, and within our general picture and our detailed knowledge of the constitution of material things.

(ii) In these reduction sentences the conditionality of the second clause is in the wrong direction. Suppose $Q1$ represents the conditions under which the occurrence of $Q3$ leads us to observe $Q2$. Then the property or state $Q3$ will be what we would ordinarily call a possible cause of the appearance of $Q2$. So the conditional sentence expressing this should have its conditionality running *from $Q3$ to $Q2$* and not the

* R. Carnap, 'Testability and Meaning', §5, *Philosophy of Science*, **3** and **4**, 1936–7.

other way. Thus, at best, a sentence that could be used to express the relation of the theoretical property to the observable property would, in this very restricted system of logic, run as follows:

$$Q1 \rightarrow (Q2 \leftarrow Q3)$$

and similarly the other member would run as follows:

$$Q4 \rightarrow (Q5 \leftarrow \sim Q3)$$

though what we make of the presence of not-$Q3$ is very far from clear. The whole question of negated predicates will be discussed in a later chapter.

To express the matter otherwise than as in the first of the above two reformulated sentences, is to make two related mistakes. It is, as was noticed in the first argument, to restrict the import of $Q3$ to a logical function of $Q1 - Q5$. But it is also to overlook an important methodological principle. This fundamental principle derives from the Stoics but it is perhaps fairest to call it Gassendi's Principle. It is that the inference from observables to hypothetical properties, states, objects or processes is proper, even though it commits the formal fallacy of affirmation of the consequent. Thus, if a certain spectrum pattern is used to obtain knowledge of the electronic constitution of something, then, the principle upon which this is based is that the electronic constitution is a possible cause of, or possibly responsible for, the spectrum pattern. So, if E then P. P is observed, but not E. So to infer E from P we affirm the consequent. In effect this principle allows theory to change without our having to abandon observations, and, at the same time, by pointing out the *propriety* of the inference, it allows one to pass from known effects to possible causes. How is this principle to be justified? First it is the principle which is used, and hence is required, if what we now think we know is to be preserved. Further, the method of analogy, to which we shall turn in great detail later under the guise of the theory of models, supplies us with the required content for the theoretical terms. Any logical system which asserts that this mode of inference is improper must be too narrow. It is a mistake of the first magnitude to abandon a perfectly *proper* principle because of its conflict with whatever logical system one happens currently to possess.

The Logical Form of the Laws of Nature and the Power of Evidence

I shall have much to say about this topic in a later chapter. At this point I intend merely to bring out two important conflicts of methodological intuition and well established practice with some consequences of the deductivist theory of science. This aspect of the deductivist theory derives

B

merely from its being deductivist. It restricts statements to sentence-forms between which deductive relations hold. It has been said frequently that 'natural laws have the form of strictly universal statements; thus they must be expressed in the form of negations of strictly existential statements.' This logical relation is the one logicians call the square of opposition. In the old style expression it asserts that a law of nature is either of the form 'All A are B', or, more rarely, of the form 'No A are B', which are the negations of 'Some A are not B' and 'Some A are B' respectively. Having made this mistaken identification of the form of the laws of nature with the universal propositions of logic, the deductivist view compounds its sins by a characteristic principle of evidence, Nicod's Criterion. It is this. If, as in the more modern form, we express the universal statement as

$$(Vx)\ (Fx \rightarrow Gx)$$

then evidence for the statement is of the form

$$(\exists x)\ (Fx\ \&\ Gx) \text{ or } Fa\ \&\ Ga$$

and the evidence against it

$$(\exists x)\ (Fx\ \&\ \sim Gx) \text{ or } Fa\ \&\ \sim Ga$$

where 'a' is a referring expression. Sometimes this is called the 'Principle of Instance Confirmation', that instances of laws confirm them and counter-instances disconfirm them, or sometimes, it has been said, falsify them. This leads to a destructive and paradoxical dilemma. Evidence on this theory is either of the form

$$(\exists x)\ (Fx\ \&\ Gx) \text{ or } Fa\ \&\ Ga$$

i.e. positive evidence, or of the form

$$(\exists x)\ (Fx\ \&\ \sim Gx) \text{ or } Fa\ \&\ \sim Ga$$

i.e. negative evidence. If we take the positive alternative, it has frequently been pointed out, notably by Hume, that inductive evidence, the piling up of favourable instances, never can confirm a universal statement. If we take the negative alternative we would be forced into the absurd position of having to claim that the only significant evidence in science would be what would falsify a universal statement. The adoption of this horn of the dilemma as a methodological principle by Sir K. Popper in his *The Logic of Scientific Discovery* (121) and *Conjectures and Refutation*, Routledge and Kegan Paul, London, 1963, has allowed him to work out *as far as possible* a consistent deductivist philosophy of science. So we are impaled on the horns of a dilemma, either with the 'problem' of induction, or with the absurdity of making out that all evidence in science is falsificatory. It seems clear again that it is the deductivist account that is at fault. We must deny, in this case, both principles. The form of a law of nature is not adequately, though it is partially, rendered by the classical formula. The form, and relation to evidence, of laws is not adequately, though it is partially, rendered by Nicod's Criterion. Other unwelcome and counter-intuitive consequences are connected with this view of the laws of nature. As we have seen, the deductivist principles go naturally, though not necessarily, with a metaphysics of events, the occasions of the occurrence of properties being the instances which provide us with favourable evidence for our laws. Both accidental and law-like, (or in a special sense, causal), sequences of events will be described by statements of the same logical form, because in the event ontology as understood by positivists, notably Hume, each event is separated and distinct. How can there be causality, that is other than mere statistical regularity, if events are distinct? How can we distinguish accidental from law-governed sequences? On the deductivist account, we cannot allow to causality a generative power, nor to laws any greater necessity than accident. And yet we *know* that there are generative powers, and we acknowledge the necessity of laws. Which are we to abandon? Common sense, if nothing else, must counsel that it is the philosophers' theories that must go, that is the deductivist account of laws and theories accompanied since Mach by a Humean metaphysics and epistemology. We shall see, in the course of this book, that this is not to abandon rationality, far from it. On the contrary, it involves a great and welcome enlargement of the principles of proper thinking.

The last unhappy consequence of deductivism to which I wish to draw attention, at this stage, is not created by the adoption of a new principle, but by the joint use of two that we have already noticed in the deductivist view. If we hold that predicates obtain their meaning by ostension of instances we are naturally led to the view that predicates which have the same instances have the same meaning. This, together with the principle that the confirmation or disconfirmation of laws lies exclusively in their instances, leads to Goodman's Paradox.* This 'problem' can be generated, for instance, by the introduction of an artificial predicate with a time dependence condition. To introduce the predicate, take some descriptive predicate P, and then define the Goodman predicate Q as follows:

$$(\forall x) \ (Qx =df. \ Px \equiv x \text{ is examined before time } t)$$

Suppose we have some confirmatory evidence, i.e. according to Nicod's

* N. Goodman (*see* 53), pp. 73–80.

Criterion, favourable instances in a set, of some set of objects bearing *P*. We have examined a certain number before time *t* and they are all *P*. Should we say of those unexamined before *t* that they are likely to be *P*, which would be the right answer according to Nicod's Criterion; or should we say, rather, that they are likely to be *Q*? The evidence that *Q* is the proper thing to say of them has accumulated exactly as fast as that for *P*, since each object which was *P* before *t*, must, by the definition, also be *Q*. But if the unexamined objects are *Q* then, by the kind of logic espoused by the deductivists, they can then only be non-*P*, since by the definition of *Q*, if something is *Q*, then it is *P* if and only if it is examined before *t*. So, if it is examined after *t*, to preserve truth-functional equivalence, *x* cannot be *P*. So it seems that the very same evidence is equally confirmatory of the hypothesis that the unexamined cases are *P*, as that they are *Q*, and so non-*P*.

There have been many attempted solutions of the paradox, within the genre, for different ways of introducing *Q*, none of which are satisfactory. A much better move is to welcome the paradox! Because it depends upon Nicod's Criterion. If the evidence for inductive projection to unexamined cases is that some predicate is correctly attributable to examined cases, then surely Goodman's paradox is viable. Of course what is required is a distinction between the accidental presence of character *P*, in each case, and the genuine scientific situation, where, in addition to the instances, we know enough about the internal constitution of the objects we are studying to have good reason to expect them to show the property or character *P* on examination, say of fluorescing in ultraviolet light. But no such reason for attributing *Q* could be found, since the time of a thing's examination does not flow from the inner constitution of that thing, so there can be no *scientific* attribution of *Q* to things. But these important distinctions will be unfolded in greater detail in other parts of this book.

The general approach I am taking can be illustrated in a couple of anecdotes, as well as found foreshadowed in the works of Bacon, Locke, Whewell and many others. I once met an elderly philosopher of great fame. The great man was maintaining that induction was an unsolved problem. There was, he maintained, no more reason for supposing that the clock on the mantel-shelf, whose face we could see, would continue to behave regularly with respect to the circumsolar motion of the earth, than that the next person passing the window would be wearing a blue jacket, as the last person had. Our mutual host then silently rose and quietly turned round the clock, which was revealed as one with glass sides and back, through which the beautifully constructed mechanism, which ensured its regularity, could be seen. He then remarked, 'Since there is no Air Force station in the vicinity I can think of no reason why we should expect a flush of blue jackets.' On hearing this the great man changed the subject.

The intuition that this illustrates, that our proper reasons for expecting some sequences to continue and others not, in many cases have nothing much to do with instances, but have a great deal to do with how much we know of the mechanism responsible for the instances. Clockwork and the bio-chemistry of inheritance, provide regular sequences. The whims of jacket buyers, and the wind blowing different length sticks and twigs off the trees in winter, do not. This is a proper intuition and must be preserved, and logic must be adjusted if it does not fit.

At Eindhoven, a gentleman amused himself by cutting circular slits in the ceiling of an upper room of his house, through which rods were suspended and on the end of which hung the planets. The whole was, and indeed is, exquisitely painted in blue and gold. A wait of an hour or so enables one to detect a little movement in the inner planets. Saturn, having a thirty-year cycle, has completed seven revolutions, or so, since this is an eighteenth century planetarium. Were we, the spectators of the show, to leave it with a few notes of the regularities observed, like modern Babylonians? Not at all. Our intuition told us that above the ceiling a causal mechanism was at work, by which the regularities were produced. For an additional guilder a ladder was revealed in a cupboard, and up we went. Sure enough our intuition was justified: there was a fantastic contrivance of wooden gears, huge wheels and enormous weights, from the behaviour of which the regularities in the movements of the dependent planets flowed. It is the view of this book that scientific research consists both in the observation of the course of nature, both as to its regularities and irregularities, *together with* the persistent attempt to uncover the mechanisms from which both flow. The deductivist view, together with the metaphysics of events, the epistemology of observations, and the restriction of the principles of rational thinking to the logic proper to mathematics, can lead only to a denial that the search for mechanisms, which I would regard as the characteristic activity of science, is anything but the effort to find a heuristic prop for poor logical thinkers. It is my intention to use the counter mythical principles of rational thinking which I have sketched, to show how the scientific enterprise is actually undertaken, and to extract from this its principles of rationality. These, we shall find, offer an enlargement of intellectual machinery much beyond that which has been found to be required for the rational reconstruction of the intuitions of mathematics.

Works of art also bear a connection with, and express in perhaps several different ways, an apprehension of nature. A certain theory about the nature of works of art follows from this general view of science, and could be developed in parallel with the account of scientific theories. Making works of art and formulating theories about natural phenomena are aspects of intellectual activity that have not always been clearly distinguished. It may well be that some of the motive to theorize may derive from its likeness to artistic creation.

The prime vehicle for thought is the statement-picture complex. For the purposes of record and for communication we 'express our thoughts' in sentences and models, diagrams and pictures. In this way the statement-picture complex is fragmented. Usually, in a scientific text, charts diagrams, photographs and so on, and in scientific practice, the continual experience of the scientific worker in dealing with subject-matter, reunites that complex so that sentences do not merely express abstract relations between terms, but appear in a distinct context, determined by the beliefs and the knowledge we have about the nature and constitution of things. The vehicles of communication are objective and sensually known. It is a fact about human beings that pleasure, and as experience widens, satisfactions of various kinds, are obtained through and in the experience of sensations and the awareness of perceptions. So the constant activity of expressing themselves, by which people record and communicate their thoughts, comes to be conducted under another sanction or control than objectivity, consistency and truth – the aesthetic sanction. Certain naturally occurring objects, properties and processes seem superior to others because they are more appealing, and certain pictures of the natural world, both as to structure and as to process, seem to have a superiority, (which can come to seem to include their truth), because they give superior satisfactions. These are illusions. Nature has no aesthetic preferences, nor does she truly differ in this manner, since nature is not the externalization of the thoughts of beings who communicate by the senses. At this point we turn to the detailed exposition of the intellectual methods and standards of the natural sciences.

By way of summary one final point should be added. The polemical remarks in this chapter, though intended only by way of introduction to the positive exposition of the profounder understanding of science which is to follow, illustrate a general point of considerable power. If it is maintained that criticisms of deductivism which hinge on the divergence between that theory and practice, are beside the point since the theory depicts an ideal form for knowledge, then deductivism as an ideal should be free from 'problems'. But it is just the very theory which runs fastest into difficulties. To hold that scientific knowledge should develop towards a deductively organized system of conditional statements, describing regularities of succession among types of events, is the *logical* ideal, leads us straight into a situation in which we have to say that in their ideal form scientific theories are not confirmable (Problem of Induction), are about regularities of sequence of types of events, indistinguishable from accidental sequences (Problem of Causality), change their plausibility relative to evidence under purely logical tranformation (Problem of Instance confirmation), that scientific laws are indistinguishable logically from accidental generalizations (Problem of Natural Necessity), are such that their component predicates are logically independent (Problem of Subjunctive Conditionals). Finally, though this is a point to which I do not devote much critical attention, the deductive structure, as has been frequently pointed out, leaves open the question of which axioms or axiom should be revised in response to unfavourable evidence, and indeed from the point of view of traditional logic a decision as to which to modify can only be arbitrary. But science has well defined rational procedures by which this choice is made, and these we shall study carefully. An ideal for scientific knowledge which effectively makes scientific knowledge unachieveable is grotesque. It is the purpose of this book to offer an ideal for scientific knowledge which is derived from science, and not from mathematics, and which is not such as to make the achievement of scientific knowledge in its image impossible.

SUMMARY AND BIBLIOGRAPHY

The main burden of argument in this chapter concerns the opposition between the view of philosophy of science which takes the methodological intuitions of scientists as data, and by subjecting them to critical scrutiny attempts to derive the general principles of scientific thinking from them, and the view which takes some system of logic as the data (it might be syllogism, or first order predicate calculus with identity) and builds, *a priori*, certain structures which are offered as ideals to which science should (must) conform. The programme of the second view has been called 'rational reconstruction'. The opposition between these views is very old, and can be found in Antiquity. But for the purposes of studying the debate it is probably better to start with the methodological controversies of the sixteenth century.

For the controversy in astronomy see

1. E. ROSEN, *Three Copernican Treatises*, Columbia University Press, New York, 1939.
2. C. CLAVIUS, *In sphaeram de Ioannis de Sacro Bosco*, Lyon, 1602, pp. 518–20.
3. NICHOLUS RAIMARUS URSUS (BAER), *Fundamentum Astronomium*, Argent., 1588, Ch. 1.
4. J. KEPLER, *Apologia Tychonis*, Chs. 1 and 2, in *Opera Omnia*, ed. Ch. Frisch, Frankfurt, 1858–71.
5. R. M. BLAKE, C. J. DUCASSE and E. H. MADDEN, *Theories of Scientific Method*, University of Washington Press, Seattle, 1960, pp. 22–49.
6. R. HARRÉ, *The Ideals of Science*, Hamish Hamilton, London, 1970. For translations of relevant passages from (3) and (4), see Chapter Three.

For the controversy in chemistry see

7. D. M. KNIGHT, *Atoms and Elements*, Hutchinson, London, 1967.
8. W. H. BROCK (editor), *The Atomic Debates*, Leicester University Press, 1967.

Both these books are worth reading in full.
The blend of positivism and deductivism that is characteristic of twentieth-century philosophy of science is not a necessary one. The opposition between a positivist epistemology and metaphysics and realist positions in the seventeenth century, cut across the opposition between deductivists and 'logical realists'. The epistemological and metaphysical opposition can be traced through

9. F. BACON, *Novum Organon*, London, 1620, II, Ch. 9.
Which should be read in conjunction with

10. F. BACON, *Valerius Terminus*, in *The Philosophical Works of Francis Bacon*, translated by Ellis and Stebbing, Longmans, London, 1905, pp. 195–9 (fragment of Ch. 11).
11. J. LOCKE, *Essay Concerning Human Understanding*, new edition, edited J. W. Yolton, Everyman's Library, London, 1961, Bk. II, Chs. 8, 21, 23, and Bk. III, Chs. 3, 6, though it must be remembered that Locke attempts to reconcile an atomistic theory of 'ideas' with a realist theory of knowledge.
12. G. BERKELEY, *A Treatise Concerning the Principles of Human Knowledge* in *Works*, new edition, edited by T. E. Jessup, Nelson, London, 1949, Volume II, particularly Principle CI to the end of the book.
13. D. HUME, *A Treatise of Human Nature*, Clarendon Press, Oxford, 1946, Bk. I, Parts i and iii.
14. D. HUME, *An Enquiry Concerning Human Understanding*, edited by L. A. Selby-Bigge, Clarendon Press, Oxford, 1962, Sections II, III, IV and VII.

In my view these books are not fully intelligible without an adequate understanding of the views of the scientists of the day, particularly

15. R. BOYLE, *The Origins of Forms and Qualities*: Vol. III, *Works*, London, 1772, pp. 35–59.

See also

16. J. H. KULTGEN, 'Boyle's Metaphysic of Science': *Philosophy of Science*, **23**, 136–41.
17. G. GALILEO, *The Assayer*, which is easily obtainable in G. Stilman Drake, *The Discoveries and Opinions of Galileo*, Anchor Books, New York, 1957, pp. 273–9.
18. SIR I. NEWTON, *Principia*, 2nd ed., reprinted, University of California, Berkeley, 1947, Bk. III, Rule III.

It is really only with the controversy between W. Whewell and J. S. Mill that phenomenalism, deductivism and positivism become explicitly associated in the manner with which we have become familiar, though that association is certainly implicit in Hume. For a survey of the history of philosophy of science from that controversy to the present day see

19. *The Encyclopedia of Philosophy*, Macmillan and Free Press, New York, 1967, **6**, 289–96, which includes a bibliography.

For contemporary unification of these themes the best sources are

20. R. VON MISES, *Positivism*, Harvard University Press, 1951.
21. C. G. HEMPEL, *Aspects of Scientific Explanation*, Free Press, New York, 1965.
22. A. J. AYER (editor), *Logical Positivism*, Free Press, Glencoe, Illinois, 1959.
23. R. CARNAP, *The Philosophical Foundations of Physics*, Basic Books, New York and London, 1966, particularly Part I and Part V.

24. P. H. NIDDITCH (editor), *The Philosophy of Science*, Clarendon Press, Oxford, 1968.
25. F. WILSON, 'Definition and Discovery', *BJPS*, **18**, 287–303, **19**, 43–56.

There is a growing body of literature in opposition to this point of view.

26. N. R. CAMPBELL, *The Foundations of Science*, Dover, New York, 1957, Part I. This is a new edition *Physics, The Elements*, Cambridge, 1919.
27. R. HARRÉ, *Theories and Things*, Sheed & Ward, London, 1961.
28. W. SELLARS, *Science, Perception and Reality*, Routledge and Kegan Paul, London, 1963, particularly Chs. 1, 3 and 4.
29. MARY HESSE, *Forces and Fields*, Nelson, London, 1961, particularly Chs. 1, 2 and 11.
30. N. R. HANSON, *The Concept of the Positron*, Cambridge University Press, 1963, which, though in some ways an unsatisfactory book, has some admirable sections, see particularly Ch. II.
31. E. F. CALDIN, 'Theories and the Development of Chemistry', *BJPS*, **10**, 209–22.

For a defence of the theory of deductivism in isolation from epistemological and metaphysical issues see

32. R. DESCARTES, *Principles of Philosophy*, translated Haldane and Ross, Cambridge University Press, 1934, II, LXIV, III, IV, XLIII.
33. R. DESCARTES, *Rules for the Direction of the Mind*, Cambridge, 1934, III.
34. H. HERTZ, *The Principles of Mechanism*, Dover Edition, New York, 1956, 'Introduction', pp. 1–41.
35. J. H. WOODGER, *The Technique of Theory Construction*, University of Chicago, 1939.
36. R. BRAITHWAITE, *Scientific Explanation*, Cambridge University Press, 1953, Ch. 1, pp. 12–21.
37. W. W. BARTLEY, III 'Achilles, The Tortoise and Explanation in Science and History', *BJPS*, **13**, 15–33.
38. E. W. BETH, 'Fundamental Features of Contemporary Theory of Science', *BJPS*, **1**, 291–301,

and for a thorough-going attack on deductivism, see the edition of N. R. Campbell, *The Foundations of Science* (26).

39. J. J. C. SMART, *Logic and Language*, edited A. G. N. Flew, Blackwell, Oxford, 1953, Ch. XII.

Much of the controversy in recent years has concerned the alleged logical identity of prediction and explanation, positivists and deductivists defending the identity (or symmetry). There are those who defend the alleged symmetry:

C. G. HEMPEL (*see* 21), 245–95.
40. P. DIETL, 'Paresis and the Alleged Asymmetry between Explanation and Prediction'. *BJPS*, **17**, 313–18,

and those who attack it, on very various grounds.
41. A. GRUNBAUM, *Philosophy of Science*, **29**,
42. N. RESCHER, 'On Prediction and Explanation', *BJPS*, **8**, 281–90.
43. I. SCHEFFLER, 'Explanation, Prediction and Abstraction', *BJPS*, **7**, 293–309.
44. M. SCRIVEN, 'Explanation, Predictions and Laws', *Minnesota Studies in the Philosophy of Science*, University of Minnesota Press, Minneapolis, 1962, III, pp. 170–230.

Many other items from both sides of this controversy will be found in the discussion bibliographies of particular topics in later chapters.

For some discussion on the identification of language and thought see

45. L. WITTGENSTEIN, *Tractatus Logico-Philosophicus*, Routledge and Kegan Paul London, 1963, translated by D. F. Pears and B. McGuinness.
46. L. WITTGENSTEIN, *Philosophical Investigations*, Blackwell, Oxford, 1953, translated by G. E. M. Anscombe.
47. EMILE MEYERSON, *Identity and Reality*, Allen and Unwin, London, 1930 (New Edition, 1964).
48. L. S. VYGOTSKY, *Thought and Language*, Harvard University Press, Boston, 1962.
49. K. BÜHLER, *Sprachtheorie*, Vienna, 1934.

For a critique of the metaphysics of events see

50. D. HAWKINS, *The Language of Nature*, Freeman, San Francisco and London, 1964, Ch. 5.
51. N. MAXWELL, 'Can there be Necessary Connections between Successive Events?' *BJPS*, **19**, 1–25.

The best discussions on the need to accept affirmation of the consequent are to be found in

C. CLAVIUS (*see* 2).
52. P. GASSENDI, *Syntagma*, Lyon, 1658, I, 68–9.

The curious problem of Goodman's Paradox has been raised in this chapter. It has generated a considerable literature. The items mentioned below have been chosen as illustrative of most of the trends in the discussion.

53. N. GOODMAN, *Fact, Fiction and Forecast*, Athlone Press, London, 1954, pp. 73–80.

The main *standard* forms of resolution, i.e. resolutions which still accept Nicod's intuition are found in

54. H. FAIN, 'The Very Thought of Grue', *Philosophical Review*, **76**, 61–73.
55. M. B. HESSE, 'Ramifications of "Grue"', *BJPS*, **20**, 13–25. See p. 25, for a useful bibliography.
56. The points made briefly in my text are established in detail in C. A. Hooker, 'Goodman, "Grue" and Hempel', *Philosophy of Science*, **35**, 232–47.

By way of summary of the opposition between the two main schools of thought in philosophy of science, the contrast between rational reconstructionist, positivist way of studying science philosophically, and the realist approach can be seen very vividly in

57. S. TOULMIN, *Scientific American*, **214**/2, 129–33.
58. E. NAGEL, *Scientific American*, **214**/4, 8–9.

There is a branch of literature, which perhaps would be aptly entitled 'science criticism', to take up a suggestion of Sir Peter Medawar's, which studies actual science in a very concrete way. Valuable insights can be found in the following:

59. A. KOESTLER, *The Watershed*, Hutchinson, London, 1961.
60. T. KUHN, *The Structure of Scientific Revolution*, Chicago University Press, 1962.
61. P. MEDAWAR, *The Art of the Soluble*, Methuen, London, 1967, particularly pp. 131–54.
62. J. D. NORTH, *The Measure of the Universe*, Clarendon Press, Oxford, 1965.
63. M. POLANYI, *Personal Knowledge*, Routledge and Kegan Paul, London, 1950, Parts I and II.
64. S. TOULMIN, *Foresight and Understanding*, Hutchinson, London, 1961.

2. Models in Theories

The Argument

1. *Definition of Sentential Model*

   T' is a model of T (T and T' are sets of sentences) if for every p in T there is a q (one and only one q perhaps) in T', such that if q is acceptable then p is true, and if p is false then q is unacceptable.

2. *Definition of an Iconic Model*

   M is an iconic model of N, if T' describes M and T describes N and T' is a sentential model of T.

3. *The Ordinary Use of the Word 'Model'*

   (*a*) Participle and proposition: ('modelled on', 'modelled by', 'modelled in', 'modelled for').

   (*b*) The verbs 'to model' = 'To wear exemplarily', 'To make in plastic material'.

   (*c*) model = type.

   (*d*) in mathematics the word is usually used for a sentential model, but sometimes for an iconic model.

4. *Distinction of a Model from a Symbol*

   This can be made by using the distinction between arbitrary and representative conventions, based upon using the object modelled as the basis for explaining why the model is the way it is.

5. *Taxonomy of Models*

   A. Subject = source; homoeomorphs

   Homoeomorphs

   Micro and megamorphs   Teleiomorphs   Metriomorphs

   (Scaling problem)

   Idealizations   Abstractions

   Capable of being used as class representatives

   B. Subject ≠ source; paramorphs

   Paramorphs

   (*a*)   (*b*)

   in relation to subject   in relation to source
   partial analogue   semi-connected
   complete analogue   singly connected
   partial homologue   multiply connected

   Step 1 in theory construction involves the creation of a paramorph.

   Step 2 in theory construction involves the hypothesis of the paramorph as a hypothetical mechanism.

   Thus a theory generates existential hypotheses.

6. *Problems arising from* (*a*) *and* (*b*)

For semi-connected and singly connected paramorphs, existential hypotheses propose entities of received *kinds*. But some abstraction may be involved, so

Metaphysical Problem 1. How much abstraction is possible without losing plausibility of existence? The theory of primary qualities can be treated as an attempt at a general *a priori* answer.

For multiply connected paramorphs, existential hypotheses involve novel kinds. (See Chapter 3.)

Metaphysical Problem 2. How are novel kinds to be incorporated into our ontology?

Answer: either through causal hypotheses, which creates the problem of the nature of causal mechanisms, which will be solved by the doctrine of powers, or by re-categorization.

7. Hypothetical mechanisms must bear, hypothetically, the same kinds of relations to phenomena as do real mechanisms.

Two such relations

I. Causal. This generates the problem of transcategorial causality, e.g. virus to disease, compared with field to particle.

II. Modal. This generates problems of taxonomical priority, i.e. which descriptions are fundamental to a classification into kinds, e.g.

(1) thoughts, acts, etc. and brain and body states (see Chapter 8),

(2) molecular swarms and gases.

8. *Plausibility Control*

For logical constructions only consistency, i.e. obedience to the principle of non-contradiction, is required; for models, consistency is required, plus a fit with received scientific ontology, either by incorporation, modification or addition.

9. *Protomorphs*

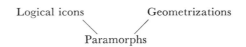

10. *Four Kinds of Hypotheses*

(*a*) Existential: are there molecules? (Viruses? Fields? Are there complexes?)

(*b*) Model Description: are molecules in random motion?

(*c*) Causal: is pressure caused by impact of molecules?

(*d*) Modal: Is temperature another way of conceiving of mean kinetic energy?

I begin my positive account of the nature of scientific thinking by setting out a new view of theories in which they are to be seen as essentially concerned with the mechanisms of nature, and only derivatively with the patterns of phenomena. I shall be considering only one component of those intellectual constructions which I have used the metaphor of 'statement-picture complex' to describe and which are actual theories *in vivo*. Scientists, in much of their theoretical activity are trying to form a picture of the mechanisms of nature which are responsible for the phenomena we observe. The chief means by which this is done is by the making or imagining of models. Since enduring structures are at least as important a feature of nature as the flux of events, there is always the chance that some models can be supposed to be hypothetical mechanisms, and that these hypothetical mechanisms are identical with real natural structures.

It might be in point here to anticipate briefly the argument of the next five chapters in which this view of theories is followed out in all its ramifications. Theories are seen as solutions to a peculiar style of problem: namely, 'Why is it that the patterns of phenomena are the way they are?' A theory answers this question by supplying an account of the constitution and behaviour of those things whose interactions with each other are responsible for the manifested patterns of behaviour. This might be a microexplanation like the explanations of chemical reactions in terms of the behaviour of chemical atoms. It might be a macro-explanation like the explanation of the tides in terms of the behaviour and powers of the heavenly bodies. To achieve this a theory must very often fill in gaps in our knowledge of the structures and constitutions of things. This it does, in ways to be detailed, by conceiving of a model for the presently unknown mechanism of nature. Such a model is, in the first instance, no more than a putative analogue for the real mechanism. The model is itself modelled on things and materials and processes which we do understand. In a creative piece of theory construction, the relation between the model of the unknown mechanism and what it is modelled on is also a relation of analogy. Thus, at the heart of a theory are various modelling relations which are types of analogy. The rational construction of models will be shown to proceed under the canons of a theory of models, to which the old logic of analogy is a crude approximation. In developing this theory we shall be looking at the way scientists train themselves in the skilful judgement of likeness and unlikeness plausability, and the like. But we can hold no hope of successful formalization of the principles of this art. At this stage we describe them. A theory cannot, therefore, be a single deductive structure. It consists, as we shall see, of at least three sets of sentences, the successful deductive organization of any of which is quite fortuitous. There will be one for the description of the phenomena for which the theory is devised. There will be another for the description of the central model, and one or more describing the material upon which the central model is based. These sets of sentences are tied together by various relations of analogy, that is by further sets of sentences whose extent cannot be discovered *a priori*. Finally, it will be shown that this is the structure which generates the most crucial of all the kinds of scientific hypotheses, namely existential ones. This happens by the step of coming to treat the central model as a hypothetical mechanism. All sorts of consequences follow from this view of theories; consequences which tie together the scientific enterprise in a rational structure which no simpler theory of theories, however elegant, can approximate.

I hope it will become clear as this book proceeds that I do not in the least intend anything specifically mechanical by the word 'mechanisms'. Clockwork is a mechanism, Faraday's strained space is a mechanism, electron quantum jumps is a mechanism, and so on. Some mechanisms are mechanical, others are not. I choose the word 'mechanism' for this use largely because it is the word most usually used for this purpose. We talk of 'the mechanism of a chemical reaction', 'the mechanism of bodily temperature control', 'the mechanism of star creation', and so on.

I turn now to a detailed exposition of the nature, use and structure of models. The first distinction to which I want to draw attention is that between sentential and iconic models. Needless confusion has been created by naively identifying the two. They can be distinguished formally as follows:

If T and T' are sets of sentences, then T' is a model w.r.t. to T if for each p such that p is a member of T, there is a q, such that q is a member of T', and when q is acceptable, p is true, and when p is false, q is unacceptable. Such a model I call a *sentential model* of T. In general, there may be more sentences in T' than in T.

If T is a set of sentences about some subject matter N, then M, is a model of N, if T' is a sentential model with respect to T, and M is a set of objects etc., such that for all q which are members of T', q is known to be true or false by reference to M. Such a model M, I call an *iconic model* of N. In this chapter, my concern is first of all with iconic models.

As another important preliminary to clear away possible misunderstandings, I shall look into the common usage and grammatical peculiarities of the noun, 'model', and the verb 'to model'. My use of the word 'model' will not be precisely that of everyday, though not unconnected with it. I use it as a technical term, and the precise technical usage I intend can be grasped, only after my exposition of the way I intend to use the term is complete. In the next section I give a general definition of model, but it is rather a sketch of certain broad conditions under which I intend to use it than a precise definition. Just to show that I do know what I am about, I introduce at this point a sketch of the ordinary usages of the term, not all of which will be reproduced in the technical usage.

1. *Participle and Preposition*

(*a*) 'The Ghanaian Assembly is *modelled on* the British Parliament.' Here the British Parliament is the source of certain features of the

Ghanaian Assembly, but it would be incorrect to say that that Assembly is a model of the British Parliament, or that it is a model Parliament.

(*b*) 'The gold lamé cocktail dress was *modelled by* Samantha.' This is a very derivitave usage, harking back to Samantha being a model, and that too is derivative, harking back to Samantha being, in some sense, ideal.

(*c*) 'The head was *modelled in* clay.' Here, by transferred epithet, the process of making a model is called 'modelling', and what it is modelled in, asks of what substance has it been made.

(*d*) 'She made a living *modelling for* the art class.' She might be an ideal shape for Venus, or for a grandmother, but she might just be something to draw. In this sense, though not in that of *b*, it is not necessary for her to be an ideal for her to be a model.

2. *The Verb 'to model'*.

Two common usages that I ignore later are 'to wear in an exemplary fashion' and 'to make something with the fingers out of a plastic material'.

3. Then there is the use of the word 'model' to designate a type. This can be exemplified by the famous apochryphal headline in *The Times*: 'Rolls Royce announce new model: radiator lowered one inch.' Sometimes this slides over, in careless talk, into reference to the token as a token representative of the type, in such expressions as 'I bought this year's model' where I mean to indicate that I bought one of this year's type.

4. In mathematics the word 'model' is used in two ways: for a set of sentences, what I have called a 'sentential model', and sometimes for a set of objects, real or imaginary, statements about which are made by means of sentences and which constitute a sentential model of the mathematical sentences modelled. This latter I have called an 'iconic model'.

I now turn to a more exact account of how I intend the word 'model' to be used. First the difference between model and symbol; I shall use 'model' so as to include pictures. This distinction has to be stated with some care, since both models and symbols depend upon a convention or set of conventions by which they become vehicles for thought about a certain subject matter. I am going to distinguish projective conventions for models from arbitrary conventions for symbols. It might be noticed in passing that Wittgenstein in the *Tractatus* tried to assimilate arbitrary conventions to projective conventions. A model can become a symbol when its source of projection is lost or forgotten as, for instance, the bull's head became *alpha*, but a symbol cannot become a model. This last clause is slightly stipulative, since some people still talk of equations as models of motions and processes. At that rate every vehicle for thought would become a model, and a valuable and interesting distinction would be lost. If anyone insists that that is just how he uses the word 'model', or that that is how the word model is used in his culture-circle, I will give him the word 'model' gladly, and mark my distinction by means of another word, let us say 'modella'. It's well to remember the old saying, if our eyes were made of green glass then *nothing* would be green.

What is the difference between a projective and an arbitrary convention? Projective conventions involve the use of the source of the model or picture in the act of construction of the model or picture. The reason why A is a model of, or for, B in one kind of case, is that A was constructed from B, so that A and B have certain likenesses, and it is these that make A a model of B. But because A and B must be unlike in certain ways (or else A is just a replica of B), and just how unlike and in what respects is at the will, and under the guide of the purpose of the model-builder, I call the projection of B into A, a convention. Arbitrary conventions have, *inter alia*, the following *differentium*: whatever may have been the source of the symbol β, no likeness or unlikeness it may bear to α its subject matter counts as a reason why it is a symbol for, or of, α. Any sign would do as well, provided a symbol convention had been agreed for it. Hieroglyphs provide an interesting case, since the marks seem to function both as models and as symbols. I understand from Egyptologists that this appearance is misleading, and that they make a clear distinction themselves between the comic strip era, and the syllabary. The only case I can think of where a word might get its meaning, or be thought to get its meaning, by a projective convention, is the alleged phenomenon of onomatopoeia. I understand linguists no longer believe in this phenomenon, that is in sound-likeness being *determinative* of meaning. If any reader still does, I simply challenge him to give the meanings of alleged onomatopoeic words in a language he does not know; for instance the following words, which were given to me by a Saudi Arabian, as colloquial Arabic, which are something like this: *tutátiko*, and *tsúktsuka*. I think enough has been said to separate at least the extremes, if there is a continuum, between models and symbols, as vehicles for thought, so I shall say nothing further about symbols, in this chapter.

A cardinal distinction, of the first importance in understanding the way models work, is that between the source of a model and the subject of a model. The *subject* is, of course, whatever it is that the model represents: that it is a model of, or model for. There are several different relations between models and their subjects. The *source* is whatever it is the model is based upon. There are again several possible relations between a model and its source. Broadly speaking models belong in two great categories, depending on whether the subject of the model is also the source of the model, the *homoeomorphs*; or whether the subject and the source differ, the *paramorphs*. For example a toy car is a model for which source and subject are identical, the toy being modelled on a car, and being a model of the very same car. However, even with toy cars, the matter is not quite so simple as it seems at first glance. Is the toy car modelled on a particular instance of the 1963 Mini-Cooper, say the one which won the Monte Carlo Rally, or is it modelled on the 1963 Mini-Cooper type? Both cases occur. Models of the particular 1963 Mini-Cooper which won the 1963 Monte Carlo Rally are sold, and models of the 1963 Mini-Cooper are also sold. The only difference that I can see is that while all models of the particular 1963 Mini-Cooper must be, in all respects in which they *are* models of *the* car, the same, the models of the 1963 Mini-Cooper can be different, in the same kind of respects as those in which 1963 Mini-Coopers differ, namely in colour, radiator grille, presence or absence of spotlights, and so on. If someone produced a model of the 1963 Buick Riviera, and declared that he wished to treat it as a model of a 1963 Mini-Cooper, we would, quite properly, have to tell him that he *could not do this*, despite the common possession of four wheels, two doors and so on. Again this shows how models differ from words. Only a certain latitude is permissible in projective conventions, and is determined by the need to eliminate certain possibilities of being misled, in particular cases. Of course, if it *is* a model, as contrasted with a replica, then there will always be *some* way of being misled, just because a model is not exactly like the subject in all respects.

Lord Kelvin's model of gyroscopes and elastic thread for a particle of the luminiferous ether is a paramorphic model since its source and subject are different. The subject is the unknown mechanism of the transmission of light (supposing light is transmitted!), the source is the known mechanical behaviour of gyroscopes and of elastic threads. Of course, a model's subject and source may differ even when the subject is some known mechanism, since models may be made, or imagined, for such conveniences as ready intelligibility, or any one of a number of reasons which would suggest that the mechanism of the thing or process, etc. modelled, should be modelled by some other kind of mechanism having suitable characteristics.

Part of the great importance of models in science derives from their rôle as the progenitors of hypothetical mechanisms. It often happens that the antecedents of an effect are well known, but the casual mechanism by which the antecedents bring about the effect is not. Consider, for instance, the long history of the use of catalysts, and of antibiotics, before the way they worked was found out. Then, in the imagination, we make a model for the unknown mechanism. Whatever *is* in the black box, one might say, could be like this. This process differs fundamentally from making a model of a subject which is already well known. I shall sharpen up the terminology a bit by using '*A* is a model for *B*' where *B* is quite unknown, and '*A* is a model of *B*', where *B* is known. Generally speaking, making models *for* unknown mechanisms is the creative process in science, by which potential advances are initiated, while the making of models *of* known things and processes has, generally speaking a more heuristic value.

Systems of classification, which are not purely arbitrary, are called 'taxonomies'. The taxonomy I am introducing in this chapter classifies models, not with respect to their intrinsic properties, but with respect to their relations to their sources and subjects. The very same entity may be a paramorph with respect to its subject, if it is a model for an unknown mechanism, or a model of a known one, and a homoeomorph with respect to its source; what it is modelled on. For instance, the Clausius model for a gas was a paramorph for an unknown structure, but an idealization and abstraction of its source, that is a swarm of material particles. It both omitted many of the properties of material particles, such as volume, and idealized those it retained, such as demanding perfect elasticity. But when it came to be believed that gases really were swarms of material particles then the Clausius model had to be reclassified, since its subject has to be made out to be one in kind with its source, since gases are made up of molecules as their parts.

An object, real or imagined, is not a model in itself. But it functions as a model when it is viewed as being in certain relationships to other things. So the classification of models is ultimately a classification of the ways things and processes can function as models. Fashion models are not a special category of humans, but fashion modelling is something that some people can do.

Homoeomorphs

I shall begin a detailed analysis with those models for which source and subject are identical. Of this category of models three kinds can be distinguished, differing in the way they are related to their source-subject. The first kind are the *micro* and *megamorphs*. These are the scale models. They differ from their subject, which is also their source, only in scale. They are created by applying to a suitable characterization of their source-subject some such instruction as 'Reduce (Increase) the main linear dimensions by 1/64'. Micro and megamorphs are to be regarded as typified by those toy steam engines which are 'perfect in every detail'. A pure micro or megamorph has every feature that its source-subject has. A solid model of an airliner used for wind-tunnel testing is not a pure micromorph, since some important features of the source-subject

are omitted, like the seats in the passenger cabin, and the pressuriza-
tion system. Similarly, fibreglass models of viruses, millionfold megamorphs, are impure, since not every structure of the source-subject is
reproduced. Even for a pure micro or megamorph the rules of transformation from source-subject to model and back again may not be uniform for all features of the subject. In model aeroplane making, the
control surfaces must be scaled down by a lesser factor than the major
linear dimensions, if the stability and manœuvreability of the model is
to match that of the plane. The exact scale factors for any features other
than the major linear dimensions may have to be discovered by experiment, if the model is going to be able not only to represent the physical
structure of the plane but also to behave like it. This is generally true of
all those micromorphs that come under the species *pilot plants*. To model
both behaviour and structure, some compromise on the details of structure is required.

The second kind of models for which source and subject are identical
are *teleiomorphs*. This name is intended to suggest that they are, in some
respect or respects, an improvement on their subjects. Two kinds of
teleiomorphs can be distinguished, depending on just how they are
related to their source-subject. There are *idealizations* and there are
abstractions. To see how these are distinguished, suppose that the task is
to model an object which has distinguishable properties $p_1 \ldots p_n$, relevant to the enquiry or task. Then a model of it may be made in one of two
ways: an idealization is created either by building a model, or selecting
something to serve as a model, having properties $p_1 \ldots p_n$, where each
property of the source-subject is matched by a property of the model,
but the model properties are distinguished by being, according to some
scale of values, more perfect than the source-subject's properties. A
fashion model, for instance, has all the characteristics of an ordinary
woman, but has them in more perfect form, according to some currently
accepted scale of values. A glance at an old fashion magazine shows
immediately that what shape distinguishes a teleiomorphic woman at
one time may not meet the standards of another age. A model pupil
presents some difficulty of classification, since, while he does all the
things a pupil should do better than his peers, he does not do some of
the things that they do; he doesn't cheat, roll in the mud, and so on. So
he may not only have in superior form the characteristics of any pupil,
his source-subject (though he is not constructed but occasionally found),
he may also have fewer characteristics. When functioning as a model he
may be, in my sense, an abstraction as well. Another point of importance emerges here too, and will be discussed later. It is not so much any
particular woman or any particular pupil that serves as the source-
subject of the model, though the model must be particular. Rather it is

some average or type which is the source subject. An ideal pupil is an idealization of all pupils, and a fashion model is similarly related logically to all women. I shall return to this point.

The other way of making or choosing a teleiomorph is by abstraction. If the source-subject has properties $p_1 \ldots p_n$, and here again the source-subject may be a particular, representing the type of all members of some class of things, or it may be a kind of average. An abstraction has properties $p_j \ldots p_k$, $l < j < k < n$, that is fewer than its source-subject. Which properties are chosen depends largely on the purposes for which the model is created. This is because the properties which are not modelled are those which are irrelevant, and relevance and irrelevance are relative to purposes. Typical abstractions include those structures in coloured wires which model the blood vascular systems of animals, and which represent some relations of connectivity and relative length and size among the blood vessels, but leave out the valves, the relations of elasticity, the capillaries and of course the blood! Since the model itself is a wire and plastic structure it has many features which are irrelevant to its functioning as a model, such as the chemical composition of the wires and the plastic coating. This point too we shall return to, but for the moment one can say that the object-model has to be understood and that, in a way, the actual model is something like the meaning of the structure of wire. But only 'something like' because some of the features of the wire structure are features of the blood system. Another abstraction teleiomorph is the gas molecules of the kinetic theory of gases, which have the mass and velocity of material objects, but in the earliest form of the theory lacked the volume and imperfect elasticity of perceived things. In later theories molecules still lack some of the properties of perceived things: they have no warmth, for instance, since this characteristic is defined only for aggregates of molecules.

So far no distinction has been made between real models and conceptual models, between everything from *objects trouvés* to things built as models in workshops, and imaginative constructions about which and with which one only thinks. Throughout this book the view is being maintained that theory can fruitfully be looked upon as the imaginative construction of models, according to well-chosen principles, and that, in many ways, the theory of 'Ideas', in Whewell's sense, is more helpful in the theory of theories and scientific method generally, than the logic of statements. If this logic is used as the exclusive tool of analysis, it imposes an unnatural structure upon theory, and conceals even some important features of the logic of statements, which stem from the involvement of models in theory construction some of which I pointed out in the resumé at the beginning of this chapter. The next kind of model exists only in the imagination.

It is the third sort of model for which source and subject are identical. I shall call these metriomorphs a typical example being the average family. The source and subject for a metriomorph is always a class, and the model represents that class, for certain purposes. For other purposes for which one might wish to represent classes, use is made of micro and megamorphs, and teleiomorphs. However, they differ from metriomorphs in that the latter *must* be related to a class, since it is from a class that they are constructed, while the former, for reasons which will be mentioned shortly, *may* represent a class, but need not. The metriomorph has just as many properties, in the sense of determinables, as a typical member of the class it represents, but the determinate properties of a metriomorph are arrived at by an averaging or similar mathematical operation on the set of similar properties, that is the set of determinates under each determinable, possessed by all the members of the class to be represented. They are abstractions in the scholastic sense. So the famous metriomorphic family of the human male metriomorph, which consists of 2·63 metriomorphic children, has characteristics which no actual human family could have, because the number of children in a metriomorphic family is some mathematical fraction of the number of children in all human families, namely $n_m = \sum_i^k n_i/k$.

How is it that teleiomorphs can represent classes? An abstraction does so because what may be omitted from the set of properties of the individual, from which it is modelled may be just the individual differences. If the purpose of the maker of the teleiomorphic model is to represent the class from which his source-subject is drawn, then he may simply omit those properties which differ from individual to individual, either by dropping out a whole determinable which fluctuates too widely, or by arbitrarily choosing a mean determinate, as in the metriomorph. Notice now that though he started with source-subject identity, the model-maker has passed to a new situation in which his source remains an individual, but his subject has now become a class, of which that individual is a member. Since abstraction always involves some loss of properties from source to model, there will always be the possibility, with an abstraction, that the subject may be switched from the individual modelled to the class of which it is a member.

Paramorphs

The second main category of models are paramorphs. A paramorph models a subject, but its form and mode of operation are drawn from a source different from its subject, not just as a class is different from an individual which is a member of the class, but a paramorph's source is a quite different individual from its subject, a quite different class if its subject is a class.

The relation of a paramorphic model to its subject may be of three different kinds. To differentiate them, it is necessary to notice that paramorphs are usually constructed with the ultimate aim of modelling processes, and this is often achieved by actually modelling equipment, or naturally occurring things, in which the processes one wishes ultimately to model are occurring. The paramorph is needed, because, to operate with or study the real process raises difficulties of various sorts. This may be because the process in which the interest lies occurs in nature by some path which is currently unknown, as for instance when genetic factors were introduced as a model for the unknown process of the inheritance of Mendelian characters. Sometimes it is because the equipment which one wishes to use is awkward to work with, so one builds a model as, say, an electrical network is set up to study a hydraulic system. Sometimes it is because the mechanism of the process being modelled is complex, as in heat transfer across phase boundaries. Sometimes a model is resorted to because it is not at all clear where to look for the mechanisms of the process, for instance in the theory of human learning. Sometimes it is too dangerous to work with the real thing. Usually some event or state initiates some process which has some other event or state as outcome. The dimensions of variety of paramorphs arise because there are a number of different relations which may exist between the model and its subject. (1) The initial and final states of model and subject may be exactly alike, and the processes by which they were reached differ, as for instance in computer simulation of arithmetical calculation by a man; (2) the initial and final states may only be similar, and that similarity may indeed extend only to the similarity between tables of numbers, as for instance in electrical simulation of hydraulic networks where what is similar is the structure of tables of numerical results derived from measurements of operations with each network. Let us call these models partial paramorphic analogues. The processes occurring in the model and subject are different as in case (1). This sort of paramorphic analogue is exemplified by Bohr's atom. The process by which it stores energy, and later releases it, is modelled on both electro-magnetic and mechanical processes, that is, our conception of it is built up by drawing upon the concepts of electromagnetism and mechanics. But since it contains a mechanical impossibility: the 'motion' of electrons from orbit to orbit with loss or gain of energy, in no time at all, the process is, at best, an analogue of whatever process does occur in radiating atoms. However, input and output for Bohr atoms are identical with input and output for real atoms, namely electromagnetic radiation. The classical studies on the tube-boiler are a case of a kind of model, which I shall call a 'complete paramorphic analogue', where *both* model-process and model input and output are analogues of subject-process and subject input and output. The model is built by using sticks of sugar to represent the hot tubes, and the process of mass-transfer to circulating water, models the process of heat-transfer in the boiler. Finally, by a suitable transformation, the profiles of the sugar sticks, after running the model, can be transformed into the heat profiles of the tubes.

Though the processes in a paramorphic model and its subject, i.e. their manner of working, would usually be different, in those cases where the initial and outcome states are analogues but not identical, there could be the case of what I shall call a paramorphic *homologue*. The model and subject would have an identical manner of working but only analogous inputs and outputs. The idea of using living nerve cells for building computers would be a case in point, since the input and output would be only analogous to giving information and receiving reports, but the process by which the cellular computer and the brain found the answer would be the same, whatever that is.

The relation of paramorph to source is similarly threefold. I distinguish singly connected, multiply connected and semi-connected paramorphs. A paramorph is constructed or imagined as operating according to certain principles, commonly drawn from known sciences and technology, though sometimes, in works of the imagination, fudged up for the occasion, and vitally important this last is for science as we shall see. A paramorph, like the corpuscular theory of gases is singly connected, because the principles of only one science, mechanics, supply the definitions of the entities and the laws of the processes which constitute the model. Bohr's atom is multiply connected, since to construct it in the imagination, one must draw on the sciences of mechanics and electromagnetism, and the principles of these sciences are not explicable, one in terms of the other. Freud's 'psychic energy' mind model is semi-connected, because, though to construct it, one must draw on some principles of energetics from physics, it also introduces processes occurring according to principles unknown to energetics, or any other science. This does not, of course, count against the Freudian theory, *a priori*, because sometimes semi-connected paramorphs are just what give us a new scientific development, by suggesting the idea of a new kind of entity, or process.

A difficulty here concerns what is to count as one source science. Only if that is clear can singly and multiply connected paramorphs be distinguished. I mentioned a weak criterion, immediately above, that the principles of the one science have not, as yet, been explained in terms of the other. There is a stronger criterion with which I propose to settle any question as to whether the entities and processes dealt with by two sciences count as two sources for a paramorph. It is simply to ask whether two individuals, one defined in terms of the one science, and the other in terms of the other, can be allowed to occupy the same place at the same time, and the one not be a part of the other. This would make electromagnetism and mechanics different sciences as sources for paramorphs, physiology and psychology different sciences, but chemistry and quantum mechanics branches of the same science.

The Problem of Real Models

Consider again the case of a teleiomorph created by abstraction, and the example of a wire structure as a model of the vascular system. There is a case for saying that the wire structure itself is not the model but represents the model: that, as I put it in introducing this problem above, the model is the meaning of the wire structure. If the model is an abstraction from the subject-source then, in the interesting case, whole determinables are not represented in the model. For instance, the wire model does not represent any of the tensile qualities of the vascular system, but it has tensile qualities of its own. We are to ignore the tensile qualities of the wire model, because the projective conventions which together constitute the modelling transform do not give them a rôle, in the modelling. One may well now ask, 'What is it that lacks tensile qualities?' It certainly is not the wire structure. The spatial organization of the wire structure models the spatial arrangements of the vascular system, and that is all. So the model lacks colour, temperature, volume, and so on: all qualities both of the wire structure and the vascular system. This tempts us to ask, 'Is the wire structure really the model after all?' Perhaps it only represents the model, which is something abstract and mysterious. This would be a mistake. There is no other object involved but the wire structure, but it has to be understood in a certain way. It must be read, like a sentence, so that it is treated as having a certain spatial configuration, and nothing else, for the purpose in hand. But what makes things, unlike words, models, is a projective convention, that is, what characteristics are read off the object as the model serve their function through physical similarity to the characteristics of the object modelled. What makes a thing a word is an arbitrary convention.

I now characterize again the Copernican Revolution in the philosophy of science, in the following ideas: theory construction is primarily model building, in particular imagining paramorphs. Imagined paramorphs involve imaginary processes among real or imagined entities. The crucial point for the understanding of science is that in either case they may invite existential questions, since unlike homoeomorphs, they introduce additional entities other than the given, provided it seems plausible to treat these as a casual mechanism. It is through imagined paramorphs and their connection with their sources, multiple, single, semi or fragmentary, that theoretical terms gain part, and a vital part, of their meaning. It is by being associated with a paramorphic model, in ways that I will go into later, that many laws of nature get their additional strength of connection among the predicates that they associate, that distinguishes them from accidental generalizations. A scientific explanation of a process or pattern among phenomena is provided by a theory constructed in this way.

Models, analogies and metaphors are closely related, though not identical tools for rational thought. At this point I want to draw a distinction, in passing, between models merely as the source of picturesque terminology, and models as the source of genuine science-extending existential hypotheses. A model for something, be it thing or process, can be described in the language of simile as a thing or process analogous to that of which it is a model. Thus electricity can be described as like a flow of something, indicating that electrical circuits have some likeness to hydraulic networks, that is, are analogous in certain respects to them. The fluid model provides an existential hypothesis, since it made sense, once, to ask, 'Is there an electric fluid or fluids?' In a case like this, we are not redescribing electrical phenomena in the language of hydraulics, we are offering the fluid as a causal mechanism to explain the electrical phenomena. This should be compared with an example like economic cycles, with its picturesque terminology of inflation and deflation, depression and boom. Somewhere in the background of this system of metaphors lies a model of the economy, and of trade, in which transactions are treated as the parts of some substance which can expand and contract, a model which no longer has an explanatory function. The model offers us nothing by way of explanation, and no existential hypotheses, but it does provide, in the system of metaphors, a picturesque terminology. Many metaphors are indeed just this, the terminological debris of a dead model. There has been a great deal of muddle about these distinctions, and some writers, particularly Poincaré, have not succeeded in keeping the cases distinct. At this point I am concerned solely with the use of models to provide hypothetical mechanisms, if they are paramorphs; or readily understandable analogues, if they are homoeomorphs. I want to turn now to a more detailed examination of the way paramorphs prompt existential questions.

Even for a singly connected paramorph the problem is not so simple. For the sake of exposition I shall take a straightforward example, the mechanical corpuscle as it figures, say, in nineteenth century models for chemistry, physics and other studies. Mechanics is the science of material things, in certain kinds of mutual interactions. Mechanics does give an account of the relations between velocities, masses, and so on of interacting bodies, but does not give an account of the interactions of their colours or temperatures. The colour and the temperature, for instance, simply do not figure among the mechanical parameters. So if a model is devised in accordance with the science of mechanics, the properties with which its entities will be endowed will be the mechanical parameters, and, in the first instance, and for a paramorph singly connected with mechanics, nothing else. But the ordinary things in the world of whose existence we can be sure have rather more properties than the mechanical parameters, though they do have those, too. The metaphysical question which has to be dealt with right at this point in the use of a single science for model-source, is this: what properties constitute the minimal set for us to be able to say that what we have is a thing existing in the world, *and forming one of the parts of which an ordinary thing is the whole, or a whole of which ordinary things are the parts*? Do the parts of a coloured thing have to be coloured? Does the whole, of which the parts bear structural relations to each other, which exhibit the relations of the law of entropy, itself exhibit that law? One answer to this question was provided in the seventeenth century with the doctrine of primary qualities, that the essential properties of things were their bulk, figure, texture (arrangement of parts) and motion, and that the properties of the whole were simply the summation of the properties of the parts. All other characteristics which things seemed to have were nothing but different ways in which the essential properties affected us. Other solutions have been given. But what we need to notice now is that without some metaphysical doctrine, even if only tacit, we cannot use the simplest kind of paramorph for its proper scientific purposes.

For multiply connected paramorphs, the problems raised by existential hypothesis are even more profound. If we suppose our imaginary model is perhaps a real mechanism, process, or thing, we have to ask whether nature, as we know it, can admit of such a thing. Sometimes this can be settled, in the negative only, *a priori*. Treated as a work of the imagination a centaur is a multiply connected model devised by drawing on both horses and men, but not drawing upon the characteristics of either completely. But a moment's thought on the parts of each that are conjoined, will show that a centaur is an anatomical impossibility. Where we do not have this easy way out we depend on what I shall call 'plausibility control'. This can be exercised through such questions as: 'Would existential affirmation of the entities and processes of the multiply connected paramorph introduce something into the world which, while not impossible, is in sharp conflict with the ideas currently held as to what there is?' The unpopular minority were right who said, of the Bohr-Rutherford atom, that it was a jerry-built contraption, on the basis of its incidentally requiring a massy particle, as the electron was then understood, to change position without traversing the intervening space or perhaps worse, to traverse it in no time at all.

This can be looked at in another way. For singly connected paramorphs there are only two cases. Either there is a full transference of properties from source to model, and this leads to philosophically uninteresting questions like 'Is there a planet between Mercury and the Sun whose influence explains the anomalous behaviour of Mercury?' Or there may be a partial transference. Here the existential questions raise more interesting issues, because the transference may omit only so many determinables without an entity losing substantiality. At some point we say that no existential question can be sensibly raised in the context, as for instance one might be tempted to say about point-atoms. In either case, we tacitly accept and operate with the standard received conception of an object, and in the latter, as I pointed out above, with the distinction between primary and secondary qualities.

Multiply connected paramorphs can lead to changes in these conceptions. The Bohr atom at least raised the question of the absoluteness of Newtonian criteria of existence. Perhaps entities of a new kind should be admitted to exist alongside those presently conceived possible, entities which did not obey the received principle, hitherto categorical, that one thing cannot be in two places at the same time. Part of the origin of the difficulties over electrons can be traced to the fact that they were multiply connected paramorphs, with, as it emerged, only *partial* transference of what seemed essential properties, from each source, the mechanical and electromagnetic respectively. The hypothesis of their existence would involve existential novelty. It would not be merely the addition of further items to a kind of entity whose existence was accepted on all hands only differing from standard examples of that kind in non-essential ways, typically only differing in size. Even though the acceptance of a multiply connected paramorph as a candidate for existence does raise problems they are nevertheless sometimes admitted. Electromagnetic radiation, viruses, quasars, electrons, the benzene ring, infant sexuality each had its conceptual origin as a multiply connected paramorph. Who would be so bold as to deny that there are such things?

Part of the difficulty of doing philosophy of science in terms of received logic, that is by concentrating on a particular aspect of the language of science, is that the principle of non-contradiction, or some similar principle or principles, serves as our only plausibility control, when we construct hypothetical mechanisms, that is in the language *genre*, define theoretical terms. Nearly all the difficulties of classical philosophy of science can be put down to this, and its associated bit of nonsense, that the sense of theoretical terms is a product of the sense of the observation terms to which they are, *à la* Carnap, reducible, plus their logical rôle as calculating devices. In short, plausibility control is exhausted, in this way of thinking, by the kind of observations already made, and the principle of non-contradiction. But in the new way of thinking, we can ask, 'How plausible is the model as a hypothetical mechanism?' even before the question of the reality of that hypothetical mechanism is raised. The plausibility question cannot be answered by any routine enquiry. Plausibility for a model is determined partly by the slowly changing general assumptions of the scientific community as to what the world is really like, partly by the way the model fits in to the particular circumstances for which it was created.

Protomorphs

My third and final family of models I shall call *protomorphs*. Two genera of protomorphs are known to me to date, but there may be others, in use in obscure corners of the scientific world. Members of these genera are related to homoeomorphs and paramorphs roughly as lampreys are to fish.

An example of the genus *logical icon*, of which the Markov diagram is a species, comes from Treismann's famous paper* on auditory unmasking. A diagram which, initially merely represented schematically the relations of various alternative auditory sensations, came to be thought of, in fact by what amounted to a misunderstanding, as a schematic diagram of broad neural pathways, and its junctions, representing the logical relation of disjunction, as switches. Neurophysiological evidence for the existence of the pathways and the switches is now accumulating. A logical icon is a diagram, concretely representing logical relations, but in such a fashion that it can be reinterpreted, or misunderstood, as a teleiomorph.

The other kind of protomorph, found fairly widely, is what I shall call 'geometrizations'. They are exemplified by Oresme's Latitude of Forms, in which the intensity of a quality is represented by a vertical line, proportional in length to the estimated intensity of the quality, located on a base line which represents the place where that particular intensity is found. Galileo's method of representing the law of descent, which he took from the Merton physicists, is a well-known example. I shall illustrate the geometrization as protomorph, and its development into a proper model from the studies of the octopus, made by Sutherland and Mackintosh.†

* M. Treismann, 'Auditory Unmasking', *J. Acoustic Soc. Am.*, 1963.
† N. S. Sutherland, 'Visual Discrimination and Orientation of Shape by the Octopus', *Nature*, **179**, 11–13.

Sutherland and Mackintosh found that an octopus could distinguish between rectangles presented to it horizontally and rectangles presented obliquely. A triangle with apex up was treated by the octopus just like a diamond with apex up, while both were distinguished from squares and rectangles presented with two sides vertical and two horizontal.

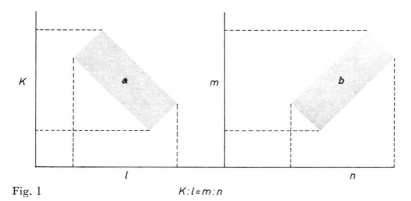

Fig. 1 $K:l=m:n$

The investigators proposed a geometrization as a protomorph for the phenomena. The octopus cannot distinguish one oblique rectangle lying along a left to right diagonal from one lying along the opposite diagonal (See Fig. 1). The shapes a and b are treated alike. But the octopus distinguishes these from rectangles, and reacts differently to a rectangle

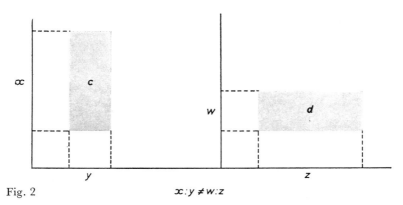

Fig. 2 $x:y \neq w:z$

with its long side in the horizontal from one with its short side in the horizontal (See Fig. 2). The shapes c and d are treated differently. The protomorph is presented in Figures 1 and 2. The mechanism of discrimination is geometrized as a first step, by supposing the shapes projected on to the horizontal and vertical axes. Oblique rectangles, whatever diagonal they are laid along, have projections of ratio $k : l = m : n$ while the rectangles of Figure 2 have projections of ratios $x : y$ and $w : z$, clearly different. This protomorph will explain the failure to discriminate diamonds from triangles, and the capacity to distinguish both from squares.

The next step is to construct (in the imagination first) a paramorph in terms of electrical circuitry, as a model of whatever neural mechanism might exist in the octopus, capable of performing the projections envisaged in the protomorph. The first of several such paramorphs, which the investigators invented, involved a square matrix of photo-sensitive cells, with a master circuit which collected and compared the total number of cells firing by rows, and by columns. In this example we have a two-step development of the paramorph through the prior invention of a geometrization. Once the paramorph is constructed, if it has existential plausibility, it is open to the investigators to suppose that it is the hypothetical mechanism, and then look within the octopus for something like this mechanism. In this particular case, the investigators did not find a mechanism like that supposed in the paramorph. They found something quite different, which banished both the paramorph and the protomorph from the explanation of the phenomena. The mechanism hypothesized simply did not exist.

I have discussed, so far, how models are related to their sources, that is how models are created. But the function of a model is to form the basis of a theory, and a theory is invented to explain some phenomena. It has a subject. I turn now to the relation between model and subject, in greater detail. An iconic model stands in for the real mechanism of nature, of which we happen to be ignorant. That is its function and that is its importance. A model, considered from the point of view of its subject phenomena, beginning merely as an analogue of the real mechanism which is supposed unknown, can become, by a switch in the attitude of its users to its status, none other than a hypothetical mechanism, and, as such, models bear hypothetically to their subject matter, just exactly the relations that real mechanisms really bear to the phenomena for which they are responsible. These can be of two kinds.

1. Some state of the mechanism is causally responsible for some state of things which can be observed. For instance a particular set up of springs, gears, and so on inside a clock is responsible for the way the hands are oriented on the face, their position from time to time and their relative motions. Or a particular arrangement of molecules in the genetic material is responsible for a particular shape or colour or chemical composition of a structure of an adult organism. Or a particular energy transaction in an atom is responsible for the colour of the light

emitted. Or a complex of repressed experience is responsible for a certain pattern of hysterical behaviour. Model states are linked to phenomena by hypothetical generative relations. In discussing the relations between the *sentences* used to describe the model and the sentences used to describe the phenomenal effects of these hypothetical states I shall talk, for this case, of a *causal transform*. There is, in fact, a partial sentential modelling between sentences describing the model and sentences describing the phenomena, but of course the mere existence of a formal modelling of the sets of sentences is neither here nor there for scientific purposes. The model, as we have seen, has to satisfy the requirements of plausibility control before it has scientific value.

2. Considered purely formally, causal transforms may not, at first sight, seem to differ from any other kind of transform, and to be able to be exhaustively expressed as simple conditional statements. But from a scientific point of view, the distinction is crucial. Where there is a causal relation the cause and the effect are independent existents, and the mechanism by which they are related can fruitfully be asked for, opening up a new dimension of explanation. In the other kind of transform, which I shall call a *modal transform*, the relation between the states of the model, considered as a hypothetical mechanism, and the phenomena, is not such as would give them independent existence. The state of the model is existentially identical with the phenomena. For instance, from an existential point of view, reflecting light of a certain wave-length and the being coloured a certain hue of a surface are identical states of the world. (This needs refinement, but the necessary conceptual machinery to handle this sort of case properly will be developed in a later chapter. The point is here illustrative only.) Nevertheless, the wave model for light is a considerable scientific advance. It does explain a great many otherwise inexplicable phenomena.

A brief sketch of some of the interesting philosophical issues raised by modal transforms will be in point here. They are connected both with the role of theory in taxonomy and the metaphysics of taxonomic theory, and with the problems raised by decisions to carry out wholesale recategorization of things in the world. Consider the discovery that crystals of common salt are cubical lattices of sodium and chloride ions. Here surely we have a modal transform between the shape of the crystal and the structure of the lattice. But to describe a substance as crystals of common salt invites one, in domestic situations, to classify the substance along with peppercorns, bay leaves and parsley, and not with other things. To describe it as sodium chloride invites one to classify it along with the electrovalent compounds, a class which quite excludes peppercorns and bay leaves. In this example the problem as to where the weight lies in taxonomy is pretty easily solved simply by referring to the context and purpose of classifying. And to the question 'What is common salt *really*?' The answer 'Sodium chloride' seems right. But if we apply the notion of modal transform, say in the province of the science of psychology, the taxonomic weight cannot so easily or so plausibly be placed upon the technically most advanced concepts. In many cases, the identification of a certain state of the nervous system as being that in which the organism hears something, say, depends upon that organism identifying its state as 'hearing something', and this applies as much to animals making auditory discriminations as to men listening to symphonies. So that the taxonomy, or general system for classifying states of the nervous system for taxonomic purposes seems to depend upon a mode of description which is the least technical of the modes involved. These questions and the associated ones about recategorization will occupy us in Chapter 8.

The total theory can be thought of metaphorically as a statement-picture complex. The iconic model is a 'picture' of a possible mechanism for producing the phenomena. It has already been explained in detail, above, how we come to know what the iconic model is like and how it works, deriving our knowledge from the parent model or models: the source of the model. By considering the sentences of the theory, abstracted from the total external expression of the statement-picture complex, that is by ignoring the model, picture or diagram of the state of the world, the following logical relations emerge:

(*a*) If the relation between the description of the model and the description of the phenomena looked at from the point of view of the complete theory, is a causal transform, then the sentences which express this relation will appear as conditionals, asserting that if the model-state obtains then the phenomenal effect of that state will come to be.

(*b*) If the relation between the description of the model and the description of the phenomena is a modal transform then the sentences expressing this relation will be bi-conditionals, asserting that if and only if the model state obtains will the phenomenal aspect of that state come to be.

In the case of the modal transform, there is no separate question as to the existence of the hypothetical mechanism and its states which the model represents, for they are the same states of the world looked at from a different point of view. But a causal transform links sentences describing quite different states of the world. Since the iconic model is introduced just because we do not know the mechanism of nature at that point, it can come to be looked upon as a hypothetical mechanism. Then the question arises, does this mechanism exist? Or one something like it? Or is what is really responsible for the phenomena something so different from the mechanism represented by the model as not to allow any modification of the original model as a version of reality? For an example consider the distinction between our being permitted to go on saying that chemical atoms exist, while continually, and drastically, modifying our beliefs about their natures, and our not being permitted to go on saying caloric exists, now that heat transfer is understood as a transfer of energy. William Odling gave a lecture towards the end of the Nineteenth Century, in which he made out, partly as a *jeu d'esprit*, that phlogiston did exist, only people had come to call it 'energy'. We shall look later into the conditions under which we make this distinction in judgement.

A theory then should prompt the consideration of four kinds of hypotheses, which are separated off by both logical and epistemological distinctions.

1. *Existential Hypotheses*; such as 'There are molecules', 'There are no pores in the septum of the heart', 'There is a planet between Mercury and the sun', 'There are complexes'. Each one of these hypotheses prompts its appropriate question, like 'Are there molecules?' the attempt to answer which generates a particular kind of experimental research. I shall emphasize later that not all questions in the form of existential hypotheses are properly answered by an experimental technique, since some questions of this form are rather invitations to change our categories than to look for new things, or properties, or processes amongst those already acknowledged to exist.

2. *Descriptions of the model or hypothetical mechanism*; such as 'Molecules are moving', 'Complexes are formed by the association of traces of repressed experiences'. The empirical pursuit of answers to the corresponding questions cannot be undertaken until questions as to the existence of the *entities* in question have been settled. Since, as will emerge, settling existence questions cannot be done in isolation from the satisfying by some entity of a certain description, some hypotheses in this class are decided along with the existential questions.

3. *Causal Hypotheses*; in such questions as, 'Is gas pressure caused by the impact of molecules?', 'Is hysteria caused by repressed experiences?' the power of a hypothetical mechanism to produce the phenomena is queried. The attempt to answer such questions gives rise to some interesting questions to which we shall return again and again, from different points of view.

4. Finally, in questions like, 'Is gas temperature really only another way of looking at mean kinetic energy of the molecules?', 'Is a slip of the tongue really an admission of guilt?' we are invited to consider *modal transforms*. Very complex issues are raised in their epistemology. Since they have the form of biconditionals, descriptions of a situation in terms

C

drawn from the vocabulary of observation or from the vocabulary used to describe the model, must be true together. Yet the model seems to introduce a new range of things and happenings, which, at least in the first instance, we cannot study independently. If all the relations between hypothetical mechanisms and phenomena were modal transforms we should be as sure of the truth of theoretical statements as of those describing the phenomena. But then explanation by adversion to hypothetical causal mechanisms would be no more than a redescription of the phenomena. We shall see that any paradoxical air which might derive from the presence of both modal transforms and causal transforms in the same theory, as in the kinetic theory of gases for instance, can be resolved in those cases where the elementary components of the hypothetical mechanism are also the elementary parts of the entity whose behaviour is being explained. Gas molecules are the elementary parts of gases as well as the elementary components of the hypothetical mechanism responsible for the holistic behaviour of gas samples.

Thus, when we turn to the sentential aspect of a 'statement-picture complex', that is, the sentences considered in abstraction from the theory, we find that they exhibit four distinct kinds, differing in logical form, appearing in:

1. Existential statements which assert, deny, or temporize about the existence of a real mechanism, identical with or like the iconic model which is used to devise a hypothetical mechanism.

2. General statements describing the attributes and modes of behaviour of the model as hypothetical mechanism.

3. Conditional statements asserting the causal relations between the states of the hypothetical mechanism and phenomena.

4. Biconditional statements asserting modal relations between the states of the hypothetical mechanism and phenomena.

In the following chapters I turn to an examination of these kinds of statements, to a discussion of their various conditions of acceptability, falsification and confirmation, and so on, and to an analysis of their internal structure. It will be continually forced upon us that no adequate account of the rôle and behaviour of the sentences used in making such statements can be obtained if we forget that they exist, *in vivo*, in association with ideas of hypothetical causal mechanisms or models of them. We must not draw sweeping conclusions about the way they are related and behave while considering them as if they were isolated individuals. To understand them, and to resolve the traditional difficulties concerning them, they must always be considered as completed in sense only when embedded in a statement-picture complex, i.e., in a theory.

The Analysis of Some Theories

The model of an unknown mechanism, whose invention is the core of the creation of a theory, is paramorphic with respect to that unknown mechanism, since in the absence of knowledge of that mechanism we can hope only for some kind of parallelism of behaviour. But the central model is homoeomorphic with respect to each of the sources upon which it is modelled. This must be shown in detail in some examples.

The phenomena which Darwin attempted to explain can be called generally 'natural variations'. This includes, but is not exhausted by, both the differences between the character of populations now living and plant populations of the past, as revealed by geologically preserved remains: and the differences contemporarily observed between similar populations inhabiting different environments. From this static picture, Darwin drew the dynamic conclusion that there was variation in the character of populations of plants and animals in nature. The mechanisms by which this happened were unknown. So in the first five chapters of the *Origin of Species*, Darwin built up a picture of a hypothetical mechanism which would explain the changes he believed occurred. In short Darwin invented an iconic model, paramorphic to the real but unknown processes by which biological change occurred. Quickly moving from model to hypothetical mechanism, he called it 'the process of natural selection'. Its sources were the known process of domestic selection by which domestic variation was brought about, and the bellicose metaphor of the struggle for existence.

Natural selection is a homoeomorph of domestic selection. It is a selection operation but without the deliberate acts of the breeder. In short, it is the kind of homoeomorph I have called a teleiomorphic abstraction. But a selection operation without some agent which diminishes the chances of reproduction of those particular plants and animals whose structure, or constitution, or patterns of behaviour is unsuited to their environment, is empty.

The deficiency was made good by Darwin by developing a second source of concepts for the natural selection model. As a paramorph, it is multiply connected. The new source was the theory of Malthus on the effect of population pressure on the division of resources, and the law that in human societies a geometrical population increase is accompanied by only an arithmetical increase in available material resources for that population, In general, therefore, there must be competition, which supplies the source for the agent conception in the Darwinian theory. And, of course, the 'struggle' conception is a teleiomorphic abstraction of Malthus' picture of human life, since, for example, the differential growth of two species of grass on a particular soil is only competitive in an abstract and metaphorical sense. These two sources combine then in the creation of Darwin's model of the actual causal mechanism responsible for organic change. Schematically the analysis of the theory can be laid out as follows:

Phenomena	Transforms	Model of Unknown Mechanism	Sources
Natural Variation	The cause of natural variation is natural selection.	Natural selection by the struggle for existence.	(1) The causal structure of domestic variation as selection.
		M	(2) Malthus theory.

M is a paramorphic model (a partial analogue multiply connected) with respect to the unknown mechanism of change.

M is a homoeomorphic model (a teleiomorphic abstraction) with respect to domestic variation and selection, and with respect to Malthus' theory.

Notice that Darwin is postulating a novel kind of process in nature, never before thought of, and up until very recently, never observed. Natural selection was as much a theoretical entity and as unobservable as the electron, for while the latter is too small to be observed, if indeed it is thing-like at all, natural selection is too slow.

Another example is the theory devised by social psychologists in which a model of rewards, costs and investments is offered as an account of various alleged social processes, such as the conferment of, or recognition of, status among people. The phenomena are the ranking of people by others as to their status, and for the purposes of the example, we will not pursue the question as to what that rather vague concept is supposed to cover. The model is constructed for the unknown mechanism which determines (if there be such) the according of status. Its source is the simple economics of costs, rewards and investments. For example, it is said by Secord and Backman (*Social Psychology*, McGraw Hill, New York, 1964, p. 297) that '*distributive justice* is obtained when the outcomes or profits of each person – his rewards minus his costs – are directly proportional to his investments'. This model is then used with respect to the phenomenon of two people according each other equal status according to the 'formula'

$$\frac{\text{'My investments}}{\text{His investments}} = \frac{\text{my rewards minus costs}}{\text{his rewards minus costs'}}$$

Here we have a case of a pure modal transform. There is no existential question raisable as to the realities of any quasi-economic process lying at the back of the according of status. The source of the model is economics. The model then is paramorphic with respect to its subject, and homoeomorphic with respect to its source. But in this case, it is not a teleiomorphic abstraction, but an idealization, since though investments, rewards and costs are genuine economic concepts, they are being used here in an idealized way. This example will serve to introduce an important new concept into the discussion. There can be no question of the model employed in this theory becoming a candidate for reality, and so a hypothetical mechanism. Instead, since *all* its connections with phenomena are mediated by modal transforms, it is a metaphor. Or to put the matter slightly differently, a metaphor is a use of a set of concepts derived as in the standard process of model-building, but which have no causal transforms with phenomena. The function of such a model cannot then be to generate existential hypotheses, so it does not form part of an explanation in the usual sense. Its function, as is clearly evident in the example I have quoted, is to illuminate the facts, to throw them into a new light, to make them more readily memorable. Its function, in short, is *literary*. This is not to denigrate the theory, but to place it in its exact logical rôle within the structure of science. For any theory with only modal transforms, we can look only for a function as metaphor and not anything else. In more formal sciences, other heuristic advantages may derive from such theories, say ease of mathematical handling.

We have noticed that the statements describing each of the components of theory bear sentential modelling relations with respect to each other, and that the transforms which give structure to the whole are each an open set of sentences. What transforms are admitted depends partly on the degree of likeness and unlikeness between homoeomorph and its source-subject which relates the model in the theory to its parent, or parents, and partly on the plausibility of the model as a putative hypothetical mechanism, that is partly on the degree to which we think it likely that the unknown mechanism for which it is an analogue is like it. *If all the transforms are modal*, then the sentences describing it become metaphorical redescriptions of the phenomena, and we commit ourselves to an effective total *unlikeness* between mechanism and model. The logical structure of the sentences describing the various components of a theory like that of Secord and Backman is now very complex, since it contains one inductive (analogical) relation between sets of sentences, namely that between the parent and the model, and one metaphorical, which comes not under logical scrutiny, but under quasi-literary scrutiny.

In theories which incorporate causal transforms, and in which the hypothetical mechanism is a possible and plausible causal mechanism, there are two inductive steps, whose validity depends upon our judgements of the likeness and unlikeness of the model and its parent, and our judgement of the plausibility of the model as a hypothetical mechanism. In neither case can the whole structure of the sentences in the expression of the theory be collapsed into a deductive system. At best it can consist of three deductive systems: that of the laws of phenomena, that of the 'laws' of the model, and that of the laws of the phenomena upon which the model is based. The sentential modelling relation between the three systems will be through a set of rules of transformation (though that itself is a modification of the real rôle of transform statements as causal hypotheses and modal transforms) which cannot be fully completed.

Models of Models

The account of theory given so far has assumed that the development of a model from a parent derives from laws of phenomena, known behaviour of known things. But other origins for models are possible. Sometimes the source of a model may be itself a model. Drude's theory of electrical and heat conduction is an example of this. In that theory the behaviour of metals as conductors of heat or electricity is explained by the help of a model which is itself based upon the molecular model of gases. A piece of metal is imagined to be a container in which electrons are in random motion in a manner like the molecules, in a container, which simulate the behaviour of gases. Conceiving of electrons in this way, and devising laws for their behaviour in such circumstances by analogy from the laws of gases, the model is found to simulate the actual behaviour of metals with respect to heat and electrical conductivity. In this case, the electron model is paramorphic with respect to the conductivity behaviour, and paramorphic with respect to the gas molecule parent, since there is no sense in which the electrons trapped in a metal are a gas. Considered with respect to the gas laws and the gas molecule, the electron model differs in source and subject. However, the gas molecules are supposed to be small particles in motion, and considered with respect to particles in motion (their source and their subject is the same) they are modelled on particles and models of particles, supposing as we now do, that a gas does consist of a swarm of particles. As a model for the electron gas, the swarm of molecules is paramorphic, as a model of a swarm of particles, it is homoeomorphic.

SUMMARY AND BIBLIOGRAPHY

Most writers on the subject of theories have held that models did play some rôle in the generation of the theory and in the giving of meaning to some of the nouns, adjectives and verbs that appear in the exposition of the theory. Controversy has turned upon two main points. Some have maintained that models have a heuristic (or pragmatic) value only, and that they can eventually be dispensed with.

N. Ursus (*see* 3), Ch. 1.

65. P. Duhem, *The Aim and Structure of Physical Theory*, Riviére, Paris, 1914, translated by P. P. Wiener, Princeton University Press, 1954, pp. 55–104.

C. G. Hempel (*see* 21), Ch. 8.

66. R. Carnap, *Introduction to Semantics*, Harvard University Press, Cambridge, Mass., 1942, §33.

67. R. Carnap, 'The Methodological Character of Theoretical Concepts', *Minnesota Studies in the Philosophy of Science*, University of Minnesota Press, Minneapolis, 1956, Vol. I, pp. 38–76.

R. B. Braithwaite (*see* 36), Chs. 3 and 4.

68. R. Ackerman, 'Confirmatory Models of Theories', *BJPS*, **16**, 312–26.

Others have maintained that models are an essential and ineliminable feature of theories, though the term 'model' has not always been used by the proponents of this view, cf. Whewell's use of 'idea', and discussions of analogy.

For an early defence of the use of iconic models as a method of discovery cf. 69 Archimedes, *The Method*, in *The Works of Archimedes* edited by T. L. Heath, pp. 12–14 (following p. 326) Dover Edition, New York, circa 1955.

70. W. Whewell, *The Philosophy of the Inductive Sciences*, New Edition, Johnson Reprint Corporation, New York and London, 1967, Bk. I.

71. W. Whewell, *Novum Organon Renovatum*, Parker, London, 1858, Bk. I, Bk. II, Chs. I, II, III, IV.

72. Mary Hesse, *Models and Analogies in Science*, Sheed and Ward, London, 1963, Part I, pp. 8–62.

R. Harré (*see* 27), Ch. 3.

73. P. Achinstein, *Concepts of Science*, Johns Hopkins Press, Baltimore, 1968, Chs. 7 and 8.

74. M. B. Hesse, 'Models in Physics', *BJPS*, **4**, 198–214.

75. E. H. Hutten, 'The Role of Models in Physics', *BJPS*, **4**, 284–301.

76. M. Black, *Models and Metaphors*, Cornell University Press, 1962, Ch. 13.

Generally speaking, those philosophers who have argued for the eliminability of models have held that the terms in a theory are all, either wholly or partially, capable of being given sense in terms of the expressions used to describe the phenomena the theory is designed to explain. For the partial interpretation view see

R. Carnap (*see* 67).

R. Braithwaite (*see* 36).

and for a critique of this view see

77. G. Schlesinger, 'The Terms and Sentences of Empirical Science', *Mind*, **73**, 394–405.

78. J. Turner, 'Maxwell on the Method of Physical Analogy', *BJPS*, **6**, 226–38.

79. P. Achinstein, 'Theoretical Terms and Partial Interpretation', *BJPS*, **14**, 89–104.

80. M. Spector, 'Models and Theories', *BJPS*, **16**, 121–42.

Further discussion of models:

    (i) of teleiomorphs:

81. M. Weber, *The Methodology of the Social Sciences*', translated by E. A. Shils and H. A. Finch, Illinois, 1949.

    (ii) of iconic paramorphs:

82. P. Achinstein, 'Theoretical Models', *BJPS*, **16**, 102–20.

    (iii) of sentential models:

83. H. Freudenthal, *The Concept and the Rôle of the Model in Mathematical, Natural and Social Sciences*, Reidel Dordrecht, 1961.
84. J. W. Swanson, 'On Models', *BJPS*, **17**, 297–311.

    (iv) of the comparative structure of a theory with and without an iconic model:

N. R. Campbell (*see* 26), Ch. VI.
85. B. Ellis, 'A Comparison of Process and Non-Process Theories in the Physical Sciences', *BJPS*, **8**, 45–56.

Metaphors have usually been discussed in close relation to models.

86. M. Black, 'Metaphor', *Proc. Aris. Soc.*, **55**, 273–94.
    M. Black (*see* 76), Ch. 3.

For a more general theory of metaphor, see

87. P. E. Wheelwright, *The Burning Fountain*, Indiana University Press, Bloomington, Ind., 1954.

And for metaphor in a specifically scientific context, see

R. Harré (*see* 27), Ch. 10.
88. Mary A. McCloskey, *Mind*, **73**, 215–33.
and also see the discussion of whether there is an independent observation language in the notes to Ch. 6.

3. Existential Hypotheses

The Argument

1. *Subject Matter of Existential Statements*

   (*a*) Individual things, processes and instances of qualities, bits of substances, powers;

   (*b*) classes of things and species, qualities in general, substances;

   (*c*) Relations, numbers, etc.; these are dependent in being, but any proof of their existence by proving the existence of the relata does not seem quite adequate.

2. *Verification of Existential Statements*

   Usually by the production of an instance, capable of being the subject of demonstrative reference. Adequate as confirmation since a particular existential statement implies a general existential statement.

3. *The Use of 'this'*

   It is proper to both ostensive and non-ostenive demonstrative reference and defines a demonstrative criterion.

   Reference as a proof of existence is incomplete without identification of kind and this defines a recognitive criterion.

4. *Demonstrative Criterion*

   (A) Ostensive
   - (i) The kind of Demonstrative act determines the kind of entity proved to exist;
   - (ii) Not all Demonstrative acts are ostensive;
   - (iii) Ostensive acts have priority.

   (B) Non-ostensive
   - (i) Introspection is considered as a mode of 'attending to';
   - (ii) The justification of attending to a cause through attention to its effect requires the Principle of Proximity of some cause to an effect.
     (The principle is not true for efficient causes.)
   - (iii) The existence of fictions: proof involves successful attending to of a *word*, and an understanding of its rôle.

   Let us call proof of a cause by its effect an 'indirect' demonstration. Other sorts of existential proof (including attention to mental states) are 'direct'. By making spatio-temporal location an ultimate criterion, we make spatio-temporal locatables ultimate existentially. Could anything but spatio-temporal location be ultimate demonstratively? Must spatio-temporal locatables be things, i.e. be characterized by qualitatively differentiated volume and endurance?

5. *Central Recognitive Criterion*

   No single recognitive criterion seems specifiable.

   C 2

Two important metaphysical distinctions can be made which yield four kinds of existents of equal status.

(i) Enduring things versus non-enduring things.

Note that this distinction is not identical with that between direct and indirect demonstration.

(ii) Independent versus dependent.

Endurers constitute, in general, the independent existents and non-endurers the dependent. But not necessarily, it seems.

6. *The Scientific Ontology*

Depends heavily on the indirect demonstration of causes and causal mechanisms, constituted by endurers, and usually the endurers are the parts of the directly demonstrable entities.

7. *Falsification of Existential Hypotheses*

	Particular	Kinds and Substances
(a) Failure of the demonstrative criterion to connect, e.g.	Vulcan	aether
(b) Failure of the recognitive criterion to be met	pores in the septum	phlogiston

(b) can lead to the disappearance of an entity from existence by recategorization, e.g. caloric into energy, substance into state.

8. *Existential Hypotheses distinguished by their Origin*

(a) a singly connected paramorph raises no problem because existential criteria for its entities are identical with the existential criteria for its source, and proofs of existence are connected with sense-extending instruments;

(b) semi-connected: whether they raise problems depends upon degree of connection.

(i) they may lead to the introduction of a new category, and hence to indirect demonstration;

(ii) therefore there may be no demand for visualizability;

(iii) *but* there may be a demand for trans-categorial causality.

Resolved by either (i) wholesale recategorization – leading to a return to direct demonstration; or (ii) a trans-categorial action can be taken as ultimate; or (iii) epistemological dualism and ontological monism may be adopted.

(c) multiply connected paramorphs *must* lead to existential novelties.

(i) the demonstrative criterion is usually indirect;

(ii) the recognitive criterion depends upon the consistency of the joint assertion of each semi-set of predicates.

The same position, i.e. that existential proofs require both demonstration and identification can be reached through a metaphysical dissection of class intensions.

The dissection is completed by the appearance of an intension related to a mode of demonstration.

In the last chapter the conclusion was reached that the sentential part of a theory consists of statements among which are four kinds of hypotheses, in the form of existential, general, conditional and biconditional statements. To pursue the study of the sentential part of theories it will now be necessary to examine these kinds of statements as they appear and function in science. Again, in keeping with the spirit of this enterprise, the major methodological intuitions of scientists about existence, about laws and about causes will be preserved and logic will be enlarged, if necessary, to accommodate them. In any case of conflict the intuition will be assumed to be proper and the principles delivered by traditional logic to be mistaken, and in need of revision or supplementation. In this chapter existential statements will be considered.

The Subject Matter of Existential Statements

This preliminary sketch of the subject matter of existential statements will be refined later as deeper considerations emerge. We commonly regard some existential questions as having to do with the existence of individual things and processes, and individual instances of properties and qualities, like a patch of such and such a colour, or a sound of such and such a pitch. All of these could perhaps be treated together, since coloured patches and sounds could be counted as rather tenuous kinds of things, since they are capable of being referred to by ostensive modes of reference, and they do have certain of the characteristics of things. Patches of colour must be spatially extended in two dimensions. Whether or not they must endure for a time, however short, seems to me uncertain. I do not think it would be self-contradictory to say that they had the temporal characteristics of instantaneous events. On the other hand, sounds, though strictly speaking they are not spatially locatable, must have a certain duration, and though we do not have to say exactly where they are, sounds are in some volume and have spatial limits. Even seemingly fleeting patches of light and colour, such as scintillations, probably need to have some duration, though it may be short. We commonly regard the question of existence as also being capable of being asked with propriety of classes of individual things, and of properties in general. 'Are there Bhuddists in Saudi Arabia?' 'Is there a colour bluer than violet?' These questions are answered affirmatively on the condition that an instance can be produced to which demonstrative reference can be made, that is, our attention can be directed to an entity identifiable as of that kind. But to say that there are Bhuddists in Saudi Arabia and that there is a colour bluer than violet is to say something different from saying that the instance produced exists, even if there is only one Saudi Arabian Bhuddist, and only one sample of the colour bluer than violet. It speaks of the existence of a class, and of a quality in general. Then there are questions about the existence of substances, and of species, which are also answered by the production of individual instances as samples or members. The fact that the relation of an instance to its class is different from the relation of a sample to a substance is not of importance in these cases. If the whole class whose existence is under consideration has but one member, or all the stuff there is, is in the bottle on the shelf that contains the sample one was shown, it is still important to differentiate the class existence question from the individual existence question. This distinction will be elaborated later.

Some entities have been thought to raise puzzling problems, when their existence is considered. Examples are relations such as 'to the left of' and numbers. Certainly an instance of a lion being to the left of a tiger can be found. But while the lion and the tiger can be pointed to and can be objects of such a common mode of demonstrative reference, the relation between them cannot, it seems, be the object of an independent act of demonstration. Is it enough to draw attention to the lion and the tiger to prove that the relation 'to the left' exists? No, because they must not only be noticed individually, but seen as in that relation. Certainly six bears can be exhibited together and the group demonstratively indicated with the help of a sweeping gesture, but does this show that the number six exists? Clearly not. Existence questions for such entities seem to lead to considerations of a different kind. Sometimes it seems plausible to deal with relations by simple grammatical transformation. For instance, an answer to 'Does motion exist?' can be got by transforming the question into 'Does anything move?' The answer 'Something moves' might be taken as equivalent to, but more perspicacious than 'Motion exists'. Hypothetical mechanisms involve relations, but though they have many parts and the parts can be counted, numbers are not among the hypothetical entities of natural science. Only the problem of the existence of relations need concern us, and that I shall treat mostly by the move of grammatical transformation.

The Verification of Existential Statements

Both general and particular existential statements are verified in the same way: by the production of an instance capable of being the object of an accepted mode of demonstrative reference by which attention is directed to it, and meeting the appropriate description. The verbal part of the verification procedure goes something like this:

'This is a positron'

The reason why the instance verifies both the particular and the general existential statement is that the truth of a particular existential statement entails the truth of the general statement. The significance of this fact for understanding the principles of theory testing is enormous. Existential facts count powerfully for a theory because only for such facts does truth of a particular statement entail the truth of a general statement, of an existential hypothesis.

The use of 'this' in the particular existential statement is incomplete if considered as a purely verbal act. It must be supplemented by the gesture of pointing, or by some other attention drawing device, that is by a demonstrative act, or act of demonstration.

In this simple grammatico-semantic fact lies the occasion for a great deal of metaphysics. Anything of which 'this' can be said, and a demonstrative act, of whatever kind, provided it is a recognized kind, a proper mode of demonstration can be performed, shall be said to have satisfied a demonstrative criterion. The satisfaction of a demonstrative criterion by an entity I shall call the demonstration of the entity. But pointing to any old thing and saying 'this' will not prove that a positron, or Atlantis, exists. The object signified must be of the right kind, or be the individual intended. So there must be recognitive criteria too, which are satisfied by the individual singled out by the act of demonstrative reference. The predicate of an existential statement must be capable of being understood in terms of the predicates appropriate to the thing or things said to exist: that is, satisfaction of them must lead to an identification of the thing demonstratively referred to as one of a certain kind. If someone says 'This is gold', then whatever it is that he is demonstratively referring to must be yellow, malleable, ductile and perhaps of specific gravity 19.6, and so on, until the sample has been identified.

Because the recognitive criterion incorporates the satisfaction of the criteria for the instance being an entity of a certain kind, the satisfaction of it by something capable of being indicated by demonstrative reference proves not only that the instance exists but that the kind does too. This is why the truth of a particular existential statement entails the truth of a general existential hypothesis. The satisfaction of the recognitive criterion by an entity I shall call the identification of the entity.

The most obvious kind of demonstrative act, ostension or pointing, directs our attention to a spatio-temporal location. The most obvious kind of recognitive criterion is met by the entity to which our attention has been drawn having certain publicly observable characteristics or properties, which are the appropriate marks of an entity of the right kind. It is quite vital to getting the scientific process of discovery right to avoid the temptation of supposing that these most obvious cases are the only kind of case or even the most typical. One must not assume that they are the basis of the only kind of existential proving that there is. Not all demonstrative reference is by spatio-temporal location, nor is all recognition by publicly observable characteristics.

(A) *Direct Demonstrative Reference.* We are liable to be misled by the obvious cases of direct demonstrative proof of existence where the entity demonstratively identified is that about which we have formed an existential hypothesis. There are two important kinds of direct demonstrative reference which have not always been accorded equality of status.

(1) *Simple Ostension,* Simple ostension, pointing with one's finger or some physical indicator, Jack Horner's criterion for the existence of plums, is frequently a way of satisfying the demonstrative criterion. Another version of the simple demonstrative act is looking in the best place. This could be called Mother Hubbard's criterion for the non-existence of bones and it is often the demonstrative act that proves something does not exist. Simple ostension can be to anything at all and can change from moment to moment. The very simplicity of that mode of demonstration can mislead us into supposing that the demonstrative criterion has no descriptive content. It could lead us to think that Russell's 'logically proper name' analysis which invites us to treat demonstrative pronouns as having no connotation, exhausted the meaning of 'this'. One might suppose, wrongly, that the use of 'this', together with a demonstrative act, merely indicated an entity, and, carried no implications as to the nature of whatever attention was thus directed to. It is true that we can point to many entities other than things and their observable properties and behaviour, even though the simple ostensive demonstrative act restricts the existent to items of public perception. These exhibit an enormous variety. There are some interesting relationships between the referents of 'this' plus ostension, since by this means a great variety of kinds of items can be indicated, such as rainbows, scintillations, musical notes, bird song, the smell in the cupboard as well as the more commonly discussed cannon balls and bric-a-brac. I shall return to a detailed discussion of some of these relationships later in the book. However, despite the variety of kinds of entities to which a finger can point, the fact that we must supplement 'this' by a demonstrative act, which may be one of many kinds of attention drawing demonstrative reference each to an appropriate group of classes of things, is not restricted to simple ostension, and hence is not without some implications as to the nature of the entity which is indicated in some way.

For instance, when indicating the blue colour of a litmus paper I may say, quite properly it seems to me, 'This shows the existence of an alkali in the soil.' Here the simple ostensive act is not to the existent, but to one of its effects. I draw attention to the cause by pointing to the effect, an important kind of case for natural science. In this case, and in the more straightforward cases in which the item can be indicated directly, the total verbal-ostensive performance directs our attention to an occupant of a spacetime location, an occupant whose presence excludes the presence of anything else, in short a thing, observable in the one case, itself unobservable in the other. But there are ways of drawing attention to entities other than things, for instance to the taste of a particular honey, and of using this attention drawing to prove the existence of a taste. The mode of demonstrative reference and the metaphysical status of the entity in question are clearly intimately related. But the variety of entities capable of being the reference of acts of simple ostensive demonstrative reference shows that this relation is not a simple one to one relation, which would put one metaphysical category in relation to each mode of demonstrative reference. Unfortunately it is not so simple as that.

However, though simple ostension is not exhaustive of demonstrative acts it does have a certain priority. It derives its importance from the fact that it is both public and immune from critical revision. We shall see that science leans heavily on a tacit theory of dependent and independent being, formally not unlike the traditional theory of being, but differing radically from it in content. Some preliminary distinctions can be drawn here which will make this point clear. The existence of chemical atoms is demonstrated by pointing to a wide variety of phenomena, the existence of which is dependent upon there being chemical atoms. Thus, the order of dependence of knowledge of existence is from perceptibles to imperceptibles: from knowledge of the existence of chemical phenomena of various kinds to knowledge of the existence of atoms. But the order of dependence of existence is from imperceptibles to perceptibles. It is only because we believe that the existence of the simply ostensible phenomena depends upon the existence of chemical atoms and certain states among them that we can use knowledge of the existence of phenomena to establish knowledge of the existence of their causes. The epistemic priority of simple ostension is connected with another philosophical myth, the theory of ostensive definition. This myth will be dealt with in detail in a later chapter. For the moment I wish only to insist that in this context, at least, ostension does not define meanings. At worst it might lead one to suppose that the meaning of a term for an existent was that existent itself; at best it would lead to no certainty that each person understood the definition in the same way. On the contrary, simple ostension identifies an instance of whatever it is a term is used for, and the meaning of the term must already be known, for it to be possible to know when, by simple ostension, we *have* indicated an instance which has satisfied the recognitive criterion. Proofs of existence and givings of meanings are not the same thing.

(2) *Non-ostensive Demonstrative Criteria*. There are important cases, variously involved in the sciences, where indication of the existent, though direct, is not achieved by an act of simple pointing, in complementation of the speech act of uttering the demonstrative pronoun 'this', followed by a predicate.

One can be asked to attend to, and thus make indicating or demonstrative reference to, one's thoughts, sensations, feelings, emotions, and states of mind. In this way the existence of certain kinds, and instances, of these categories of entities are shown to exist. The method is commonplace in the teaching of art appreciation. The pupil is confronted with an object, or asked to listen to a certain sequence of sounds, or taste wine from a certain set of bottles, and then asked to pay attention both to the sensations and feelings he has, and to the thoughts and emotions he has. The teacher tries to encourage one kind of mental reaction and to inhibit others. So too, in becoming aware of an illusion, say that brought about by crossing the fingers and running them along an edge, one is asked to direct one's attention to a sensation: that of feeling two edges. In this case there is a direction of attention, but in the nature of the case, no ostension is possible. Here too there are characteristic mistakes about the use of this criterion. One of the errors of early Logical Positivism, a philosophy much under the spell of simple ostension, was to suppose that where simple ostensive acts could not be performed no other demonstrative act had any virtue as part of a proof of existence. Thus, in the early days of the movement, Logical Positivists were led to deny the existence of states of mind, thoughts and so on, as independent entities over and above pieces of behaviour because they could not be the subjects of simple acts of ostension, performed in public. The fact that an act of reference to thoughts, feelings and so on is possible, makes nonsense of simple materialism, but the fact that it is a different kind of act of demonstration from simple ostension leads to both philosophical and scientific problems raised by the fact that the organ of thought, the brain and central nervous system, is a simple ostensible object. A solution to these problems at least for scientific contexts will be advanced in a later chapter.

Demonstrative reference by simple direction of attention, in non-ostensive cases, particularly associated with an alleged realm of mental entities, and so of importance to the science of psychology, runs into an important difficulty. The act of demonstration, turning one's attention to one's state of mind, is not a public act of demonstration.

Compare the standing of a claim to have experienced and thus proved the existence of a new kind of feeling, and that of a public demonstration, and thus proof of the existence, of a new kind of colour, or a new species of animal. There can be no doubt whatever that for conclusive proofs of existence, which will bring an entity into a scientific community, one person's assertion that he has experienced, or had experience of, that entity is usually not enough. But must demonstrative criteria be satisfied in public? May it not be enough that the criteria are not satisfied unipersonally?

It is important to see that publicity – that is in principle, *all* person participation in the same act of demonstration – is not required. With people of proven public reliability we are prepared to accept one-person assurance of existence when the act of demonstration is private. However, when a new kind of entity is in question, rather than a new example of a well known kind, the repeatability of the experience by others is paramount, and it is this which has sometimes been confused with the unnecessary requirement of publicity. The necessity of *any*-person participation has been confused with the contingency of *all*-person participation. The enhancement of colours under the influence of L.S.D. is something that occurs with many persons who take it, and since some of those who did so before its dangers were known were persons in good standing in the scientific community, the effect must be presumed to exist even by those who have not taken the drug. This feature of non-ostensive demonstration is, as we shall see, of crucial importance to the possibility of psychology as a science, and in its relations to the physiological study of the brain, the central nervous system and organs of perception.

However, it is just as important to see that publicity of demonstration which ostension always achieves, does have a superior though not exclusive claim. In a certain sense, it has an absolute priority in proofs of existence. Public ostension of an entity is once and for all, as Professor Challenger so clearly understood, at least in principle. I add this last qualification because there is the phenomenon of public hallucination, if that is the right word, in such cases as the Angel of Mons, and the flying saucers. However, even in these cases it is not so much the public act of ostension that is fallible, but the alleged satisfaction of the recognitive criteria. No doubt some phenomenon existed which was the occasion of the existence claim, but in neither case was it subjected to the plausibility control exercised by current science on what the existent was. There was something over Mons, and something did plunge into the sea by the liner, and so existents were demonstrated – but – the plausibility control, operating through the recognitive criterion, would demand that we assert rather that there was a cloud over Mons, and that a meteorite had fallen in the neighbourhood of the ship. We shall see that this priority of the public act of ostension, operating in the theory of dependent and independent being, is deeply embedded in current science.

(B) *Indirect Demonstrative Reference*. Simple ostension of an effect is used to show the existence of its cause. Ordinary medical diagnosis proceeds by exactly this method. It is one of the contentions of this book that this is indeed, and has always been, one of the major methods of science, namely, the proof of the existence of causes by the ostension of their effects. Most causes operative in nature are not easily observed, if observable at all. Must the causal substructure of nature be dismissed as a myth? Of course not, and this is one of the methodological intuitions of practising scientists I shall do much to defend. Part of the seductiveness of the case for dismissing the inner workings of nature from serious consideration comes from conflating this case with another which is really not in the least like it.

If one is asked about the existence of Pursewarden's sister, one goes to the work of fiction in which she supposedly figures and finds in it an expression which, were the characters real people, would refer to her. For mythological creatures and persons, for fictions and the like, one answers questions as to their existence, such as it is, by proving the existence of an appropriate *expression* which would refer to them, did they exist. And, in a sense, all that the existence of such creatures amounts to, is that of their referring expressions. It has often been argued by deductivists, from Ursus to Hempel, that hypothetical causes are fictitious, and that only the theoretical terms which would refer to them, did they exist, are worthy subjects of study. This attitude we have seen in Chapter 1, fully developed in the theory of reduction sentences. It is part of my purpose to know that it cannot be the case that all such entities are fictitious.

The denial to indirect Demonstrative Reference of any real significance apart from the introduction of fictions has been a central feature of certain philosophies of science, at least since the sixteenth century. Fictions do not *exist*, and the proof of their 'existence', through the ostension of expressions which, did they exist, would refer to them, does not shake this important truth. The fictionalist account of science, so often associated with deductivism, invites us to treat methods of indirect showing of the existence of causes through their effects as special cases of the same general kind as the verbal introduction of pseudo-referring expressions associated with the entry of fictions, and indeed insists that we suppose the referents of neither kind of verbal reference to exist. This last 'exist' is, of course, linked with the direct demonstrative reference to some of the entities in question. Proponents of this view have sometimes denied virtue to any but ostensive references, so restricting their conception of existence that if we cannot point to an instance of a kind of entity, they do not exist. However, two points of distinction can be drawn between the cases, which are adequate to re-establish the intuition of the scientific community as to the independence of its method of indirect Demonstrative Reference, at least *prima facie*.

Genuine fictions differ from hypothetical entities in two ways. No claim for the occupancy of spatio-temporal regions can be made out on behalf of fictions. If it *is* made out, as it sometimes is for the personages of mythology, this is one important way of opening up a possibility of demonstrating that they are not fictions. The location of hypothetical entities in space and time is often achieved, and this is because of the second way in which they differ from fictions. The states of, and changes in and among hypothetical entities, are introduced for the purpose of supplying the causes for observed effects. To prove that there is an aeroplane in the stratosphere, by pointing to the vapour trail, depends upon our belief that the passage of aeroplanes causes (produces) vapour trails. To prove that a certain gene is present in the germplasm of certain animals by pointing to an observed structural peculiarity (say a fifteenth rib in the offspring of a particular bull), depends upon our belief that the presence of the gene causes the development of the structure, in the sense that it probably both initiates and controls its development. This important principle which we could call that of the Proximity of Some Cause to an Effect has not been much discussed by philosophers. For the present it is sufficient to notice that it has been used to locate hypothetical entities in space and time by reference to the proximity of these entities to the effects they are supposed to cause. For instance, this principle lies behind the reasoning that puts forward an invasion of micro-organisms in the patient, as the cause of the syndrome of that disease in *that* patient. Part of the difficulty *we* have with the African and Australian Aboriginal spells theory of disease is that in the light of the principle we try in vain to think of a spatio-temporally proximate cause, itself an effect of the spell, which is neither spatially nor temporally proximate. But even in our thinking the principle is not without its qualifications. Obviously the attribution of causes in practical affairs recognizes that for practical or judicial or moral purposes *the* cause may have to be chosen from among remote ancestors in the complex of influences which generate events. Nevertheless, for practical purposes, a proximate cause too is recognized. It may be the drunkenness of the driver that is the proximate or immediate cause of an accident, at the same time as one might blame a weak man's friends for persuading him to drink some time before, and express this by saying that it was his meeting these friends that was *the* cause of the accident. Probably one might say this because of a feeling of inevitability about everything that subsequently happened. Only *this*, we might feel, could have been different. But in the scientific context, for those states or presences of hypothetical entities which are functioning as proximate causes in time, the principle of spatial proximity holds too. Whatever cause is next prior in time is next in space. In this way by pointing to an effect we point to a cause of it. By and large the sciences do not recognize, and never have recognized, action at a distance. It has seldom been accepted as more than a temporary and unwelcome expedient, at the best a metaphor for ignorance of causes. It is often just when a gap of space and time intervenes between two parts of a process that we invent hypothetical entities or processes to satisfy the principle of the proximity of some cause to an effect. Freud discovered that what happened to a child made a difference to the way that person behaved when adult. His machinery of repression, the unconscious and the complexes, was introduced to bridge the gap. Methodologically it was just the right move, though criticisms of the particular hypothetical mechanisms he proposed have often been made, probably justly.

In all the cases considered, public ostension to material things and the states and processes in them, and to any other publicly perceptible point of spatial reference, plays a central rôle. Even in non-public direction of attention to our own thoughts and feelings it is a spatially locatable and temporally enduring embodied person who does the directing, and they are his thoughts and feelings, thought and felt in definite places and at more or less definite times. It is this fact which ultimately determines the structure of our theory of dependent and independent being, and from it we can derive the desiderata for a central demonstrative criterion, and its referents.

The Central Demonstrative Criterion

An entity exists, if (but not only if) it is the referent of a public act of demonstration, and the characteristics of the referent satisfy the recognitive criterion for an entity of that kind.

(1) The adoption of that criterion leads to some problems which deserve some comment. Must the basic entities, whose existence is proved by their meeting the demands of this criterion, necessarily be perceptibles, like things? For example must they be interpersonal objects of demonstrative reference? Sensibles, intrapersonal objects of demonstrative reference like sensations, can sometimes be the objects of public demonstrative acts of ostension. In some cases sensations are publicly

demonstrable and so satisfy at least a part of the requirement of a demonstration of existence. A doctor is not committing any philosophical infelicity when he asks a patient 'Where does it hurt?', and he can understand perfectly the patient's gesture indicating the back of the knee. There is a public act of demonstrative reference possible for many sensations. But only he who has them can tell whether a particular description is satisfied, that is, whether they satisfy the recognitive criterion. This latter fact does not, of itself, take sensations out of the public domain, because there is a public language of descriptive terms with which to describe sensations, in terms of which recognitive criteria are indeed specified, but seeing that they satisfy that criterion is not public. But as I pointed out before, publicity of identification is but one method of satisfying the demand for reliability of identification. It follows from this that sensations exist in exactly the same world as do cars and cannon balls. They could not exist without people, but then neither could clothes.

(2) But of all possible objects of acts of demonstration and identification, only things and processes seem to have the necessary qualifications to be ultimate and fundamental, since they are both spatially locatable and temporally enduring. I have assumed that spatio-temporal location must have a fundamental rôle in demonstration, because the direction of attention, whether ostensively or not, whether direct or indirect, must be to somewhere at some time. We shall see in a later chapter how the endurance of things and the possibility of a spatio-temporal framework are connected. From this one is inclined to slip into thinking that things and processes in things must be fundamental. But we shall also see that we may, in the ultimate analysis, have to abandon things as fundamental. The possibility of verbal reference to something out of sight which can yet be located at moments other than those of a demonstrative act of reference, is made viable by relating the thing or process to which verbal reference has been made, to a more or less enduring structure of things, by which the continuity of places, intermittently occupied by things, can be determined. Of course to prove the existence of a thing, process, quality or relation, we need merely point to where it is exemplified, but for it to go on existing it must endure among other things. This latter requirement of permanence, at least of relative permanence, clearly does not belong in the demonstrative criterion, since existence is proven once and for all by one successful act of demonstrative reference.

Is there a Central Recognitive Criterion?

To prove that something definite exists it is not just sufficient to point to anything at all. The thing pointed to must have the characteristics of things of that kind. The queer fish found in the net must be identified as a coelocanth. The referent of the demonstrative act must satisfy the appropriate recognitive criterion. Entities differ in important ways, some of which show up in the variety of modes of demonstration, but others are marked by broad differences in the recognitive criteria. We shall follow them up.

Public acts of demonstrative reference can be made to at least the following categories of existents:

(a) *Things and Substances*. Here the recognitive criterion requires permanence of a certain sort, and drives one to distinguishing primary from secondary qualities. Things must endure.

(b) *Events*. They need not all be changes in things. Events may endure. There is a use of 'event', often noticed, for what ought, in philosophy, to be called, short, internally organized processes.

(c) *Properties, Powers and Sensible Qualities*. These are two of the kinds of dependent beings, since properties and sensible qualities are usually attributed to one or other of the kinds of entities of categories (a) and (b). But it is not quite so simple as this. For instance sensible qualities can be attributed to such non-things as the sky, and properties can be attributed to other properties and to sensible qualities. Properties can be attributed too to such non-things as the universe, and to space, at least they have been so attributed. This brief sketch shows how extraordinarily various are the entities capable of direct, ostensive demonstration. It is in their identification that their differences appear.

Broadly speaking, recognitive criteria can be classified in two dimensions: (1) according to whether they require, or do not require, or positively exclude duration from the characteristics demanded of an entity; this seems to separate things from other categories of beings; (2) as they characterize an entity as having dependent or independent being. It emerged above that the central demonstrative criteria forced upon things, the anchor points of the spatio-temporal frameworks of referents and the bearers of qualities, sufferers of change and possessors of powers, a rather special rôle in demonstrations of existence. Demonstration led to them, and other forms of reference seemed to be dependent upon them, though in more complicated ways. This could be expressed picturesquely by saying that things, that is the enduring referents, have independent being, all else that exists is dependent for its existence upon things.

However, sometimes, and this is a point of central importance for the philosophy of science, the things which are held to be permanent are not the referents of simple ostensive acts, nor are they the perceptible things of the public world. They are the ultimate entities as conceived by scientists. This is why the order of recognition of existence, as one might put it, is importantly different from the order of dependence of being. In the seventeenth century perceptible things and qualities were, as they are today, the terminations of simple acts of ostension, and thus had, and still have, primacy in the order of recognition of existence. In that century, however, the ultimate and really permanent elements of nature upon whose arrangements the observable characteristics of the perceptibles depended, were the material corpuscles. The proofs of their existence, as cited for instance by Robert Boyle,* proceeded by that non-ostensive mode of demonstration I have called attention to as the indirect method, to the validation of which we shall frequently return. However, the special rôle of things as referents of simple ostensive acts does affect the view we take of what characteristics any ultimate entities as constitutive of the world are supposed to have. Ultimate entities tend to be modelled on public things. Classical atoms were streamlined things: teleiomorphs of which the source was ordinary perceptible things. Fluids too were teleiomorphs of perceptible liquids and gases, substance-like in many respects. It seems as if the sciences tacitly assume the principle that the endurers are, in general, also the independent existents, and external, non-enduring entities, the dependent existents. But must it be this way? Could the order of existence exhibit different priorities?

There are three obvious possibilities which are worth just noticing at this point. (1) Perhaps sensations could be existentially prior to things. The idea that sensations should be prior, the basis of phenomenalism, seems to have involved a slide from an arguable priority in knowledge since it might be held that we know about things only because we have sensations, to a most dubious priority in existence which would ignore the public character of the demonstrative criterion. (2) Perhaps events could be prior to things. Could not we think of a permanent seeming thing as a kind of internally scintillating flux of external events? (3) Perhaps some entity of quite a different category, only dimly glimpsed in the perceptible world, and our ideas of which are constructed in the imagination through a paramorphic model, could be the ultimate existent. Each of these positions has been maintained.

Assuming that the perceptible thing has been at least the main model for the ultimate independent existents, certain broad ontological principles are discernible at work in our thinking. Demonstration by simple ostension of the perceptibles is related to the class of material things and by simple ostension of the perceptibles taken as effects we generate

* R. Boyle, *The Origins of Forms and Qualities*, *Works*, Vol. III, London, 1699–1700, p. 112.

the class of hypothetical entities. Science is concerned with a variety of ontological kinds, and though material things and substances, including hypothetical entities, whose being in certain states constitutes them causes, have claims to our attention, it would be unwise to let their importance obscure other categories. The order of knowledge is from things and their qualities and relations, through effects, to causes and thus to ultimate entities. The order of dependence of being is exactly the reverse. We learn about the nature of things through the study of their sensible qualities, in which some of their powers are manifested but there might easily have been a universe in which men did not exist: one in which there would be no sensible qualities.

Internal Connections of the Scientific Ontology

Hypothetical causes are dependent upon there being the class of hypothetical entities, which constitutes the ground and mechanism of the causes. In most of science, phenomena are treated as effects and their causes hypotheticized. This will be regarded as the typical movement of thought, and its principles, the central principles of scientific method. Sometimes what were originally hypothetical causes and entities are shown to be among the perceptibles. For this to happen, not only must they be made the termination of simple ostensive acts of demonstration, but recognitive criteria for them must be satisfied. How do we know what these are? The answer to this we have already seen in the last chapter, since hypothetical entities are none other than paramorphic models treated as candidates for existence.

Hypothetical causes have actual effects. Actual effects are modifications of things, either as to their number, their relations or their properties, any of which may lead to a change in their sensible qualities. Thus actual effects are dependent upon their being actual things with powers.

The circle is closed by *the* great scientific hypothesis – that the parts of actual things and their internal constitutions, *are* the entities, which looked at from the point of view of the order of knowledge, are hypothetical. In this way the hypothetical entities assume actual priority in being, and both they and their congeries, perceptible things, substances and so on, are of equal ontological status, *the* independent beings. The rest of this book will, in a way, be the working out of the consequences of this hypothesis.

The Falsification of Existential Statements

In each section of this study the conflict between the principles of scientific thinking and received traditional logic has had to be resolved.

The study of existential statements is no exception. In formal logic, as it has so far been developed, existential hypotheses are not strictly falsifiable at all. In traditional logic falsification would have to be by instances, and no instances of the non-existence of a particular hypothetical entity seem, at first sight, to be possible. How, for instance, do we tell the non-existence of one kind of entity from the non-existence of another kind? It has indeed been held traditionally that all empty classes are identical, so one kind of non-existence would not differ from another. But falsification of existential hypotheses does take place. So even though the hypothesis 'There are M's' is not falsifiable by instances, it cannot follow that it is not falsifiable at all. What does follow is that the falsification of existential hypotheses does not occur by instances, at least, not in any simple way. This result is perfectly consistent with the position that we shall find ourselves obliged to take with regard to general hypotheses too. In the case of statements of the form 'All M are N' it is their *confirmation* which is impossible by the standards of traditional logic. But they *are* confirmed, so we shall have to conclude that their confirmation does not occur by simple instances. Broadly speaking then, there are two modes of disproof of existence, the empirical and the conceptual.

For particular existential hypotheses, say that concerning the existence of the planet Vulcan, empirical falsification can occur in either of two ways:

(a) The hypothesis of existence is held to be false because the demonstrative criterion is not met: that is the act of demonstration, in particular the simple ostensive act, fails to link up with any object at all. One points, but nothing whatever lies at the end of the pointing finger; one looks, but the cupboard is bare. This is in fact what happened with the planet Vulcan. Telescopes were directed to where it would have been, had it existed, and nothing whatever was seen in that region. Compare the discovery of the planet Neptune, where a telescope did lock on, ostensively, to something in the place indicated by the theory.

(b) The hypothesis may be held to be false because the recognitive criterion is not met. An act of demonstration links up with an object all right, but it is not what was expected there. The falsification becomes conclusive if the nature of what is demonstrated excludes the existence, in the same place, of a thing of the kind we were looking for. This might be because both were material things, and two different material things cannot exist in the same place at the same time. An example of this would be the discovery that the pores in the septum of the heart did not exist, because the septum was a continuous muscle. Another striking case is the disproof of the existence of the Aristotelian crystalline spheres. Their existential *coup de grâce* was given by the discovery of the unobstructed, trans-solar system, movements of the comets.

These techniques can be generalized. Demonstration, for instance direct demonstration by simple ostension, must be performed with respect to a limited spatio-temporal region. Let this be called the 'demonstration-region'. To falsify a general existential hypothesis two conditions on the demonstration-region must be met:

1. The demonstration-region chosen must be that in which the general background of science, knowledge of conditions and the like, lead to the hypothesis that if there are any such entities, this is where either they themselves, or their effects, will be found, or are most likely to be found.

2. Either the demonstration-region is empty, or what is more usually the case, the demonstration-region is occupied by something not meeting the recognitive criterion, which is identified as an entity whose existence excludes the possibility of the existence of the entity in question.

A wholesale falsification of existential hypotheses took place at the end of the eighteenth and the beginning of the nineteenth centuries. At this time various demonstration-regions were outlined which turned out either to be empty or their occupants were found not to be fluids, or the effects of fluids. I shall consider two examples.

Phlogiston was a substance, occupying volume and having the characteristics of substances of a more familiar kind, though the determinate manifestations of such characteristics as mass, in negative weight, did not fit easily into the Newtonian scheme. The evidential facts for phlogiston were chemical phenomena, for which its existence seemed to provide an elegant explanation. Copper, heated in air, turned to calx, losing phlogiston, while copper calx, heated with coal, was reduced to copper. Coal, supposed rich in phlogiston, because of its readiness to burn and its leaving little ash, provided the necessary ingredient to combine with copper calx to yield copper. The parent of this model was the supposed substance, fire, and as the matter of fire, an ancient 'element', its history was of respectable duration.

Phlogiston was shown not to exist. Another and different substance was demonstrable in those chemical situations where phlogiston was most likely to be found, and this new substance, whose existence could be demonstrated, served as the basis of an explanation of at least equal adequacy. The existence of oxygen was proved by existential demonstration. Transactions in it explained combustion and it was metaphysically commonplace. There were no further grounds, in unexplained chemical phenomena, to make it necessary still to espouse phlogiston exchange as a hypothetical mechanism. Finally, oxygen as identified satisfied a recognitive criterion that excluded the co-presence of anything meeting the recognitive criteria for phlogiston. But in its last stronghold in physics phlogiston was eclipsed by a different process of disproof which will lead to the last facet of the logic of existential statements that I propose to explore; what I shall call a conceptual mode of disproof.

In physics, phlogiston was not shown not to exist by something else being demonstrable where it was thought most likely to be. It was, instead, transmuted into a different category. Phlogiston was a substance and as a substance it explained, under the title 'caloric', many of the simpler thermodynamic phenomena. But it was transmuted from a substance, existing alongside other substances, into a quality of substances, energy. Heat, as a manifest quality, had been explained by treating it as the manifestation of a substance, additional to the substance which was hot. The model of explanation was that of the explanation of the wetness of a sponge, by the postulation of the additional substance, water, with which the sponge became permeated. In the energy theory heat as a manifested quality is explained as the manifestation of a qualitative difference in the internal constitution of whatever is hot, associating increase in manifested heat of anything with increase in the average kinetic energy of its constituent corpuscles, and associating that with the primary quality, motion. Thus caloric underwent a category transformation. It passed from a substance into a quality: in fact a relational quality. Caloric, as a substance, ceased to exist.

In just the same way, the electric fluids were recategorized as electrical energy, and then recategorized again as particles, and then recategorized again as fields, a conceptual oscillation with which we have not yet done. But unlike phlogiston in chemistry the essential extra-material electricity has not been displaced by a commonplace substance. Instead, it has survived, undergoing transmutation from category to category, from substance to quality, from quality to corpuscles, from corpuscles to fields.

It is not too much to say that each great leap forward in scientific theory takes place by a complex of moves, an important step of which formal logic denigrates, namely, the falsification of existential hypotheses, either empirically *à la* Mother Hubbard or conceptually by recategorization.

The Practical Formulation of Existential Hypotheses

Apart from mere catalogues of facts, the positive content of science consists primarily of confirmed existential hypotheses of various kinds, and along with these, determining recognitive criteria, go hypotheses as to the behaviour and characteristics of those things, properties and processes the *existence* of which has been established.

The basic principle, upon which all attempts to confirm or falsify theories, by trying to prove or disprove the existence of the entities, properties or processes-postulated in the theory, is this:

A Hypothetical Entity is Derived from an Iconic Model

An iconic model is built or imagined, and we ask, 'Is there anything in nature like this, really doing the job which this was invented to simulate?' A hypothetical entity is derived by considering that an iconic model might possibly be real, so the kinds of hypothetical entities will be a reflection of the kinds of paramorphic models, since, with the exception of the odd teleiomorph, it is paramorphic models which are constructed to fill in gaps in our knowledge of natural mechanisms, serving in the first instance as models for unknown processes, and mechanisms. I shall speak of a hypothetical entity being derived from a paramorphic model but the process of 'derivation' must be understood to lie in the changing *attitude* of scientists to the model.

Two cases can be distinguished, depending upon whether the hypothetical entity is derived from a singly connected paramorph, or a multiply connected one. The former are those models which are built upon the basis of the entities of some single science, such as microorganisms, and gas molecules. Once the early teleiomorphic stage is passed, these are supposed to be material particles of certain kinds, and each is supposed to have the essential characteristics and obey the essential laws of their parent, that sort of thing they are modelled on. Since they are the same *kinds* of things as the things they are modelled on the problems of proving their existence are technical only. The most interesting kinds of model from an existential point of view are the semi- and multiply connected paramorphs. The latter are those models which are built by utilizing only some of the essential characteristics from several sciences, as Bohr's atomic model drew on only some of the mechanical and some of the electro-magnetic attributes of matter. The importance of these kinds of models is that they introduce entities, which if they did exist, would be of a novel kind.

Case 1. A hypothetical entity derived from a singly connected paramorph has the same existential criteria as the actual entities of the parent science, upon which the model is modelled. In short, it is just another thing of the same kind. That there are entities like the sun responsible for the points of light seen in the night sky enters into science first as a singly connected paramorph, competing with such others as that they are holes in a firmament, behind which burns a perpetual fire.

So if they can be located and identified as bodies distributed in space, and having the leading characteristics of the sun, they can be said to exist, that is, other suns exist, because they have satisfied both the demonstrative and recognitive criteria for suns. But to make this step a telescope is required. That there are entities like bacteria, responsible for otherwise inexplicable diseases, suggests a hypothetical entity, the virus, derived from a singly connected paramorph, whose parent is the bacterium. And this is itself derived from a model (the germ theory of disease) whose ultimate parent was perceptible organisms. To prove the existence of bacteria the optical microscope was required, not just to locate the entity, to make demonstrative reference to it, but to tell if it is of the right kind, to satisfy the recognitive criterion.

These examples have in common the fact that the ultimate parent in each case was perceptible. It is, therefore, to be expected that instruments of that special kind which extend the power to perceive should be central to the demonstration. They allow us to make simple ostensive acts which, as a merely contingent fact, we could not otherwise do. With instruments of this sort, the theory that explains their powers can be changed, without upsetting their existential deliverances. The same is true of the whole range of such instruments, that is, of stethoscopes, X-ray machines, mechanically magnified probes, and so on. Why, incidentally, has no one made an instrument for increasing one's powers of taste and smell? Is it impossible, and if not, why not? There are ways of detecting the agents responsible for tastes and smells, but they are chemical, and so work, by reference to secondary effects. But one might imagine a kind of a concentrator, which would be a nose or palate extending instrument.

In this way the basic ontological class from an epistemological point of view is, and must be, the class of perceptibles. It is to that class that the entities revealed by the extended *senses* belong. They are one with the entities revealed by the senses alone. But as has been noticed several times, priority in knowledge does not necessarily imply independence of existence. In fact we have been driven to the position that the only fully independent existents are the basic entities which are the parts or form the inner constitution of perceptibles, or are their causes, where the perceptibles are sensible qualities and the like. So in this case existence revealed by effects creates no problems if the hypothetical entity is derived from a singly connected paramorph, for it introduces entities of a kind already sanctioned by the sciences, and whose existential criteria, even if through effects, have been legitimised into the body of science. For instance, a well known entity, known only through its effects, is energy. By using a singly connected paramorph we develop an idea of electrical energy from our known concept of mechanical energy. It too is revealed through its effects, and produces them by a different causal law, but it is an entity of exactly the same kind and its existence is proved by satisfaction of precisely the same criteria.

Case 2. A hypothetical entity derived from a semi- or multiply connected paramorph. In neither case does the entity derived have the full complement of essential properties of any known kind of existents, because in the modelling process the paramorph is constructed either by drawing on less than its parents' full set of essential properties, or drawing upon two or more parents, and usually drawing less from each than either alone would provide.

(a) *A semi-connected paramorph.* A field is less than a substance but is extended in space, and passes influences from one perceptible to another, revealing time dependent and perceptible and detectable effects. A mechanical atom is less than a thing, it has no colour nor temperature, yet it is supposed to resist penetration, to move, to transmit momentum, and to acquire and lose mechanical energy. All depends in these cases and others like them, on *how* like their parents are the models from which the entities are hypothesized, how semi is the paramorph. If the characteristics preserved play a large rôle in the recognitive criteria of the parent entities, like impenetrability does with things, and if the demonstrative criteria involve the same mode of reference, then there is no metaphysical problem involved, and this is probably the case with mechanical atoms. There may be technical problems about making the things reveal their presence, or scientific problems about their capacity to cause the effects they are alleged to, but there are no metaphysical problems provided that their share of the full set of characteristics of their perceptible parent is that part of the parent's characteristics by which *their* existence is determined. This sort of case, where a gradual extension of an ontological class takes place, is very typical of the sciences, particularly in those periods which Kuhn calls 'normal'.

What happens, however, when the demonstrative criterion is by effects, and the recognitive criterion is derived from an essential set of characteristics, unlike the effects, unlike perceptibles? There are intermediate cases, between this and the previous case. The extreme case of difference in category is the most instructive. The claim for the existence of a field in empty space poses just such a problem. All other states of strain are states of strain of a medium. All other smiles are on faces. What is a state of strain existing by *itself*? Or a grin after the cat has vanished? At least in the case of the state of strain, though we are required to use our imagination, we do not have to visualize any such thing, for the existence of the field is revealed demonstratively by its effects. Nevertheless we are required, if we hold to the independent existence of the field, to accept a causal relation under very peculiar circumstances. Briefly, to anticipate the account of causality to follow later, a causal relation obtains when one state of things or substances generates or produces another, and some generative or productive mechanism can be found, or imagined under the licence of science, by which the production or generation is achieved. The problem of trans-categorial causation is this: if the field is unlimited, insubstantial, etc., and the things it affects limited, substantial, etc., of what kind is the mechanism by which the unlimited and insubstantial affects the limited and substantial? If field-like the categorial problem arises at the production end of the relation between the mechanism and the effect, if thing-like at the stimulus end, between the cause and the mechanism, later to yield the effect.

Actual science resolves this in three different ways:

1. Faraday, Boscovich and Kant resolved the difficulty inherent in this very example by adopting an immaterial view of things, as mere points whose being consisted in their force fields, which explains, as we shall see in a later section on matter, their material characteristics without making them substantial. In short, in Faraday's resolution one category swallows up another. Like Newton's mechanism of the aether, Faraday uses this device to resolve any crucial difficulty, particularly in his correspondence with Hare.*

2. The relation of the states in each category is taken as a basic interaction, and so no mechanism is required, because none could be forthcoming, without a complete conceptual revision. This was the way Locke† resolved the difficulties about how mind and matter could interact.

3. Each separate state is understood as being actually only one aspect of the one state. So the correlation of states which led to the hypothesis of causality between them is explained by making their differences epistemological, that the two 'states' are but two ways of knowing what is in fact only one state. This is the basis of the most promising solution to the difficulty of explaining the apparent trans-categorial causation between mind-states, and body and brain-states. This is explored in detail in Part 1 of Chapter 8.

In each case the resolution seems thorough-going and the problem fully resolved. The problem, it seems, is only one of making the right choice of mode of resolution. A Faraday-type resolution merely obfuscates the mind-body problem, and so on.

(b) *Multiply connected paramorphs.* An apparently problematic situation

---

* Discussed by L. Pierce Williams in his *Michael Faraday*, Chapman and Hall, London, 1965, pp. 309–11.

† J. Locke (11), Bk. IV, Ch. 3, § 11.

has arisen in the philosophy of physics through the difficulties that have arisen in that science by the widespread use of multiply connected paramorphs, with parents which, taken each in the round, are mutually incompatible. Many physicists failed to see that an entity which was derived from a mixed wave and particle model was an existential novelty, and provided that its share of the essential properties of waves and particles together made up a consistent set, the incompatability of the parents was irrelevant. Sub-atomic entities have the recognitive criteria of neither of the parents of the model from which they are derived. They have their own recognitive criteria, derived from the characteristics and powers with which they need to be endowed in order for them to be able to perform their causal functions. The demonstrative criterion for them operates through their effects, particularly their power to ionize gases, and their power to affect screens and photographic plates in ways which have visible effects. Using the principle of the proximity of causes to their effects, it is not difficult to prove that they satisfy a perfectly proper demonstrative criterion. Their powers to produce those effects being just those required by their recognitive criterion, their *existence* can surely not be in doubt. An electron is identified as that entity which has the power to affect a photographic plate in a certain specific way, and so on. Of course, *what* we shall ultimately say they are is another thing. There is no circularity in this method of proceeding, because in attributing powers to produce particular effects the way is open to develop theories about the natures of the entities in virtue of which they have the powers in question. The lack of specificity of the attribution of a bare power is but the sign of the temporary termination of our enquiries. Whatever we are forced to attribute mere powers to, become the basic entities of the world, as science currently and ephemerally sees it.

Existence, though, seems in many cases to require more still. Duration, persistence in being, seems to be required of thing-like and substantial entities, as opposed to certain classes of events, for whom only instantaneous being is required. I have tried to show that the satisfaction of the demonstrative criterion leaves it open whether the existent endures or has only momentary being. Whether or not an entity has to endure to exist is contingent and must be part of the recognitive criterion. This serves to distinguish recognitive criteria into two grand kinds, making a deep metaphysical differentium in modes of being.

There is another way of reaching the position outlined in the chapter. I have enlarged upon it in other places,* but for the sake of completeness I shall sketch it here again. In the definition of any thing, sub-

* *Matter and Method*, Macmillan, London, 1964, pp. 48–52.

stance, event, property, or process, two groups of characteristics can be distinguished. There are those which express the categorial nature of the entity, and those which serve to differentiate it within the category. The former group of attributes is not usually mentioned explicitly in defining a kind of entity because we are usually working within a tacit categorial framework. Chemistry books do not have to say that oxygen is a substance rather than a quality before they go on to define what kind of substance it is. Dictionaries, on the other hand, being eclectic in the category of their entities might, on occasion, make explicit mention of category.

The categorial differentium and the particular differentium can be looked at as the intensions of classes. The former I shall call an 'ontological intension', and the latter a 'descriptive intension'. Then the total definition of an entity is the intersection of the intensions of a class of each kind. Existence is settled by the identification of a member of the product class. When an identification is attempted it is seen to involve two acts, each linked logically to one of the intersecting classes. There is a demonstrative act, the nature of which is determined by the intension of the ontological class, and a recognitive act, the nature of which is determined partly by the intensions of each class. Suppose the entity is a {four-footed, ungulate} × {animal} then the intension of {animal} tells one what kingdom of entity one is to search, though in this case we have not reached a metaphysical division, while {four-footed, ungulate} picks out the species or genus.

By further dissection of {animal} we reach something like this: {animal} = {organic} × {mobile} × {material thing}. A certain sort of demonstrative procedure is required to identify a material thing. It is in fact a public act of simple ostension. That this is the appropriate demonstrative procedure follows from the general specification of what it is to be a material thing, that is, to be at some place for a certain interval of time, and to exclude the possession of the place by other material things. An act of pointing, which must be to some place at some time, thus identifies a material thing. It does not, of course, follow that everything pointed to must be a material thing. {Material thing} then, is the ontological intension, and {four-footed, ungulate, organic, mobile} is the descriptive intension. If by following the pointed finger I see something which is four-footed, ungulate, organic and mobile, I thereby satisfy myself that sheep exist. If we can all see them, etc., then sheep exist. In just the same way as one is inclined to overlook the metaphysical and categorial loading of 'this' plus the pointing finger, so in most definitions the metaphysical part is left understood, and the ontological class of an entity, property or process, is seldom mentioned. From a categorial point of view the definition of sheep

D

simply as four-footed ungulates is quite incomplete. Are these the sheep of Cyclops, or the sheep of Dan Archer, or the sheep of Norfolk? Are the entities mythological, fictional or material? Are they events, properties, things or processes, relations or qualities? Which sort of reference is appropriate will depend upon which group the entities are in. In physics one is always running into difficulties by the tacit involvement in much of that science of the metaphysical prescription for the Newtonian object incorporated through the identifying procedures for physical things. Some, at least, of the early attempts to understand the nature of the products of the disintegration of matter, and the despairing retreat into the positivism of the Copenhagen Interpretation can be put down to the feeling of contradiction arising from the attempt, tacit again, to operate with concepts in which the descriptive intension was such that it could not be conjoined to the ontological intension of Newtonian objects. The mode of demonstrative reference by which a wave is identified is quite different from that by which a material thing is identified. For instance, the inter-action between 'particles' and slits, which is required to get the correct probability for a 'particle' in the two-slit experiment, seems to require a kind of particle whose characteristics conflict with the Newtonian requirement that one thing cannot be in two places at once, i.e. with a referential requirement. This principle is not only the basis of the application of the classical probability rules, which depend on the assumption of the independence of each thing, to particles passing through slits, but is also the basis of the legitimacy of the identification of one object by pointing at one time to one place, i.e. of the legitimacy of the demonstrative criterion. In my discussion of the metaphysics of matter I shall return to the question of categories, and to the meta-physical, as opposed to descriptive, distinctions among the features, parts and structures of the world.

The examples I have used for expository purposes in this chapter have been almost exclusively Things and Substances. But it is, in practice, equally important to consider the proofs of existence of the properties and powers attributed to things. Sometimes model building involves the invention of novel properties and powers, the possession of which by hypothetical entities is used to explain the manifested phenomena. Properties too show the dichotomy into the novel and the familiar that I noticed above for things and substances. A novel property (or power) is exemplified by the microphysicists' 'strangeness' as opposed to their use, though modified, of the familiar property 'spin'. In a detailed analysis even 'spin' can be seen to be not quite the same characteristic as say the mechanical whirling of a top. It is not at all clear that the thing-like model of the electron which spins is the only way in which a left and right handedness distinction can be introduced. The proof and disproof of the existence of properties and powers follows similar lines to the proof and disproof of the existence of things and substances. That is there is both a demonstrative and a recognitive criterion to be satisfied. In many cases too the existence of a property or power is established through the existence of its effect. I shall return to this topic again in the critique of matter and the theory of primary qualities, later in the book.

The Existence of Events

In that sense of 'event' in which an event is an instantaneous happening, nothing of endurance enters. But the machinery of demonstrative and recognitive criteria for things, substances, properties and powers is so set up that the transfer of its use to the establishing of the existence of events is quite smooth. Remember that endurance is introduced through the recognitive criterion. The demonstrative act need be no more than the instantaneous direction of attention, so it can be used without modification for referring to events.

In this chapter, three important principles of scientific thinking have emerged.

1. The Principle of Proximity of some cause to every phenomenon or pattern of phenomena considered as effects.

2. The Principle that the most Permanent Entities should be regarded as the ultimate components of all that exists.

3. The Principle that an entity can be made to occupy the same position, i.e. the same place at the same time, as another entity, by transmutation of category.

All these principles can be seen as at once methodological and metaphysical. None can plausibly be regarded as a law of nature. Their justification, as with the other principles of scientific thinking which will be brought to light, ultimately rests upon the degree to which these principles allow the formulation of a rational methodology of science which justifies as much as possible of scientific practice.

SUMMARY AND BIBLIOGRAPHY

Consideration of what is involved in showing that an entity exists has led to the conclusion that existential proofs require both demonstration and identification. An existential hypothesis, such as a scientist might entertain, is not just that *some*thing exists in a certain place and at a certain time, but that whatever does exist there has certain characteristics, in virtue of which it possesses the causal powers which are required to produce the phenomena its existence explains. It is from ordinary cases that we derive a paradigm of proofs of existence. I have tried to show that the same paradigm can be used in cases where demonstrative reference is to effects rather than to causes by assuming that in the end these cases do not really demand another and dubious paradigm of proofs of existence. This is shown through the analysis of such intermediate cases as the proof of the existence of bacteria as the cause of the symptoms of a disease. At first we say such things as 'This is a streptococcus infection' speaking of the cause though indicating its effect, i.e. drawing attention to the symptoms. The technical innovation of the microscope allows this to be split into two applications of the original paradigm. 'This is an s-type infection' claims the existence of the symptoms, while the microscopist with 'This is a streptococcus baccillus' claims the existence of the causal agent. Thus some cases of the proof of the existence of causes through the ostension of their effects can be turned into a pair of simple proofs of existence. It is the continuing possibility of this that makes it reasonable to accept the tentative claims for existence of entities as remote from observation as the positron.

In this chapter, I have considered existential hypotheses, their content and the conditions under which they are confirmed and falsified. The key idea has been the connection between mode of reference and ontological kind or category. The main force of the argument has been directed at trying to show that the ordinary ways of settling existential questions can be made to yield a paradigm which can be applied successfully in scientific contexts.

Two main problems about existence have occupied the attention of philosophers; the problem of the existence of universals, classes, numbers and the like; and the problem of the justification of existential hypotheses about the unobserved, of which the referents of theoretical terms in science are a case in point. While I am not really concerned in this book with the first problem, some important points about the representation of existence statements in appropriate logical form have been made, for instance in

89. L. S. STEBBING, 'The Philosophical Importance of the Verb "to be" ', *Proc. Aris. Soc.*, **18**, 582–9.
90. W. V. QUINE, *From a Logical Point of View*, Harvard University Press, Cambridge, Mass., 1953, Chs. 1, 6, 7, 9.
91. C. LEJEWSKI, 'Logic and Existence', *BJPS*, **5**, 104–19.
92. F. SOMMERS, 'Types and Ontology', *Phil. Rev.*, **72**, 327–63.

The more specific topic of the logic of demonstrative reference has been extensively discussed. Most of the leading points can be found in

93. J. L. AUSTIN, *Philosophical Papers*, Clarendon Press, Oxford, 1961, Ch. 8.
94. P. T. GEACH, *Reference and Generality*, Cornell University Press, 1962, Ch. 3.
95. P. F. STRAWSON, *Individuals*, Methuen, London, 1959, Part II.
96. S. C. COVAL, 'Demonstrative without Descriptive Conventions', *Philosophy*, **40**, 334–43.
97. I. Scheffler and N. Chomsky, 'What is said to be', *Proc. Aris. Soc.*, **59**, 71–82.

On the question of the empirical *criteria* of existence, both in non-scientific and scientific contexts there exists a considerable literature. The following items pick out some of the many points which have been made:

98. E. NAGEL, *The Structure of Science*, Routledge and Kegan Paul, London, 1961, pp. 146–52.
    R. HARRÉ (*see* 27), Chs. 4, 5 and 6.
99. F. DRETSKE, *Seeing and Knowing*, Routledge and Kegan Paul, London, 1969, Ch. II, p. 4.

The problem of the existence of fictions has been discussed in

100. M. SCRIVEN, 'The Language of Fiction', *Proc. Aris. Soc.*, Supp. Vol. **28**, 184–96.
101. C. CRITTENDEN, 'Fictional Existence', *APQ*, **3**, 317–21,

and in

102. A. PAP, *An Introduction to the Philosophy of Science*, Free Press, Glencoe, Ill., 1963, pp. 354–6, the view that hypothetical entities should be treated as fictions is advocated.

The confirmation and falsification of existential hypotheses is discussed in

103. J. W. N. WATKINS, 'Between Analytic and Empirical', *Philosophy*, **32**, 114–30.
104. G. BUCHDAHL, 'Sources of Scepticism in Atomic Theory', *BJPS*, **10**, 120–34.

The question of the existential priority of things, or of their independence as existents is discussed in

P. F. STRAWSON (*see* 95), Ch. I.

The classical attempt to substitute events for things is

105. A. N. Whitehead, *Concept of Nature*, Cambridge, 1964, Ch. IV and VII.

For a general discussion of 'reality', see

106. J. L. AUSTIN, *Sense and Sensibilia*, Clarendon Press, Oxford, 1962, Ch. VII.
107. M. K. MUNITZ, *The Mystery of Existence*, Appleton-Century-Crofts, New York, 1965, Ch. V.
108. J. F. BENNETT, ' "Real" ', *Mind*, **75**, 501–15.
109. D. W. THEOBALD, 'Observation and Reality', *Mind*, **76**, 198–207.

4. Laws of Nature

The fundamental cause of the development of a thing is not external but internal.
'The Works of Chairman Mao'

The Argument

Three classes of general statements can be distinguished according to their subject matter. Each group can be organized by the degree of immunity to falsification of its members.

1. *General Theoretical Statements*

   (*a*) The *logical power* to imply true particulars is not sufficient as an acceptance condition:

       (i) it removes distinction between laws and accidental generalizations;

       (ii) it does not permit unequivocal falsification.

       Logical power can be treated as a limiting condition, but

       (iii) consequences in conflict with fact can be explained away, or the fact can be explained away;

       (iv) some consequences are better approximations to the facts than any other theory gives.

   (*b*) *Material conditions.* Theoretical statement 'D' about iconic model M, is acceptable if, there is a law 'L' about P, some field of phenomena, and $M \gtrless P$, 'D'$*$'L', i.e. M is an iconic model of P, and 'D' is a sentential model of 'L'.

   Comment. Sentential modelling of 'D' in 'L' is insufficient. There is no plausibility control, and hence infinitely many 'D's can be modelled on any 'L', any one of which might be the correct theory. This is why we need the concept of an iconic model over and above that of the sentential model.

2. *Laws of Nature*

   (*a*) *Acceptance of 'L' as a Law*

       (i) the requirement that there be favourable instance statistics must be met;

       (ii) the requirement that there is some reason other than the instance statistics for accepting 'L', must be met.

       (ii) is achieved by turning the material conditions of 1(*b*) round, so that the existence of 'D' and M support 'L'.

   (*b*) *Reasons for this view*

       (i) Deducibility is not enough for attachment of a law to theory, since there are infinitely many possible theories in deducibility relations with any putative law.

       (ii) The view we are advocating has the power to resolve classical difficulties.

3. *Causation*

   (*a*) Causal laws are distinguished from other kinds of general statement associated with some conception of a generative mechanism.

(*b*) Causal reports are distinguished from causal laws.

(*c*) Three aspects of causal laws:

    (i) *Universality*: achieved by the power to explain away counter instances, therefore the basis of acceptance cannot be just instances;

    (ii) *Necessity*: derived from the nature of the associated generative mechanism;

    (iii) *Assent*: generative power must be justified by reference to some generative mechanism.

(*d*) *Proof of Causal Law*

Prima facie criterion – some non-random pattern is observed, e.g. regular concomitance among events, regular structure among things, e.g. a crystal.

Definitive criterion – the existence of causal mechanism has been established, or an acceptable model created.

(*e*) *Critique of Hume on Theories*

An 'Idea of Connection' can be supplied by creative theory construction.

Proof. By consideration of examples it is shown that regular concomitance is not identical with causality, let alone a necessary condition for a causal relation, but a mechanism of connection is adequate.

Hume's plausibility derives from his correct identification of the *prima facie* criterion.

(*f*) *Causal Reports*

Derived from causal laws, concentrating upon the point in the whole set-up where human agents can interfere. This fits with Hart's 'making-a-difference' theory.

(*g*) *Stratification of the Causal Account of Nature,*

    (i) is a necessary consequence of this theory;

    (ii) ultimate connections are *prima facie* causal and leave room for the later satisfaction of the definitive criterion. The doctrine of powers resolves the problem of the ultimate without *ad hoc* methodological hypotheses.

4. *Conditionality*

(*a*) The conditionality of laws of nature is not adequately expressed by material conditionality, since that captures merely the statistics of truth;

(*b*) Goodman's attempt to provide an increase of strength by the notion of projectability and entrenchment is still subject to the old inductive doubts;

(*c*) Conditionality only makes full sense if the antecedent of a conditional is treated as describing the stimulus required for some generative mechanism to be activated.

5. *Natural Necessity*

Import of 'Necessity'

(*a*) *Absence of Alternatives.* Generative mechanisms provide restrictions on empirical possibility and so lead to a conception of natural necessity.

Measure

    (i) Non-probabilistic case:

$$N = \frac{1}{n_a} \text{ where } n_a \text{ is number of alternatives}$$

(ii) Probabilistic case:

$$\mathcal{N}_k = \frac{p^{k+1}.\,\ldots\,p^n}{n \times p^1.\,\ldots\,p^k}$$

(b) *Inviolability* of laws is only relative for natural necessity, but absolute for logical necessity.

6. *Subjunctive Conditionals*

Why does the inferability of a subjunctive conditional from p show p to be a law statement?

(a) *Subject Matter of Conditionals*
    (i) Indicative: possibilities including those actualized;
    (ii) Subjunctive: possibilities excluding those actualized;
    (iii) Accidental: possibilities excluding those not actualized;

(b) *Truth Grounds*
    (i) Indicative: (A) favourable instances, and (B) a reason for extending the application of the law to possible cases, i.e. a hypothesis about, or knowledge of a generative mechanism.
    (ii) The truth grounds for the subjunctive conditional are confined to (B).
    (iii) The truth grounds for the accidental conditional are confined to (A). But (i) a conditional statement is law-ish only if (B) is satisfied. Satisfaction of (B) permits the inferability of (ii), the subjunctive statement therefore if (ii) is inferable (i) law-ish.
    For (iii) there are no (B) grounds, therefore it cannot be law-ish.

(c) *Remarks.* In the future tense:
    (i) tense expresses possibilities;
    therefore
    (ii) subjunctive future expresses remoter possibilities.

(d) *Problem of Belief in Particular Subjunctive Conditionals*
Individual instances are translated into specimen instances.

7. *Hempel's Paradox*

(a) Three conditions:
    (i) Nicod's criterion: confirmation is by instances;
    (ii) Equivalence condition: whatever confirms L confirms all logical consequences of L to the same degree;
    (iii) Hempel's Intuition. Some logical consequences of L are not confirmed to the same degree, by the same evidence.

(b) *Solution*
    (i) If L is a law-ish statement, then certain logical consequences (A-consequences) (term negation) are not law-ish because of the loss of the generative mechanism which made L law-ish, therefore Condition (ii) needs modifying. (Condition (i) is defective anyway as shown by subjunctive conditional problem.)
    (ii) B-consequences (proposition negation). Here Condition (ii) is preserved, since the subject matter remains the same and the generative mechanism is preserved, but Condition (iii) collapses.

If we distinguish statements by differences in the status of their subject matter, then the sciences contain three main groups of statements, other than existential hypotheses. There are those which describe, and are sometimes treated so that they can be said to lay down, the modes of behaviour, the essential characteristics, and so on, of hypothetical entities and hypothetical mechanisms. These are pure theoretical statements. Then there are those which describe or lay down the relations between states of hypothetical mechanisms and phenomena. These have been variously described as bridge statements, transformation statements, reduction sentences, dictionary entries, and so on. Then there are those which describe phenomena. This schema is less simple than it looks, particularly because of the problems that arise in trying to give any general prescription as to what is to count as phenomena, but at least central cases of each kind of statement are not difficult to identifiy. It is clear that a statement about the number of electrons in the K-shell of beryllium is a theoretical statement. It is clear that a statement about the relation between a gene pool and the distribution of characteristics in a population is a bridge statement, and it is clear that a statement about flocculent white precipitates is a statement about phenomena.

Within each group, statements can be roughly ordered by the relative strengths of belief which scientists accord them, from those which are given up very readily to those which are treated as practically immune to falsification. It is a marked feature of scientific work that theory is sometimes used as a powerful corrective to naïve acceptance of fact. As Dumas once put it, 'The main utility of theory is to help us decide which "facts" are false.' There are many, many striking instances of this. It was theory, for example, which allowed physicists to conclude that Miller's apparent evidence for the aether drift was false.

Those theoretical and bridge statements which are, at some given time, treated as most immune from revision, I shall call 'principles'. The importance of this way of treating statements is so great that I shall reserve its discussion for a separate chapter. It will become clear that statements which are being treated as principles are not absolutely immune from revision, only relatively so. Because a statement is being *treated as* a principle it does not thereby acquire a different logical structure from other empirical statements, i.e. its status is not conferred upon it by its being analytic in logical form, but by its being treated by scientists as inviolable for the time being. Differences in inviolability I shall call differences in *strength*. Of course, it is difficult not to accord inviolable status to a statement one believes to be analytic, but it is just a mistake to think that all statements accorded inviolable status must be analytic.

Statements which describe phenomena are also differentiable in degrees of strength. Following customary usage, I shall call those with the greatest strength, laws of nature. Those with the least strength I shall

D 2

call accidental generalizations. This latter class of statements must be distinguished in principle from a class of general statements which, for reasons I shall develop in this chapter, lack the strength of laws, but which might acquire that strength. These I shall call 'protolaws'.

(a) *General Theoretical Statements*. The reason for regarding as satisfactory, those statements which describe models, and subsequently, sometimes hypothetical mechansims and the inner constitutions and powers of things, cannot be (because of observation of the state of affairs they purport to describe) that a theory may be held to be satisfactory long before there can be any certainty as to the existence or non-existence of a hypothetical mechanism based upon the model. Deductivism led to a serious muddle about the acceptance conditions of these statements. It arose in this way. If one makes the initial mistake of accepting the Cartesian myth and attempts in accordance with its tenets to impose a simple deductive structure on statements occurring in a scientific theory, the general theoretical statements will tend to occupy, relatively, the position of premises. Statements describing possible or actual observations will tend to occupy the position of conclusions, particularly when they are used for stating a prediction. There will, of course, be observation statements among the premises, which describe initial conditions. Observation statements, since they are particular, can be simply true or false within some conceptual system. If we follow the deductive method of organizing knowledge, theory seems to make contact with reality at the level of those observation statements which appear at the end of deductions, as conclusions, since such statements look as if they are 'right up against the world'. The theoretical statements are, from a logical point of view, the most removed from reality, so on this view it would look as if truth and falsity were fed into the system at the bottom, so to speak. It was made clear by Descartes in *Principles*, XXII, and has been tacitly or explicity accepted by all other deductivists, that acceptability of theoretical statements hinged solely upon their power to imply true predictions, and, should a prediction be falsified, something in the theoretical deductive structure must be wrong; either a step is invalid, or a premise false. Since, in many cases, theoretical statements do not themselves describe observable states of affairs, it might look as if their independent confirmation was impossible. If we accept this line of argument we seem to have a problem because some way must be found of accounting for the degree of confidence that scientists actually have in their theories. But the confirmation of the consequences derived by logical reasoning from the theoretical statements and description of initial conditions is not enough. It is obvious that, using classical logic, the truth of consequences cannot be transferred to their antecedents, because on the deductive model, infinitely many theories could be constructed to imply the same consequences, only one of which might be true. The *reductio ad absurdum* of deductivism comes from seeing that, in accordance with the canons of received logic, the only relation theory can have with fact is that a theory or hypothesis is falsified by the discovery of states of affairs the description of which contradicts conclusions from the theory. This creates the other horn of a dilemma, since infinitely many theories can imply the same consequences; falsifying any one can give no guidance in the rational choice of its successor. Another, less objectionable, version of deductivism has it that general statements function as principles of inference among observation statements rather than as some of the premises from which they are deduced. On this interpretation of the rôle of theoretical statements, truth and falsity would be confined to particular statements, because only particular statements are ever premises or conclusions of inferences about observable matters of fact, e.g. the reading of a particular instrument at a particular time. There is a great deal to be said for this view, provided it is not combined with the further doctrine that the general statements are merely conventions, or instruments for inference, or something of that sort, since this would make them immune from correction, and hence would not explain how they are in fact corrected. We owe to Whewell, and to N. R. Campbell, the insight that the acceptability of a theory has only what one might call 'limiting' connection with its power to imply consequences which turn out to be correct predictions of the facts, that is, to imply, with other statements, particular statements which, because they describe possible observations, can turn out to be true or false.

Both these men, in their own way, made the point which is central to my treatment of laws of nature. The acceptability of theoretical statements is a product of two factors. One is their power to imply true predictions, but not necessarily by simple deduction. This is a mere limiting condition because many theories can imply the same facts, i.e. it is at best a necessary condition for a theoretical statement or hypothesis to be acceptable. The other condition is that the relation between the parent model and the iconic model involved in building up the picture of nature which the theory expresses should ensure the plausibility of the iconic model as a hypothetical causal mechanism. These conditions can be put as follows:

A theoretical statement, *D*, is acceptable if

1. *Limiting Conditions*

(i) Any true observation statements which conflict with the consequences of the theory containing *D* can be ignored or explained away, preferably by reference to some feature of the hypothetical causal mechanisms involved in the theory of which D is a part.

(ii) There are some consequences of the theory containing D, drawn from the theory together with some statements of initial conditions, which provide a closer approximation to the facts than do the consequences of any other theory.

2. *Material Conditions*

D is about some iconic model M, and there is an acceptable law of nature L about some phenomena P, which form the parent model for M (written $M \rightleftharpoons P$ for short) so that D is a sentential model of L (written $D * L$ for short), and there are states, etc., in P for which L is a law of nature, and these are modelled by like or identical states of M, described by D, and D forms part of the explanation of some phenomena N, since some of the laws of N, e.g. T, are such that $T * D$.

What this condition (2) amounts to is that *it is not enough* that D be a sentential model of some L, there must also be the modelling relation between the subject matter of D and the subject matter of L, so that D describes an iconic model whose parent model is described by L. In general, the phenomena which are, of course, the behaviour or natures of things and processes described by L will be different from the phenomena described by D. Only thus can what D describes, by a growing attitude of acceptance by scientists, become a hypothetical mechanism for the phenomena that D explains.

Thus, out of the infinitely many possible theories from which statements describing the known facts can be deduced, only a very few command serious attention, and rarely more than one is picked out as meriting tentative acceptance. Those which command serious attention are just the theories which describe models which might be hypostatized to be the actual mechanisms of nature and the real structures of things. The old way of resolving the difficulty of the weakness of the deductive limiting condition, that we accept the simpler theory, is unsatisfactory, since apart from the notorious difficulties of defining simplicity of theories, one could well argue that for two theories, considered just as sets of sentences, to be truly equipollent (capable of implying, with the same initial conditions, exactly the same facts), they must, in the end, be the same theory, and only show superficial or merely notational differences. Genuine differences must hinge on differences in what the sets of sentences *mean*, i.e. on the plausibility, as hypothetical mechanisms, of the iconic models they describe. Faraday's *strain-line* model derived from the theory of continuous media under strain becomes the *line of force* hypothetical mechanism, and it is the explanatory power of a description of this hypothetical mechanism which gives it priority over the equally deductively successful Ampère theory of action at a distance between poles.

(b) *Laws of Nature*. To reach the required logical power, that is, to be capable of implying indefinitely many instances, given some initial conditions, that is, to have the power to lead to predictions, a law of nature must be general. But mere generality is not enough, because there is also a class of statements which are general in form but lack the superior strength of the laws of nature. These are accidental generalizations. They, like the laws of nature, have favourable instances, but no consequences for other parts of science would follow should an accidental generalization turn out to be false. Our confidence in it goes no further than its instances. This distinction is in accordance with the intuition of scientists, and characteristically appears when some non-randomness in phenomena has been spotted, and it has yet to be decided whether the pattern has any significance, whether it is the mark of a real connection, or produced by some causal mechanism. In the first instance we can assert only an accidental generalization summing up the facts we know, but not going beyond them. For instance, people get better when they have been treated for a nervous condition by an analyst. Is this an accidental regularity or is the treatment in some way responsible for the cure? The studies made by Eysenck* suggest that it is an accidental regularity, because he found that people who have no treatment at all recover at the same rate as those who have analysis, i.e. it makes no difference to the prognosis of the disease if the fact that the patient is being treated by an analyst is included in the evidence or left out of consideration altogether.

The tendency of philosophers of the empiricist and deductivist turn of mind is to suppose that the only favourable evidence there could be for laws of nature would be favourable instances. This supposition has created problems of enormous difficulty. Instance-statistics, that is, a count of favourable and unfavourable instances, are just the same for laws as for accidental generalizations, as has been pointed out, at least since Hume. So, if instance-statistics were all there is to go on, sceptics would be right to conclude that all generalizations partake of the character of accidental generalizations, and that there is no class of general statements which are of superior strength to be marked off as laws. Rationalists, working from the same basic and erroneous position about the bearing of factual considerations upon laws, concluded that the laws of nature, which they saw did have superior strength, must derive it by being provable by some sort of *a priori* method, without recourse to instance-statistics. If we call the superior strength of laws,

* H. J. Eysenck, *Uses and Abuses of Psychology*, Penguin Books, London, 1953, pp. 196–8.

their natural necessity, the rationalist move would be to make natural necessity coincide with logical necessity, which is a most dubious proposal. Neither side paid much attention to scientific practice.

In practice, and I shall show that there are excellent reasons for accepting practice, the criteria for accepting a general statement as a law of nature, are two-fold.

1. The law has instances, that is, there are particular sequences of events, or states, that is, there are regular patterns of co-existing properties or other non-random patterns in nature, such as are described in the statement of the law. Thus the logic of instance-confirmation will be shared by laws of nature, accidental generalizations, and those statements of uncertain status I called protolaws. For instance, facts which run counter to examples of these kinds of general statement are *prima facie* evidence for the falsity of the statement, though only conclusive evidence under very special conditions. I shall defer the detailed discussion of this until the section dealing with protolaws.

2. There is some reason, *other than the instance-statistics*, for supposing that what a law asserts, holds generally. This reason, it will emerge, is to be found among the material conditions, as set out above for theoretical statements.

In science the second condition reduces to the belief in the existence of generative mechanisms, structures and powers in the internal constitution of matter, pervasive fields and so on, which are responsible for the patterns and regularities observed in nature and noted in the instance-statistics. Attaching a general statement to a theory is one of the ways of satisfying this requirement, for in that act, the picture of nature which is the core of the theory, is made to depict, perhaps only hypothetically, the mechanism which produces the phenomena in whatever sequence and pattern they are found to be. For example, the Wiedemann-Franz Law which relates thermal and electrical conductivity, under moderate conditions, by the simple relation

$$\theta/k = T \times \text{const.}$$

where θ is the thermal conductivity, k is the electrical conductivity, and T the absolute temperature, can be shown to be just the law of nature that it would be proper to expect, *assuming the nature of metals* and their inner constitution, described in Drude's theory of conductivity, where the 'free' electrons are pictured as behaving analogously to the molecules of a gas.

Two preliminary points should be noticed. A limiting case of the connection of a putative law to a theory is provided by the deducibility of the statement of the generalization expressing the law from some set of hypotheses. This is the minimal case of producing a reason for accepting the statement as a law. If the generalization is deducible from some set of hypotheses, and *if* the hypotheses describe a model (or at an advanced stage a hypothetical, or even a real mechanism of nature) there is a much stronger reason for accepting it as a law, that is, for according it the kind of generality which endows it with superior logical power, i.e. we can use it to infer unexamined cases. So deducibility is *prima facie* evidence of law-likeness, but only *prima facie*, since the hypotheses must have the character of a theory (that is, be descriptive of the underlying mechanisms of nature) and not merely be one of the infinitely many hypotheses and sets of hypotheses from which the generalization can be deduced. This point goes back to at least the sixteenth century, when it was pointed out that mere deducibility of 'All *A* is *B*' from the theory 'All *M* is *B*' and 'All *A* is *M*' is insufficient to support the lawfulness of 'All *A* is *B*', since the deduction will be valid, *whatever M is*, and in general, the likelihood is that with most interpretations of *M*, 'All *M* is *B*' and 'All *A* is *M*' are both false. A party did exist in the late sixteenth century for taking the *FFT* syllogism as the proper form for theories, but they were put to rout, after enjoying considerable vogue, by Kepler and others.

The general theme of this book is that laws of nature are abstractions from what is, in epistemological and psychological actuality, a statement-picture complex. The picture is at least a hypothetical depiction of the nature of the actual world, and only in relation to this, with its depiction of the inner constitution of things, can the way laws of nature function and their powers in our intellectual economy be understood. Accidental generalizations are statements describing patterns in nature for which no picture of a possible generative mechanism can be formed. It is this insight which will resolve the classical difficulties associated with the philosophy of laws of nature, for with it I will forge (*a*) the two-criterion theory of causality, which resolves the difficulties of distinguishing causal laws from other descriptions of successive states, (*b*) a theory of nomic necessity, which explains the kind of necessity laws of nature have, in which it will be shown exactly how the superior strength of the laws of nature arise, (*c*) an explanation which accounts for our intuitions about the connection of subjective conditionals with the laws of nature, and finally (*d*) a resolution of the confirmation paradoxes from which will flow the correct theory of the confirmation of generalizations.

a. The distinction of causal laws from other kinds of laws is the first task for this theory of the laws of nature. It is my view that the aim of science is to try to find the structures, states and inner constitutions from which the phenomena of nature flow. It is a task of molecular biologists to find the structures of molecules in the nuclei of cells which are responsible for the development of various characteristics by organisms. It is, in short, to look for the causal mechanisms of which the patterns and regularities of phenomena are the effects. It is also, though this is sometimes confused with the previous task, to find the inner aspects of things of which the phenomena are the outer aspects. It is to try to find what is happening in a metal when it is conducting electricity for example. In many cases where the spatial metaphor of in and out is misleading, scientists can be described as trying to find the internal and unobserved states of things of which the external states are observable phenomena. Sometimes both a cause and its effect are observable. In such a case, they might both be events, or the cause may be an event, the effect a permanent or semi-permanent state, and so on. Of course, once a stable state has been reached the fact of its endurance has no need of a cause. It is from the observed cases of the action of one thing upon another that we get the idea of causality, which permeates the whole of our scientific enterprise. It is simply a mistake, arising from preconceptions of logic and from ignorance of scientific practice, to suppose that the only knowledge sought by scientists is knowledge of regularities in observable phenomena. Science is by no means just the collection of the statistics of events. Indeed, it will follow from my account that to distinguish laws of any kind, causal or otherwise, from accidental generalizations, a distinction denied only by the most diehard empiricist, in fact requires the full theory of statement-picture complexes, of theory as an account of causal mechanisms, to which this book is devoted.

b. *The analysis and implications of causal laws.* Causal laws are those which state the permanent or enduring conditions under which a certain kind of phenomenon will occur, in short they describe the modes of generation or mechanisms of production of phenomena. Throughout this book, I assume that what are taken to be phenomena are picked out of a more complex reality under the influence of previous knowledge, perceptual schemata and so on. Even so, we can distinguish those happenings which are causes, and which *bring about* other phenomena. These are described in what I shall call 'causal reports', which state the initiating or responsible kind of event for another kind of event or state to occur, and are to be distinguished from causal laws. Causal reports are unintelligible without the idea of a real or imaginary agent. They are anthropocentric, in that they analyze natural happenings from the point of view of a human manipulator.* Examples will make this clear. It is a causal law that the cutting of magnetic lines of force by a conductor generates a current. The Law of Mass Action is a causal law since it describes the conditions which determine the rate of a chemical

* H. L. A. Hart and A. M. Honoré, *Causation in the Law*, Oxford, 1959.

reaction. The Theory of Natural Selection is a causal theory because it describes the way and the means by which new varieties (and perhaps new species) are generated. But to announce that the cause of combustion is ignition is to give a causal report, because it assumes the mechanism of the production of flame, and draws attention to that element in the production of burning that is typically under human control. Causal reports are required by law courts, causal laws by science. Causal reports are derived from causal laws by concentrating upon what was, will be, or might be imagined to be affected by an agent, supposing all other factors in the situation constant.

Species under the genus *cause* include 'produce', 'generate', 'necessitate', 'make', 'lead to', 'bring about' and others, but are to be distinguished from other species of order among phenomena in nature, such as 'accompany', 'is concommitant with', 'mutually varies with' and so on. Science was, is, and always will be, concerned ultimately with the causal genus of order. But fear of anthropomorphism, of injecting unwarranted agents and agencies into nature, brought about perhaps by a long standing muddle between causal laws and causal reports, has led to an overt suppression of at least some of the causal vocabulary, and in the often naive methodological pronouncements of scientists, a disclaimer of the search for causal law.

Essentially, laws have three aspects. (i) Their being laws requires that they aim to describe what happens universally, provided the conditions are right, and not just what happens usually or for the most part. To show that a statement is a causal *law* it is necessary, but not sufficient, to produce some instance-statistics favourable to the hypothesis, that is to show that there is a regularity, or order in the pattern of cause and effect of whatever sort they be. For instance if we want to advance the hypothesis that the selection of micromutations leads to the evolution of new species, as a causal law, there must be some observations of the alleged phenomenon to back it up, that is there must be some favourable instance-statistics. To show that that order is inviolable, that selection of micromutations must lead to evolution of species, is a further step, and no amount of instances will give us the wherewithal to explain, and hence explain away, apparent counter instances. For the special case of events, a criterion of regularity of sequence of events of the cause type with events of the effect type, must be satisfied. This provides some evidence for the generality of the form of that sequence of events.

(ii) But a law is mandatory. How we rule out the possibility of alternative effects under the same cause will emerge from the account of what it is to be a *causal* law, and lead on to an account of natural necessity. I shall postpone the account of the necessity component of causal law, until the concept of natural necessity has been made clear.

(iii) Causal laws are saying ultimately how a cause generates its effect. To establish this element in causal law the proponent of the law can be called upon to describe the manner and mechanism of the generation of the effect by the cause. To do this he must advert to the inner constitutions, structures, powers, encompassing systems, and so on, of which natural generative mechanisms are constituted, and of which the connection between cause and effect usually consists. Sometimes such a generative mechanism can be observed, as for instance when, by dissection, the afferent and efferent nervous systems are laid bare and the outlines of the mechanism of certain kinds of muscular response to stimuli made clear. Sometimes this can be done only with the extended senses, when, for instance, by microscopic studies, the life history of a parasite is followed and the mechanism of infection thus discovered. Very frequently neither of these possibilities is open. Then the proponent of causal law has the whole techinque of model building and theory constructing to fall back upon, and he constructs a picture of the causal connection which he may one day be lucky enough to observe by developing a model of those generative mechanisms he has not yet observed.

A causal law *asserts* then the generative connection of cause and effect. It will follow that in the right circumstances the presence of the cause must always lead to the coming about of the effect. The *proof* of causal law is the satisfaction of two criteria. There must be favourable instance-statistics, and it must be possible to describe, and better to produce, a plausible generative mechanism. The fact that statements of causality (mentioning only events, states and so on) are associated with and indeed only make sense in conjunction with beliefs about the existence of generative mechanisms, was forgotten by philosophers, and so the source of our beliefs in natural necessity and generative connection was lost. The legitimacy of the claim of natural science to discover necessary generative connections was questioned and indeed scouted. Some, misunderstanding the nature of scientific thinking, and concentrating only on the superficial form of isolated statements of law, and forgetting the associated picture of the mechanisms of nature without which a law loses its force, recommended the restriction of the meaning of causality to a mere regularity of similarity of type among sequences of events.

I turn now to examine critically the theory from which this view as to the meaning of causal laws derives.

There is a powerful tradition, springing from Hume's discussion of causality, of reducing the causal relationship from an internal connection between cause and effect to a summary expression of an external relation between events, in which the cause is simply the invariable antecedent to the effect. I propose to argue against the view that the meaning of the relational predicate '. . . causes . . .' is '. . . is the invariable antecedent of . . .', by (i) showing that Hume's arguments against 'connexion' are not strong enough to carry the sceptical conclusion he derives from them, and by (ii) showing that we cannot succeed in identifying the invariable antecedents of an event from the previous state of nature, i.e. from among its total temporal antecedents, unless we employ some method other than the recognition of constant conjunction among event-pairs. It will turn out that constant conjunction is only *prima facie* evidence for asserting that there is a causal relation between event-pairs, and, at best, can be a necessary condition for causality only.

Hume's attack on the claim that in the assertion of the holding of the causal relation we refer to a real connection between the events which are the terms of the relation runs, in summary, as follows: 'After he [an observer] has observed several instances of this nature [i.e. conjunction of event-pairs] he then pronounces them to be connected. What alteration has happened to give rise to this new idea of *connexion*? Nothing but that he now *feels* these events to be connected in his imagination, and can readily foretell the existence of the one from the appearance of the other' (Sect. VII, Pt. ii, *Enquiry*). That is to say we have not learned any new empirical fact other than the conjunction of the event-pairs, upon which we base the pronouncement that the events are causally connected. All we have is a new feeling about them derived from a new habit of mind, that of expecting the one on the occurrence of the other.

A 'cause', then, is simply 'an object, followed by another, and where all objects similar to the first are followed by objects similar to the second' (*loc. cit.*); and, 'if the first object had not been, the second never had existed'. To see why this will not do we need to examine Hume's illustrative example. 'The vibration of this string is the cause of this particular sound' is to be understood, following the above explication, as 'that this vibration is followed by this sound, and that all similar vibrations have been followed by similar sounds: Or, that this vibration is followed by this sound, and that upon the appearance of one the mind anticipates the senses, and forms immediately an idea of the other' (*loc. cit.*). This, Hume contends, must be the correct analysis since we can form no idea of the connection between the vibration and the sound. But the theory and experiments of sonic physics and neuro-physiology give us a very good idea of the connection between the vibration and the sound. We all know nowadays of the train of pressures in the air, the operation of the ear-drum, the cochlea and so on, and we now know something of the train of electrochemical happenings between the inner ear and that part of the brain identified as the seat of audition. Furthermore, to explain what we mean by 'the vibration causes the sound', rather than something else, typically involves, I contend, reference to the intervening mechanism which links the vibration in the string to the sound we hear. The vibration of the string stimulates a mechanism which then acts in such a way that we are stimulated and hear a sound. Some people of Hume's persuasion concede this point but argue that analysis of the intervening mechanisms reduce ultimately to Humean conjunctions. I will consider and dismiss this reply below.

A word we commonly use for the causal relation is 'makes'. For instance to ask, 'What makes the wheels go round?' is to ask a question to which a causal answer is demanded. Suppose we give the answer 'Explosions in the cylinders'. No matter how often we showed that there were explosions in the cylinders when the wheels went round we could not secure conviction that this was the cause, and this is partly because, though we can put the gears into neutral, it still remains true that what makes the wheels go round is the explosions in the cylinders. We secure conviction that this is the cause by describing the mechanism of pistons, connecting rods, crankshaft, gear-box and final drive, which *connects* explosions in the cylinders with the movement of the wheels. Incidentally this allows us to put the gears into neutral with the engine running, and to coast downhill with the engine off, without either of these cases counting as falsifications of the candidature of the explosions in the cylinders for being the cause of the wheels going round. In fact, since explosions in the cylinders neither invariably accompany motion of the wheels nor are they always absent when the wheels are stationary, to say that the explosions cause the wheels to go round can not mean that explosions are the necessary and sufficient conditions for the wheels to go round.

Why then has Hume's analysis so often secured conviction? The answer, I believe, is that it correctly describes a test for causation; that is, it lays down one of the criteria the satisfaction of which *prima facie* entitles us to assert as a hypothesis the relational predicate '. . . causes . . .' of two events X and Y, or two classes of events of the type of X and of the type of Y. If 'X causes Y' means 'X generates or produces Y', then it follows that a test for whether X and Y are related in this way will be to see whether whenever X occurs, Y follows, and whenever X is absent Y is absent. The motoring example shows that failure always to satisfy this test is not an adequate reason, given a mechanism linking cause and effect in a generative or productive way, for denying the causal connection; though if Hume were really explaining the meaning of causation it would have to be. Similarly it is not difficult to find cases, for instance astronomical conjunctions, where, though certain events always occur together, and in the absence of one the other does not, as a matter of fact, occur; that is where the necessity and sufficiency criterion is satisfied in the highest degree, we should still be most reluctant, and rightly, to say that a causal relation holds. If this is the case, even though we may say that the conjunctions we observe are the visible appearances of quite different trains of happenings, 'X causes Y' cannot mean 'X and Y are invariably conjoined'.

To see how causal reports and causal laws are related, it is necessary to consider the matter from a slightly different point of view. 'The cause of an event is that set of antecedent conditions which are necessary and sufficient for the occurrence of the event.' Can this be construed as 'The cause of an event is that set of antecedent conditions in the absence of which the event does not occur and in the presence of which the event invariably follows'?

'The cause of the explosion was an electrical fault.' The electrical fault is here an instance of ignition. An analysis might run as follows:

'If a gas/air mixture is not ignited it does not explode', and
'If a gas/air mixture is ingnited it does explode'.

We are tempted to say that the necessary and sufficient condition for a gas/air mixture to explode is that it be ignited. But this will not quite do. It is also a necessary condition for explosions that the mixture be in the gaseous phase and that it be at a sufficient pressure. But these further conditions are not events which generate or produce explosions. They could not figure as items in a causal report unless they were states which we suspected had been brought about by some guilty party. We do rule them out from the causes of the explosion by making them the conditions of the ignition – explosion sequence. That is they are descriptive, in part, of the mechanism which is responsible for explosions following ignition. Adopting the production/generation view of causality, it is very easy to see why these conditions, though necessary, are not part of the cause of the explosion. That they hold, ensures that the appropriate mechanism to produce an explosion on ignition is present. By distinguishing between those conditions necessary for there to be the appropriate production or generation mechanism present, and the event or state which activates this mechanism to produce or generate the effect, we can distinguish the cause more or less uniquely. By searching out the necessary and sufficient *conditions* for an event we invariably get more than the cause, either because we can be driven to include the whole state of the universe just prior to the appropriate moment, or because we can be driven to include the whole history of the universe up to the appropriate moment, since if anything in these two extensions of the conditions had been different it is at least on the cards that the event would not have occurred just as it did. By adopting the view that the causal relation is generative this difficulty is automatically removed, for we can use the idea of the generative mechanism to distinguish the conditions for an event, that is those conditions which ensure that the appropriate mechanism should be present, from the causes of an event, that is *the* event sequence which activates just *this* mechanism to produce or generate just *this* effect.

A certain degree of plausibility can be given to the constant conjunction analysis of the causal relation by concentrating on a special case of causality which is found in all the sciences. There are interactions, for example, the final products of the analysis of intervening 'connections' between events, which we would be prepared to call 'causal' but for which we cannot describe the connection; that is for which we have no model of the hypothetical link by which we suppose the cause to produce or generate the effect. It will emerge that the existence of these causal relations, far from undermining the general case of this chapter, reinforces it.

As an example of this consider Newton's mechanics, which admits two such interactions, impact and gravity, represented by the relational predicates '. . . collides with . . .' and '. . . is attracted by . . .'. These relations are specified very exactly by the laws of impact and by the law of gravitation. The causes of all changes of motion are impact and gravity. Furthermore we are quite prepared to say that one body, colliding with another, causes the second body to move, and that the sun and the moon cause the tides. But from the concepts of Newtonian science we cannot supply an account of the links between bodies which explains what happens on impact, or why the sun and earth exert a mutual attraction. We can only describe impacts and gravitational systems. However the 'cannot' in 'cannot supply an account of' finds its way into the discussion, not because of some feature of the causal relation in general, but in virtue of the place these particular causal relations have in the logical structure of Newtonian mechanics. If they are the basic interactions then, *a fortiori*, interactions more basic, in terms of which a model for a generative mechanism can be constructed to explain them, cannot be found in Newtonian mechanics. This is shown by the fact that by giving a different formulation to mechanics we can supply the link that is absent in Newton's version. In a mechanics based upon the transfer of energy, the link between impacting bodies, in virtue of which the first causes the second to move, is provided by the mechanism of the transfer of energy which now becomes fundamental and inexplicable. The laws of impact are then not just descriptive of what happens in collisions, but can be reinterpreted as laws governing the transference of energy. That is, they now describe the causal link in virtue of which the first body generates (since energy and not motion is transferred) motion in the second. Newton himself, dissatisfied with two basic interactions, suggested that the cause of gravity be looked for in some contact-like interaction between the ether and material bodies. In this he tried to make impact basic.

It follows from these considerations that the relation '. . . causes . . .' can be asserted of the components of the basic interactions, since one of the criteria for the assertion of the relation is satisfied, namely constant conjunction of cause and effect, but this is not to say that even in the basic interactions this is what '. . . causes . . .' means. We can use '. . . causes . . .' of the basic interactions just because, in some other theory either alternative or more basic, a model of the generative or productive mechanism could be supplied. To say that the basic inter-actions are causal is, as it were, to indicate some unfinished business.

Hume's theory has sometimes been defended in the following way: the vibration-sound example is only apparently defeated by the supply-ing of the connection between vibration and heard sound in scientific terms. All that has been done is to shift the causal problem back a step from the relation between the vibration and the sound to the relations between the components of the intervening mechanism; and to these Hume's sceptical doubts of discovering connectivity still apply. The answer to this objection can be found by applying the idea sketched in the last paragraph more generally. When we begin to investigate some phenomenon of nature we know, by having discerned some non-random pattern like constant conjunction, that it is likely that a causal relation is present. To supply an account of this causal relation we describe a mechanism which is such that on the occurrence of events of the type of the cause, it produces or generates events of the type of the effect. Let us call the cause event 'a', the effect event 'b' and the mechanism 'M'. Now the mechanism M consists of certain components which interact according to certain laws, and we find or propose certain conjunctions of phenomena, say p, q and r, within M, which account for the way in which when it is stimulated by a-type events it produces or generates b-type events. Causal relations between these are filled out by describing other mechanisms, M_1', M_2', . . ., which are such that p stimulates M_1' to produce or generate q, and q stimulates M_2' and so on until r is produced or generated; and a produces or is identical with p, and r produces or is identical with b. To such questions as, 'How does p produce or generate q?' we answer by describing yet more mech-anisms M_1'', M_2'', . . ., until we come to a pair of components, in a mechanism, which are related by the basic interactions of the science we are using. Any scientific enquiry with which the reader is familiar can provide instances of this procedure, e.g. physics supplies the mechanisms for chemical phenomena, chemistry for many physiological phenomena, etc., etc. Far from the discovery of constant conjunctions being identical to the discovery of causal relations, they are taken as the signs of causal relations, which have to be filled out by describing the mechanism which accounts for the constant conjunction. Compare 'Why did he faint?'; *Ans*. 1 'The room was over-heated, and over-heated rooms make people faint' with 'Why did the over-heating make him faint?'; *Ans*. 2 'Lack of carbon dioxide in his blood automatically restricted the blood supply to his head'; *Ans*. 3 'The molecules composing the walls of capillary blood vessels shorten in the absence of carbon dioxide', and so on through *Ans*. 4, *Ans*. 5, etc., until the resources of physiology, chemistry and physics are exhausted. To give *Ans*. 1 is to supply the framework for a causal account which answers 2, 3, 4 . . . progressively fill in. No end to this process of discovery and invention can be foreseen, i.e. this paradigm of causal explanation can be reapplied indefinitely often.

Genesis and genetic connection is, then, always understood as the stimulus and response of some enduring 'mechanism' in nature. The solution to the problem of how the stimulus causes the mechanism to act, and the subsequent state of the mechanism causes the required effect, is simply to add two further generative mechanisms. So natural knowledge, organized as causal laws, is stratified. The deepest stratum, to which I have only briefly referred above, will be discussed in detail in Chapters 10 and 11. Physiology provides some of the most striking examples of this stratification. We know that the body is capable of maintaining a constant temperature. When the ambient temperature is raised various effects occur, one of which is the increase of blood flow to the skin, so the cause of increased blood flow to the skin is increase of ambient temperature. At the next stratum, or level, the mechanism of the diversion of blood to the surface of the body is described and involves the description of part of the nervous system, of the heat core of the body, and so on. The increase of the ambient temperature effects the temperature regulation mechanism, and has the ultimate effect of increasing the blood flow to the skin. Then we ask, 'How does the ambient temperature stimulate or affect the mechanism?' To explain that, the effect now becomes the state of the mechanism, say the nervous impulses from temperature sensitive elements in the body surface. Our task then becomes that of describing or inventing the mechanism of heat detection, and then of the transmission of nervous impulses, and so on.

The discovery of the generative mechanism hypothesitized in a model confirms a causal hypothesis which might originally have been formulated on the *prima facie* evidence of some favourable instance-statistics. We shall see that it is the existence of the possibility of discovering generative mechanisms, or failing that, of building or imagining plausible models for them, that provides greater strength than instance-statistics are able to afford, and which distinguishes, as we have already seen, laws of nature from accidental generalizations and protolaws. It explains too, how, frequently, very few instances are required to confirm a law of nature. In those cases it turns out that the connection criterion is highly satisfied. For accidental generalizations we have no idea of the generative mechanism, or even whether there is one, so we have no reason for believing that sequences or patterns of merely accidental origin will persist. Are there *really* accidental sequences, or is it just our ingnorance that makes them seem so? This we shall turn to next.

Conditionality. The laws of nature are general statements, and their content, at least in part, seems to be expressible by a statement in the form of a conditional. Indeed, since the laws of nature are involved in describing the conditions that have to obtain for a certain phenomenon, event or state, to occur, they naturally take on conditional form when they are detached from the theories which describe the generative mechanisms responsible for the relations among phenomena they describe. An example of a statement of general form is 'All A are followed by B', and if the obtaining of A is supposed to be a condition for the coming to be of B, then that much of the content of the general statement can be expressed as 'If A obtains, then B will come to be'. But clearly the form 'All A are followed by B' would do just as well for an A and a B which were not such that the obtaining of A made B happen, but which merely as a matter of fact followed one another, and the generality of which is merely accidental. In that sort of case we would not be entitled to assume that the obtaining of A was the condition of the coming to pass of B. We would just not know what was required to bring about B, whether it was something connected with A or not. So the sort of conditionality relevant to causal explanation is a less general but stronger relation than is expressed by 'All A are followed by B'. A similar distinction runs through the analysis of the form 'All A are B'. If everything that is A is also B this may be because their being B is connected with their being A, or it may just be an unconnected accompaniment, or it may be consequent on all those things having something else, C, in common, which is separately responsible for their being A, and their being B.

To get to the heart of the muddle which deductivism has bequeathed us it is necessary to disentangle a serious conflation of kinds of conditionality which have greatly obscured the subject. There is the conditionality that holds in reasoning, between premises and conclusions, in which the truth of a premise is a condition for the truth of the conclusion. This conditionality can be strikingly reduced to the material conditionality of truth-functional logic, which expresses, as it were, the instance-statistics of truth. There is *also* the conditionality that appears in nature, between conditions (causes) and effects. The event-as-changes myth has led to the attempt to treat this conditionality as the instance-statistics of instantaneous events, and to use precisely the same logical connective to express the conditionality among instantaneous events or states, as was used to express the conditionality that obtains among propositions, in particular propositions related as premises and conclusions. But the conditionality of propositions treated as material conditionality, is exactly equivalent to accidental or mere generality, within which the superior strength of laws can not be distinguished. So having raised a dust by analysing statements of lawful conditionality by the use of the relation of material conditionality appropriate to the barest logical relation, the deductivists complain of an intolerable difficulty in distinguishing accidental and law-like general statements, which, under this theory, are now conflated.

As a preliminary to the proper analysis of conditionality, the Goodmanite heresy must be dealt with. Goodman and his followers are prepared to admit, where Hume was not, that there is a distinction of some scientific propriety to be drawn between accidental and law-like generalizations; those which I shall say are used to describe Regularities of Type *A* as distinct from those used to describe Regularities of Type *B*. Being committed to deductivism, and to the event mythology amongst other things, the Goodmanites persist, like so many deductivists, in using only material conditionality as an analysis of the law-like conditionality of causal and other scientific statements. They then try to distinguish the accidental from the lawful regularities, Type *A* from Type *B*, by a condition on the predicates employed in the statement of the law. Goodman introduces two concepts to try to express his point of view. A 'projectible' predicate is one about which we can have some confidence that it will be correctly predicable in similar circumstances to those in which it has been properly predicated before. Predicates which are to be used in law-like statements are called by Goodman, 'entrenched predicates'. Entrenched predicates are, it seems, those that have been projected in the past. But if this is what entrenchment means, then traditional inductive scepticism shows that by this criterion of entrenchment or non-entrenchment, no absolute distinction of any traditional logical validity between law-like and accidental generalizations can be made out. (Traditional logic sets the standards chosen by the deductivist.) Of course if entrenchment is extended to include incorporation of statements containing predicates to be entrenched, within theories, provided a simple deductivist account of theories is avoided, some more plausible criterion of entrenchment might be given. But, more is the pity, the Goodmanites cling, too, to the deductivist myth, so that the only grounds, for them, for holding theoretical propositions to be true, are ultimately the instance-statistics of their logical consequences, and to the use of these, inductive scepticism applies with undiminished force. Entrenchment, if this is reduced either to immediate or mediate confirmation by instances is surely no logical ground for projectability.

We have already seen how causality is established by the satisfaction of two tests, the *prima facie* test of whether there is any discernible pattern in the field of phenomena under investigation, such as a regularity of sequence among kinds of events, and the definitive test, of whether there is a generative mechanism at work; we noticed that in many cases we have to be content with only a model for that mechanism. A causal law then, gets confirmation both from empirical evidence, through satisfying the *prima facie* test and through an account of a plausible generative mechanism. The material conditionality of formal logic reflects only the instance-statistics criterion, the mere obtaining or non-obtaining of the purely statistical relations of truth and falsity of statements corresponding to the antecedents and consequents of conditionals. And, as we shall see, such statements do not express the *content* of antecedent and consequent adequately even. It is the existence of the knowledge of, or the hypothesis of the model of, the generative mechanism that distinguishes the law-like conditionality of scientific statements. That is, what makes the difference is the picture of nature within which the statement is usually embedded. Extract the statement, and consider it by itself, and immediately its law-likeness, its superior force, vanishes.

The generative mechanism is the basis of natural necessity. The general notion of necessity has two components, which have different prominence in different cases.

1. That which is necessary is without alternative. A necessary condition, for instance, is one of those conditions, which, if it is not met, and some alternative state of affairs obtains, the desired outcome will not occur. What one wishes to indicate by the use of the expression 'natural necessity' is the absence of possible alternative outcomes given the occurrence of certain conditions. Given the antecedent conditions, and the structures and powers of the things and materials involved, we can see in certain cases that there is only one possible outcome. To say that there is necessity in nature is to say that in certain conditions, only one outcome is possible. Empiricists of the Humean variety argue that so far as we can deduce from the instance-statistics of past events, any phenomenon could be the outcome of any set of conditions. And, given

their myths, they are, of course, perfectly right. Again, of course, the absurdity of their conclusion should make us suspicious of the grounds upon which it is based. We know perfectly well that there are many sets of conditions from each of which only one outcome can be expected, where there are no alternatives. Even when outcomes are given only with probability, there is a definite structure to the distribution of probabilities amongst the possibilities. Clearly, since instance-statistics yield the empiricist and deductivist conclusions, they must be supplemented by quite different considerations. These other considerations are to hand, in the nature of the generative mechanisms by which the obtaining of the conditions produces the subsequent phenomenon.

Looked at more formally, the problem can be put in terms of the relations between the antecedent and consequent of a conditional statement. Given two predicates ϕ and ψ, which represent (can be used to ascribe) two properties which have as a matter of fact, appeared together, say electrical conductivity and thermal conductivity, are we entitled to conclude that if a metal is a good conductor of electricity then it will be a goodish conductor of heat? The more we know about the mechanism of heat and electrical conduction, the more narrowly we can circumscribe the subsequent possibilities of behaviour, and the more we are justified in believing that the presence of these properties is not accidental. In short anything which tends to reduce the number of alternatives on the consequent side of a conditional adds a mite of necessity to the relation which is supposed to obtain between the obtaining of the antecedent conditions and the consequent phenomenon. And typically this reduction is achieved by an understanding of the mechanism of the connection.

A measure of natural necessity along these lines is relatively easily devised. Suppose nothing whatever is known about the generative mechanism involved in the production or generation of a phenomenon. Then we must say that, strictly speaking, any subsequent state is equally possible. The subsequent possibilities are infinite. Let $\mathcal{N}$ be the measure of natural necessity, given by

$$\mathcal{N} = 1/n_a$$

where n_a is the number of alternative consequences possible. If $n_a = \infty$ then clearly $\mathcal{N} = 0$, so on a Humean, deductivist view there is no natural necessity, since, given any state of affairs, it is logically possible that any other whatever may succeed it. Suppose on the other hand that we have a pretty complete understanding of the generative mechanism, that for instance by which covalent compounds are electrolyzed, then in a given set of conditions, we can see that only one possibility is open. In this situation $n_a = 1$ and natural necessity has a measure $\mathcal{N} = 1$.

A rather more elaborate measure is required for intermediate cases, because the usual thing is that our knowledge of generative mechanisms does not rule out *all* possible alternatives, but reduces them drastically. Our knowledge of these mechanisms enables us to weight the various possibilities as more or less probable. If there seem to be two possibilities $\mathcal{N}$ may not be simply 0·5, but rather higher, say because one possibility is very much more likely than the other. A measure of $\mathcal{N}_k$ for the kth possibility that seems plausible is

$$\mathcal{N}_k = \frac{p^{k+1} \ldots p^n}{n \times p^1 \ldots p^k}$$

where n is the number of possibilities, and their probabilities $p^1 \ldots p^k$ $\ldots p^n$. One must remember too, that, in general, $p^1 + \ldots p^n \neq 1$, since nothing at all may happen in any given set of circumstances, whatever we may have by way of antecedent knowledge.

Logical necessity and logical possibility are limiting cases, having, the values of 1 and 0. If p necessitates q logically, then the only truth-alternative to q, namely not-q, is not a possible conclusion from p, because, of course, p & not-q is a contradiction. So there is no alternative to the truth of q as the conclusion, if p is true. Thus logical necessity has $\mathcal{N} = 1/1 = 1$. Logical possibility, by parity of reasoning, gives a measure of $\mathcal{N} = 1/\infty = 0$, since if q is logically possible in the presence of p, then neither p & q, nor p & not-q, are contradictory, and so ruled out. There is an old problem of measurement here. If we treat not-q as the alternative to q on a par with q itself, then we can be led into supposing that there are just two alternatives, q and not-q, and so on giving an implausible measure $\mathcal{N} = 1/2 = 0·5$ for the measure of necessity. But of course, not-q must always be supposed, unless specifically restricted, to be infinitely rich. Not-q, if treated as the denial of a certain predicate of an object, must be analyzed as a potentially infinite conjunction of the attributions of all positive predicates which are not q, i.e. if a thing is not red, there are infinitely many colours that it might be.

2. There is another aspect of logical necessity, not found with natural necessity. This is the absolute inviolability of logically necessary truths. Our beliefs as to what are the conditions which naturally necessitate an effect are revisable with further experience. The laws of nature, causal laws, are not absolutely inviolable. Nor are attributions of natural necessity to sequences of phenomena beyond revision. Of course logically necessary statements are inviolable since their form is such that they can have no non-necessary consequences. Logically necessary statements offer neither alternatives, nor the possibility of falsification. Natural necessity reflects only the single-possibility feature of logical necessity, not its absolute unfalsifiability. But there are statements in

science which scientists treat as unfalsifiable. They are treated as un-falsifiable by the choice, tacit perhaps, of the scientific community. They are not laws of nature. They will be described and their rôle made clear in Chapter 8.

Imagine the drum of a seismograph. The chart upon which the pen runs represents all possible earthquakes at successive times. A law of nature is like the drum of a seismograph. The line the pen traces marks the inexorable course of quakes which carve actuality out of all possible tremors. But in another way a law of nature is not like the drum of a seismograph. Looked at in another way a law of nature describes the necessities among possibilities. It sketches out a network of causal con-nections between kinds of phenomena, so that if one of the possibilities is realized we can tell with some degree of certainty, what *must* sub-sequently happen. Our knowledge of what is necessary in nature is provisional, but in logic, each move can be reduced to a step or sequence of steps. Failure to take one of these steps would result in self con-tradiction. So our knowledge of what is necessary in reasoning and discourse is, with certain reservations, complete.

I want now to turn to two of the hardest problems that arise in the traditional deductivist view of laws of nature, to see how far the views set out in this book can deal with them. The Copernican revolution in the philosophy of science consists in bringing models into the central position as instruments of thought, and relegating deductively organized structures of propositions to a heuristic rôle only, and resurrecting the notion of the generation of one event or state of affairs by another. On this view theory construction becomes essentially the building up of ideas of hypothetical mechanisms. Remember too that my aim is to preserve the persistent intuitions of scientists, as far as possible, against the philosophers. If received logic runs counter to an important in-tuition then I take it as a *prima facie* hypothesis that received logic is inadequate in some way. The two problems I shall handle now are both concerned with law-like general statements, their logical con-sequences and their relation to evidence. The first problem concerns the relation between subjunctive conditional statements and indicative statements, and the justification of subjunctive conditionals. The second is the problem we might call 'Hempel's Paradox' which is concerned with the possibilities of inference from general statements, and their relation to evidence.

The problem about subjunctive conditionals can be put in a number of ways, but I shall consider it by discussing the idea that a *prima facie* case may be made out for the law-likeness of a general assertoric state-ment concerning a certain subject matter, if subjunctive conditional statements about the same subject matter seem to be acceptable. I shall proceed by trying to show that only if some idea of a hypothetical mechanism is associated with a generalization will that generalization warrant subjunctive conditional statements about the same subject matter. So, if subjunctive conditionals *are* warranted, then the general assertoric statement with which they are related, must be associated with some idea of a hypothetical mechanism. Earlier it was shown that that association is the condition for a general statement being a law statement. So, if a general statement is of the strength of a law, it must always be understood as abstracted from a statement-picture complex, that is the ideas of what hypothetical mechanisms are operative in the field must not be overlooked, in considering its logical powers. I set out first some obvious truths about the kinds of statements under discussion.

A. The subject matter of general subjunctive conditionals is possibilities, *excluding those known to have been actualized.* 'If any specimen of sodium were brought into contact with water, hydrogen would be evolved.'

B. The subject matter of general indicative conditionals is possibilities, *including those actualized.* 'If any specimen of soduim is brought into contact with water, hydrogen is evolved.'

C. The subject matter of law-like general statements is identical with the subject matter of general indicative conditionals, in which the same predicates appear. 'All specimens of sodium in contact with water produce an evolution of hydrogen.'

D. The subject matter of accidental generalizations is actual cases, *excluding possibilities.* A characteristic grammatical form for these statements, at least in the past tense, is 'When this, that and the other specimens of sodium were brought into contact with water, hydrogen was evolved.'

The truth-grounds for indicative conditionals are (i) the possibilities that are actualized, that is true particulars, known cases of sodium that have reacted with water in the way mentioned in the statement; (ii) *plus* attendant theory which supports the tacit induction to possibilities not actualized. In my view (ii) is different from (i), since (ii) involves the idea of a hypothetical mechanism, represented by a plausible model of the actual mechanism of the process, so that the reasons for accepting the statement expressed in an indicative conditional are not just the true particulars.

The truth grounds of subjunctive conditionals are restricted to (ii), in effect to our being capable of forming some idea of the hypothetical mechanism likely to be responsible for the kind of phenomena under investigation, because the subject matter of subjunctive conditional statements does not include the actualized possibilities which, taken by themselves, would constitute favourable instances. So if I am right, that is if (ii) does constitute independent grounds for accepting a statement, then there is a source of grounds for accepting subjunctive conditionals, which are among the grounds for the sibling indicative conditionals, though the exclusion of actualized possibilities from the subject matter of any subjunctive conditional statement excludes true particulars from their support, in any simple way. However support of kind (i), that is true particulars of the indicative general statements, always add their mite to (ii), our confident belief in a certain hypothetical mechanism, represented by a model, so via, *and only via*, the confirmation of attendant theory do true particulars lend some support to subjunctive conditionals about the same subject matter. This is because that theory would apply, and that hypothetical mechanism would be supposed to work, in those cases which are mere possibilities too. The truth grounds of accidental general indicative statements are restricted to true particulars, so they do not have associated ideas of plausible mechanisms, so they do not have associated subjunctive conditionals with some degree of acceptability. Only if a general statement is a law or law-like, that is part of a statement-picture complex of the proper sort, will it sanction subjunctive conditionals whose subject matter is confined to possibilities. And therefore if a general statement does seem to sanction a sibling subjunctive conditional we do have a *prima facie* case for supposing it to be law-like: a case that can be made water-tight by describing, and justifying through modelling, the hypothetical mechanism that is supposed to be at work in both cases. This theory would explain our intuition that the acceptability of an associated subjunctive conditional statement is a mark distinguishing law-like from accidental generalizations, since, another way of making that distinction, is by whether true particulars are backed up, for induction, by attendant theory, that is by ideas of plausible generative mechanisms.

All future tense conditionals partake of the logic of subjunctive conditionals because, *a fortiori*, their subject matter is possibilities, and since the future is of necessity not yet actual, future tense statements must exclude actualities from their subject matter. At first sight it might seem right to say that all conditionals in the future tense ought properly to be in the subjunctive mood, so that 'If I do stay to supper I will miss the train' would be ungrammatical. The proper form might be thought to be 'Were I to stay for supper I would miss the train.' But 'If I do stay to supper I will miss the train' does seem to be perfectly proper and to express a hypothetical relation among possibilities. One would expect a corresponding form to express a past possible relation, and the form 'If he did kick her then he deserved obloquy' does seem proper too. This could lead us to think that for the general, scientific cases we need only one category of possibility, roughly rendered as 'plausible but not actual'. But we need at least two categories of possibility for singular, non-scientific cases. Thus, on this view, 'If I do stay . . .' expresses a 'near' possibility, while 'If I were to stay . . .' expresses a more remote one.

We can get some help in finding a rationale for particular subjunctive conditionals of a quasi-scientific kind, from these ideas. The cases that have been thought difficult are on the model of 'If Caesar had not crossed the Rubicon, Rome would have remained a Republic.' The next step is to distinguish between specimen-instances, like a piece of butter, and individual-instances like Caesar. At first sight it looks as if the rationale I have sketched will do perfectly for subjunctive conditionals about particular specimen-instances, but not for those which are about individual-instances. Searching for reasons for believing a statement about possible fates for the butter, directs our attention both to actual fates of other instances, and to the picture of digestion which sets the scene for remarks about possible fates of the instance under discussion. Neither adversion seems proper in Caesar's case. There may be no other instances. There may be no theory. Do we then have any reason whatever for believing the subjunctive conditional about Caesar? Is there *any* rationale for statements of this sort? I think people do proceed in cases like this by trying to reconstrue the individual-instance as a specimen-instance. And in so far as they are successful in this, in so far as Caesar's crossing the Rubicon comes to be seen as just another preliminary to seizure of power by a right-wing revolutionary Officer's Council, or whatever, so our knowledge of the mechanism of revolution and the conditions of its success and failure can be brought to bear. Such knowledge is often encapsulated in a paramorphic model of the political currents in a state.

The second well-known problem, with which we are now equipped to deal, is due to Carl Hempel (141). Its essential features can be brought out by expressing the problem in terms of three principles:

Nicod's Criterion. A general statement is confirmed to some extent by finding that it has instances, for example, that there are things which have the properties as described in the general statement. Thus if the general statement under consideration is 'All emeralds are green' this is confirmed to some extent by finding individual instances of things, acknowledged to be emeralds, which are found to be green. Confirmation is by instances. We recognize this, of course, as part of the mythology of deductivism. Disconfirmation is also by instances, and our general statement would be disconfirmed by the finding of a red emerald. The facts of the colour of rubies, i.e. non-emeralds, would generally be held to be neutral with respect to the confirmation or disconfirmation of 'All emeralds are green'.

E

Equivalence Condition. If some general statement, *G*, is confirmed by some instances, all statements logically equivalent to *G*, are equally confirmed by those instances. In particular, for the purposes of the paradox, a statement and its contrapositive are held to be logically equivalent, that is 'All emeralds are green' and 'All non-green things are non-emeralds' are held to be logically equivalent. And indeed on traditional logical principles they are equivalent.

Hempel's Intuition. If we consider some statement like 'All emeralds are green' then it seems clear that the discovery of a green emerald confirms that statement rather better than would the production of a red ruby. But since a red ruby is a non-green non-emerald, it confirms 'All non-green things are non-emeralds' to exactly the same degree as a green emerald confirms 'All emeralds are green'. And since, by ordinary received logic, 'All emeralds are green' is logically equivalent to 'All non-green things are non-emeralds' the red ruby should, via this equivalence, confirm 'All emeralds are green' just exactly as well as would the production of a green emerald. But this runs counter to our intuition that the discovery of a green emerald confirms 'All emeralds are green' better than finding a red ruby would.

Wonderful have been the attempts to solve this puzzle within the framework of received logic!* I shall treat the puzzle as yet another *reductio ad absurdum* of the idea that received logic is adequate for dealing with even such a simple part of the reasoning involved in the sciences as the relation between general statements and instances.

The equivalence condition is clearly the most dubious, and I turn to that first. If I am to accept the statement that all emeralds are green as having mandatory force, as being of the nature of a law, and so of being capable of supporting predictions to the colour of unexamined instances, I must not only have knowledge of specimen-instances, but I must also have some idea, it may only be through a model, of the generative mechanisms at work in the differential reflection of light by gem stones. That is I must have some idea why it is that something which has the defining properties of an emerald, that is a certain chemical composition and crystalline structure, is green. When the contrapositive of the original statement is formed, it relates, not emeraldness and greenness, but non-greenness and non-emeraldness. To make that equivalently law-like, we should have to produce some sketch or model of the mechanism by which, whatever it was that constituted the non-greenness of things, produced their non-emeraldness. And of course, since negated properties are not really attributes at all, the question of the nature of the mechanism by which one comes to be associated with the other is otiose. So contrapositives are not equivalent for con-

* See the bibliography to this chapter, Items, 139–47.

firmation to their originals, since in the process of contraposition the mechanism which lies behind the lawfulness of the original general statement is just lost. This is so even of a very simple world. Suppose that there are only two kinds of things in the universe, emeralds and rubies, and only two properties which anything may have: being red or being green. Contraposition of 'All emeralds are green' yields 'All red things are rubies'. So a red ruby must, it seems, be equally confirmatory of both. Yet once again a moment's reflection shows that the connection between chemical nature and colour which would have to be sketched to supplement the instance-statistics of the emerald statement, will not necessarily do for the ruby statement. In this case, where the sorts of things in the world are restricted to the same natural kind,* we might argue that the same sort of mechanism operated in both cases. But if the world consists among other things, of crows and shoes, which can be either white or black, then the mechanism by which crowhood and blackness are associated is genetic and biochemical, while the reason that white things are shoes will surely be a matter of fashion and so call for a sociological explanation, or some such.

For any statement which purports to state a law or a strong connection between states, properties and so on, contraposition, or any other logical move that involves the use of negated predicates, is forbidden.

Clearly Nicod's criterion is no less objectionable than permissiveness over contraposition. There is no doubt that if a general statement is to have predictive power, it cannot be certified merely by instance-statistics. If it is to be anything more than an accidental generalization, if it is to give us any ground whatever for expecting the phenomena it describes to be found in similar relations and juxtapositions, in the future, to those we know now, then our confidence in it cannot be based only upon the cases we have come across so far. Nicod's Criterion is, at best, a criterion for a general statement to be a protolaw, and protolaws are general statements which are not yet associated with knowledge of generative mechanisms to give them law-like status.

Hempel's Intuition only seems plausible when we think of the positive particular as a favourable instance, and as an exemplification of a generative mechanism in action, in the appropriate sort of case. The contrapositive particular, though a favourable instance, must be an exemplification of a different mechanism in action, or just no mechanism at all. This is why it seems so much weaker than the positive instance, it just exists in the limbo of accidents, until something more can be discovered. The reason, in short, why Hempel's intuition holds, that is, why a positive instance seems to give greater confirmation than a contrapositive one, is just exactly the reason why contraposition is forbidden.

* *and* the predicates are determinates under the same determinable.

The argument, so far, assumes that the contrapositive is formed by the negation of the terms of the original hypothesis. From a statement about emeralds and greenness a statement about non-emeralds and non-greenness was deduced. In short, if term-negation is used to form the contrapositive, the argument shows that mechanism change is involved in contraposition, and so the logical equivalence condition fails, since the contrapositive does not express the same law as its original. In these circumstances Hempel's Intuition is preserved.

To complete the argument, the other horn of a dilemma must be constructed. Consider now contraposition by propositional negation. Suppose the law-like generalization is 'If anything is an emerald, then it is green'. This can be contraposed by propositional negation into 'If something is not green then it is not an emerald', and the subject matter continues to be emeralds and greenness. Just the same generative mechanism would be adverted to in support of this statement as of its original. But it should now be perfectly clear that the discovery of a green emerald supports both original and contrapositive to exactly the same degree. A green emerald supports the contrapositive just as well, because it goes to show that emeralds *are* green, and not any other colour. This horn of the dilemma preserves the equivalence condition, but at the cost of abandoning Hempel's Intuition. In neither case are the propositions, which, taken together generate the paradox, in fact jointly assertable.

The argument of this chapter has established the value of the realist conception of what a law of nature is, a conception that can be summed up by seeing how it is connected with the concept of 'power'. The concept of 'power', together with the concepts of 'capacity', 'liability', 'tendency' and others will become of great importance in later chapters of this book. Arguments will be developed to show that the differentiation and identification of things and materials in terms of their powers, rather than their qualities, leads to an epistemology and metaphysics closer to the realities of scientific thought. To say of a thing or material that it has a power to do, or to be, or to effect something or other is to say specifically what would happen under appropriate conditions, *and* to say that these effects occur in virtue of the nature of the thing or material, whatever that may be. It is to say specifically what a thing can do but only unspecifically what nature it has, because it is to say only that it is in virtue of *whatever* nature it has that it can affect things, materials and observers the way it does. To say that acidic solutions have the power to turn litmus paper red is to say specifically what they will do, and to say that it is in virtue of their acidic nature, whatever that is, that this effect occurs. A powers statement opens up a kind of empirical question different from the

empirical studies which lead to knowledge of how things behave, because it directs our attention to the problem of the nature or constitution of the things and materials involved in a reaction, an interaction
and so on. It directs our attention not only to the problem of how acids
behave, but to the problem of their common chemical constitution, in
virtue of which they behave the way they do.

It is the view of this chapter that a law of nature states explicitly that
a certain pattern of events, or of structure, occurs in nature, and our
according this description the status of a law tacitly associates this
pattern with a mechanism, more or less permanently existing, which
is responsible for the pattern in the phenomena. To accord a statement
the status of a law is to treat it as describing a process having *an* associated generative mechanism, but not to describe what that mechanism
is. Thus we have the same specific and explicit, tacit and unspecific
structure we have just identified as the logical form of a powers statement. On this view the laws of nature can be seen as describing the
powers (and capacities, liabilities and tendencies) of things and
material substances. These may be powers to act upon each other,
or to affect observers and instruments in various ways. It is just because
these effects are the effects of powers, that observation and experiment
can inform us of the natures of things and materials. To this we will
return in a later chapter when I discuss the development of a sophisticated and technical vocabulary for the description of observations and
experiments.

It has often been pointed out that laws of nature and other more
humble generalizations about phenomena are accompanied in practice
by a *certeris paribus* clause. The function of this clause is to confer a
certain degree of immunity upon the generalization to preserve it from
pseudo falsifications, and too easy rejection.

All generalizations, whether accorded the status of laws of nature or
not, describe what can take place, but only against an assumed background of stable conditions. The totality of such conditions can never
be finally discovered so the application of a law in particular circumstances is always liable to be upset by the vaccilating of some condition
which we had not taken into account. The 'other things being equal'
clause is from a logical point of view, an open set of hypotheticals to
which we must be prepared to add. The viewpoint of this chapter
would demand that *certeris paribus* qualifications be of two kinds, those
which qualify the maintenance of the powers of things and materials,
and those which qualify the stability of the circumambient conditions
in which some happening occurs or structure is found. Compare 'The
musket will go off provided the powder is dry' where a condition for
the maintenance of the power to explode is stated, with 'The musket

will go off provided there is a flint in the lock' where a condition for the obtaining of the conditions to produce detonation is stated. In general, the maintenance of circumnambient conditions is easier than the maintenance of powers. There is a widespread but mistaken assumption that by maintaining the stability of external conditions it is always possible to maintain the powers of things and materials. It was this assumption that led to the efforts to alter the rate of radioactive decay by altering all known external conditions. The failure of that enterprise simply shows that the tendency to decay is to be described in the logical form of a power (in fact a 'liability to decay'), rather than in the logical form of an unqualified hypothetical statement outlining the conditions. There are no conditions. Somewhat less extreme than this case are the innumerable phenomena of human behaviour which show a striking dichotomy between the *certeris paribus* of intrinsic natures of people, and the *certeris paribus* of the conditions to which they are subjected.

Once again we turn to the characterization of a scientific explanation. In order to give a scientific explanation every happening must be looked upon as due to the workings of some mechanism, which may be proceeding in isolation from its environment, as does a clock or a radium atom, or may be in various degrees dependent for stimulation upon changes in the circumnambient conditions, as is dynamite or litmus paper. Classical theories of nature can be roughly characterized as assuming that the mechanisms in nature remain quiescent unless stimulated. Some mechanisms are like that, but I can see no grounds for supposing *a priori* that this must be true of all. Giving an explanation then, consists in identifying the change in circumnambient conditions, if there has been one, which is to identify the stimulus to which the things or materials involved have been subjected, and then to describe the nature and constitution of the things involved which makes clear to us what mechanisms have been brought into operation. A tenuous gas is made to carry an electric current and it glows with light of a characteristic wave length. By describing the atoms of the gas according to the Bohr model we give an explanation of the specificity of spectra by coming to form an idea of what the mechanism of spectrum production is, or might be. The logical form of an explanation must then be a conjunction of a hypothetical statement linking conditions to behaviour and a categorical statement describing the mechanism responsible for that behaviour. It follows that a power statement is a schematic explanation.

One recognizes a pattern in nature, an organization which is not random: for example the cubical crystal of common salt, or the resemblance between parental organisms and their offspring, or the patterns of motion in the sky. One explains such things by, for example, referring to the mutual powers of sodium and chloride ions which one imagines to be at the vertices of the crystalline structure, and sees that if they are as imagined, then that structure must be a cube. Scientific explanation consists in finding or imagining plausible generative mechanisms for the patterns amongst events, for the structures of things, for the generation, growth, decay or extinction of things and materials, for changes within persisting things and materials. It seems natural to speak of a generative mechanism responsible for a thing coming to be in a certain state, as we think of a collapsing field generating a current in an appropriately placed conductor. It does seem to be somewhat odd and indeed perhaps an illegitimate extension of the idea of generation to speak of a generative mechanism responsible for a thing continuing to be in a certain state. However, it is, I think, quite certain that explanations of persistence in states are sometimes required, and that when they are, and the conditions under which they are will be discussed in a later chapter, we advert to mechanisms by which the state is maintained. In general, such mechanisms are adverted to when there is reason to suspect that there is some influence which tends to destroy the state. There has to be a mechanism for the maintenance of bodily temperature since left to itself any warm body cools off, or heats up if the circumnambient temperature is higher than the temperature of the body. If there is a tendency for a moving body to continue to move in a straight line then there must be an influence affecting it when, as in the motion of the planets, it moves in a curve.

Scientific knowledge consists of two main kinds of items of information.

1. Knowledge of the internal structures, constitutions, natures and so on of things and materials, as various as atoms and galaxies, for these are what persist.

2. Knowledge of the statistics of events, of the behaviour of the persisting things and materials. In this way we discern patterns amongst events. In an explanation we show how the patterns discerned amongst events are produced by the persisting natures and constitutions of things and materials. Taken together these two kinds of items of knowledge amount to knowledge of the powers of things and materials.

Of the three classes of general statements identified at the beginning of this chapter, I have discussed two. General theoretical statements were seen as best understood as descriptive of the iconic model at the core of the theory. Laws of nature have been understood as statements describing patterns and regularities in and amongst phenomena, backed up by association sometimes tacit, sometimes explicit, with some conception of the generative mechanisms responsible for the phenomena being the way they are. There remain the bridge statements asserting causal or modal relations between states of the model treated as hypothetical generative mechanism and the phenomena it is imagined to produce. Clearly statements of this sort can neither be backed up by instance-statistics nor can they be rendered plausible by reflection on the sources of the model upon which they depend, for as yet they have none. On what grounds do we accept them? I think the answer is that they are not accepted on independent grounds at all. They are constitutive of the theory and they stand or fall with its logical power to explain the facts and its plausibility as a description of a generative mechanism. But it is very important to realize that the causal bridge statements making up the causal transform do offer openings for further theorizing and further model construction. Since like all causal general statements they are of the form of powers statements, 'Collision of the molecules with the walls of the vessel produces pressure in virtue of the transfer of momentum', 'Genes produce adult characters in virtue of their biochemical nature', etc., they automatically generate empirical questions as to the nature of causal mechanism by which they produce their effects, questions as to the physical basis of the powers of the supposed components of hypothetical mechanisms and the effect that their states have in the world, which is of course the pattern of phenomena under scrutiny. The modal transform cannot be handled so easily, and we shall return to its further discussion in Chapter 8.

SUMMARY AND BIBLIOGRAPHY

In this chapter, a generative account of causality has been developed, which, it is argued, is very close to scientific practice. A conception of law is derived from it. It is shown that statements selected according to a criterion derived from this conception of law do imply subjunctive conditional statements, and fall under restrictions on logical operations upon them which resolve Hempel's Paradox. Finally, a scientific explanation is characterized as one in which some conception of the mechanism of generation of the phenomena to be explained is incorporated, and this is associated with a consonant treatment of causal powers.

The idea of simplicity as a criterion for choice amongst competing theories all with deductive relationship with a law is rejected. For an extensive study and rejection of simplicity, see

110. M. BUNGE, *The Myth of Simplicity*, Prentice-Hall, N.J., 1963, Part II, Chs. 4, 5, 6, 7.

For a more general attack upon simplicity criteria see

111. J. KATZ, *The Problem of Induction and its Solution*, Chicago University Press, 1962, Ch. 5.

The attack on the deductivist school of the sixteenth century can be found in

C. CLAVIUS (*see* 2).
J. KEPLER (*see* 4).

Three main controversies about causation have occurred.

A. Is causality as an empirical, scientific concept nothing but regular sequence?

The classical arguments in favour are in

D. HUME (*see* 13).
D. HUME (*see* 14).

An excellent summary of modern regularity theories is in

112. J. L. MACKIE, 'Causes and Conditions', *APQ*, 2, 245–64.

Against the regularity theory

113. M. BUNGE, *Causality*, Harvard University Press, Cambridge, Mass., 1959, 3.3.
114. M. GRENE, 'Causes', *Philosophy*, 38, 149–59.
115. M. FISK, 'Causation and Action', *Review of Metaphysics*, 19, 235–47.
116. P. G. MORRISON, 'On Partial Identity of Cause and Effect', *BJPS*, 11, 42–9.
117. W. SELLARS, 'Counterfactuals, Dispositions and the Causal Modalities', *Minnesota Studies in the Philosophy of Science*, University of Minnesota Press, Minneapolis, 1958, Vol. II, 225–308.

B. There has been a considerable discussion of causality as determination, particularly in the philosophy of physics. A certain view on this question is assumed in my text in supposing that we may say that some causal mechanisms can work in only one way. Most of the broader issues in this discussion are touched on in

M. BUNGE (*see* 113), passim.
118. D. BOHM, *Causality and Chance in Modern Physics*, Routledge and Kegan Paul, London, 1957, though there is an enormous literature on the subject.

C. The problem of whether there is a distinguishable and irreducible mode of causation in self-regulating mechanisms and organisms has also been widely discussed. Good discussions of these issues are

E. NAGEL (*see* 98), pp. 401–28.
119. C. TAYLOR, *The Explanation of Behaviour*, Routledge and Kegan Paul, London, 1964, Ch. 1.

Within the general approach of this chapter, this is another pseudo-problem, since any difference between teleological and non-teleological causation is referred to differences between kinds of generative mechanisms, while there remains only one kind of causation.

The attempt to identify laws of nature via the notion of 'entrenchment' of the predicates used in formulating them begins with

N. GOODMAN (*see* 53), Ch. IV.

For a useful summary and a valuable bibliography of the various attempts to find satisfactory criteria for 'scientific law' see

120. E. J. JOBE, 'Some Recent Work on the Problem of Law', *Philosophy of Science*, 34, 363–81.

Natural Necessity: Two ways of trying to give sense to this notion have been tried.

A. Natural Necessity; related to Universality.

121. K. POPPER, *Logic of Scientific Discovery*, Hutchinson, London, 1959, pp. 420–41.
122. K. POPPER, 'A Revised Definition of Natural Necessity', *BJPS*, 18, 316–24.
123. J. HINTIKKA, *Nous*, 1, 33–62.

E 2

124. S. Kripke, *Zeitschrift für mathematische Logik*, **9**, 67–96.
but this has been attacked by
125. W. Kneale, 'Universality and Necessity', *BJPS*, **12**, 89–102.
126. G. C. Nerlich and W. A. Suchting, 'Popper on Law and Natural Necessity', *BJPS*, **18**, 233–5.

B. Contingency has been seen to lie wholly in initial conditions and natural necessity seen to lie in the nature of causal mechanisms. This is essentially the view of this book. By suitably modifying the notion of 'possible worlds' the two points of view can be reconciled.

C. The inconceivability of the contrary was used by Whewell to justify the treatment of certain very general theoretical principles as necessary, though these could be modified by progressive knowledge of fact. See

127. R. E. Butts, 'Necessary Truth in Whewell's Theory of Science', *APQ*, **2**, 161–81.
128. C. J. Ducasse, 'Whewell's Philosophy of Scientific Discovery', *Phil. Rev.*, **60** 61–9, 214–17.

For a general discussion of issues involved in the attempt to separate out a non-logical sense of necessary truth see

129. S. Toulmin, 'A Defence of "Synthetic Necessary Truth" ', *Mind*, **58**, 164–77.

Further discussions concerned with the necessity and the inviolability of principles will be found in the bibliography to Chapter 8. The attempt to use the derivability of a subjunctive conditional as a mark of a law of nature has led to two main avenues of discussion.

(i) Can subjunctive conditionals be analysed using only traditional logical operators? And, if not, how are they to be treated?

130. R. M. Chisholm, 'The Contrary-to-Fact Conditional', *Mind*, **65**, 299–307.
131. J. W. Watling, 'The Problem of Contrary-to-Fact Conditionals', *Analysis*, **17**, 75–80.
132. P. Alexander, 'Subjunctive Conditionals', *Proc. Aris. Soc.*, Supp. **36**, 185–97.
133. B. Mayo, 'Conditional Statements', *Phil. Rev.*, **66**, 291–303.

(ii) Can the intuition that laws imply subjunctive conditionals be justified?
N. Goodman (*see* 53), pp. 39–57.

134. D. Pears, 'Hypotheticals', *Analysis*, **10**, 49–63.
135. W. Kneale, 'Natural Laws and the Contrary-to-Fact Conditional', *Analysis*, **10**, 121–5.
136. R. M. Chisholm, 'Law Statements and Counterfactual Inference', *Analysis*, **15**, 101.
137. P. Alexander, *Proc. Aris. Soc.*, Supp. **36**, 197–200.
138. Mary Hesse, 'Subjunctive Conditionals', *Proc. Aris. Soc.*, Supp. **36**, 201–14.
There is an enormous literature on this subject, of which the chosen items are a representative sample, raising between them most of the important issues.

Hempel's original paper on the paradox of confirmation appeared as

139. C. G. Hempel, 'Studies in the Logic of Confirmation', *Mind*, **54**, 1–26, 97–121.
It has been extensively reprinted, cf. C. G. Hempel (*see* 21), 3–51.

The main 'solutions' of Hempel's Paradox within the traditional assumptions, are outlined in

140. J. L. Mackie, 'The Paradox of Confirmation', *BJPS*, **13**, 265–77.
141. W. M. Baumer, 'Confirmation without Paradoxes', *BJPS*, **15**, 177–95.

The equivalence condition has been attacked by

142. J. SCHOENBERG, 'Confirmation by Observation and the Paradox of the Raven', *BJPS*, **15**, 200–12.
143. H. SMOKLER, 'The Equivalence Condition', *APQ* **4**, 300–7.

and defended by

144. W. C. SALMON, *The Foundations of Scientific Inference*, University of Pittsburgh Press, 1969.

and Nicod's Criterion by

145. P. R. WILSON, 'On the Confirmation Paradox', *BJPS*, **15**, 196–9.
146. D. STONE, 'On Logical Definition of Confirmation', *BJPS*, **16**, 265–72.

Again, an extensive literature exists of which the articles chosen here are a representative sample only. Points very similar to the ones I wish to make are to be found in

147. W. W. ROZEBOOM, 'New Dimensions of Confirmation Theory', *APQ*, **35**, 134–155.

Throughout the discussions of this chapter, I have made considerable use of the notion of possibility. For discussion of this important concept cf. bibliography to Chapter 7 and

148. I. HACKING, 'Possibility', *Phil. Rev.*, **76**, 143–68.

The theory of explanation advocated in this chapter can be usefully considered in the context of two general papers on the nature of explanation.

149. J. W. YOLTON, 'Explanation', *BJPS*, **10**, 195–208.
150. A. W. COLLINS, 'Explanation and Causality', *Mind*, **75**, 482–500.

5. Protolaws

The Argument

1. *Definition*

   (*a*) Protolaws are generalizations which are not backed up by an account or picture of a causal mechanism;

   (*b*) We seek causal mechanisms for the patterns of regularity or structure described in protolaws.

   Problem. How are protolaws assessed, with only instances as a basis?

2. *Form of Protolaws*

   (*a*) Examination of standard, formal analyses;

   (*b*) Accepting standard forms and hence traditional logic leads to a dilemma: Either evidence is favourable, but useless, or evidence is unfavourable, and hence falsificatory but stultifying;

   (*c*) The true form of protolaws distinguishes universality over three 'dimensions', substances, conditions and space-time.

   In actual science successive hypotheses are rationally connected, though connections different in each 'dimension'.

3. *Taxonomic Generalizations*

       (i) the extension of general words occurring in them is without order;

       (ii) some 'potentially' necessary, i.e. those which might be chosen as part of a definition;

       (iii) Taxonomic quantification presupposes no order in the extension of the predicates.

   (*b*) *Non-taxonomic Generalizations*

       (i) the extensions of predicates are ordered sets;

       (ii) numerical values are a typical ordered set.

4. The traditional square of opposition is too crude a device for real science.

   Actual Effects of Counter Instances

   (A) Taxonomic Generalizations

   Case 1. A predicate F, not originally part of the differentia, is found not to hold of some members of a species:

       (i) F must be excluded from the differentia and original hypothesis is false, and was *not* potentially necessary;

       (ii) F can be included in the differentia, and the original hypothesis is modified, and apparently saved, and was potentially necessary.

   Problem. Is it the same or a different hypothesis?

   Case 2. F is part of the differentia of a species.

       (i) a single case of an entity having all the differentia except F, is not a counter instance;

(ii) many cases of entities having all the differentia except *F* leads to either

    (*a*) *F* is expunged from differentia, *or*

    (*b*) *F*, and a *G* which excludes *F*, are differentia of two subspecies.

The considerations governing these cases have to do with the meaning of *F*. *Problem.* If cogency of reasoning depends upon meaning-considerations, should these considerations figure in logic?

(B) Non-taxonomic Generalizations

Case 1. Counter-instances have different effects depending upon their place in the range relative to original instances.

(i) *Limitation.* Counter-instances are used to define the limits of holding;

(ii) *Modification.* Counter-instances are combined with the old law to form a new law;

(iii) *Idealization.* Counter-instances are used to define a teleiomorph of which the original law is true.

Case 2. Favourable instances extend the range of holding, but within the known range of holding are useless.

(C) Inference-Patterns

    (i) Substance-dimension of generality: taxonomic;

    (ii) conditions-dimension of generality: non-taxonomic.

  For problems of spatio-temporal generality see Chapter 10.

(D) A formal logic of non-taxonomic inference

    (i) definitions and principles;

    (ii) an associated numerical confirmation theory.

(E) Accidents

    (i) Defined (as traditionally) as the product of two or more relatively independent causal mechanisms;

    (ii) A qualification is required due to the necessity of the connection between phenotype and genotype, though production of given essence (e.g. given gene composition) may be accidental in the sense of (i);

    (iii) Examination of the principle of the independence of predicates, and its dismissal;

    (iv) Consequential theory of essences and natural kinds;

    (v) Truly accidental general statements must be a finite conjunction of particular statements.

Some generalizations have favourable instances but are not associated with a picture or a description of a plausible mechanism by which the phenomena they describe could be caused. Darwin would never have said that all blue-eyed white cats are deaf had he never seen a blue-eyed, white, deaf cat. He had no idea why this concomitance was to be found, that is, he had no idea of what mechanism might be responsible for the phenomenon. *Prima facie* empirical generalizations share this lack of theoretical background with the profoundest principles of natural science, but unlike the latter, which describe the basic interactions of the things, properties, states, and so on of the universe, known at any time, such generalizations are the most superficial of scientific statements. Basic principles lack background because they are the background for everything else; empirical generalizations lack background because of our ignorance. I believe that most non-random patterns of events and things which we identify in nature are caused, and so any general statement describing such patterns and for which there are favourable instances, ought to be capable of coming to have the status of a law, that is the generative mechanism of the pattern might be found. For this reason, I want to call those generalizations which are accepted only upon the evidence of instances, 'protolaws'. We might formulate a whimsical law of life for them, a law which expresses their epistemological status.

> A protolaw strives to become a law, by seeking a theory to which to attach itself. In this way it becomes associated with ideas of a generative mechanism which explains the phenomena it describes.

A protolaw is a parvenu in search of a place in the society of laws.

No one is ever satisfied with a mere generalization, even if it is associated in some way with the body of scientific knowledge. The fact that the chemical elements arranged in order of atomic weight showed a periodicity on the base 8 was once such a generalization. An explanation for the fact it expresses is required, and that, as has emerged, cannot be achieved merely by formulating some other propositions from which the general proposition in question follows logically. A set of propositions is only an explanation if it describes the generative mechanisms responsible for the phenomena, in our example, the electronic structure of the atom which was found to be responsible for chemical behaviours. In short, whether a set of propositions is an explanation or not is determined by their content, not their form. As Campbell long ago pointed out: that a set of propositions imply some proposition does not mean that that set explains the other. Deducibility is not even a necessary condition for explanation. There is explanation by analogy for instance. As we have seen, deducibility is characteristic of a rather special case. We have noticed how the myths of deductivism have served to blind the deductivists to this fact, and have led them to elevate what is in fact a rarity into a norm.

Before an explanation is found, how does a protolaw recommend itself to our consideration? It is by the appearance of instances of the phenomena it describes and the absence of exceptions to the pattern. Only later when the protolaw has acquired a background of theory will it gain the power that comes from having the wherewithal to explain away unfavourable instances. Here we turn to what is traditionally called 'inductive logic' which, in the deductivist's view, of course, exhausts the study of the confirmation and falsification of hypotheses, since, for them only instances of the facts mentioned in generalizations are evidence for such generalizations.

What is the proper form for a protolaw? Much of the philosophy of science has been done under the assumption that it is the Aristotelian form, in either its ancient or modern versions. In its ancient version, a protolaw has the form 'All *a* are *b*' where *a* and *b* are terms, and terms are the formal expression of common nouns. Aristotle himself distinguishes terms, generally, as the 'limits' of propositions, by which it seems he means simply the verbal beginning and end of the sentence expressing the proposition. Aristotle distinguishes between individual terms, like 'Cleon' and 'Callias', which cannot be predicated of anything but which can have something predicated of them, and universal terms which can be both predicated of something and have something predicated of them. In short, individual terms cannot be terms of convertible propositions while universal terms can, since they can be in either the subject or the predicate position. So far as I can see, the only grammatical element of modern English that can reasonably be said to meet these conditions, that is to be both universal and able to appear as either subject or predicate is the common noun. 'Men are rogues' and 'Rogues are men' exhibits the kind of convertibility Aristotle seems to have had in mind. A further refinement in the same spirit would suggest that adjectives like 'white' and 'black' should be replaced by common nouns like 'whites' and 'blacks'.

In modern times the Aristotelian general proposition has been understood in two distinct ways:

1. As a statement about the relations of classes, as to how the subject class, αs, are related to the predicate class βs. In the protolaw the predicate class includes or is identical with the subject class, represented

$$\alpha \subset \beta$$

or as expressed in Boolean algebra, this is equivalent to the assertion that the class of αs, and non-βs is empty, i.e.

$$\alpha \times \bar{\beta} = \Omega$$

2. As a statement about the copresence of properties. A protolaw then is taken to assert that the subject property ϕ is always found associated with the predicate property ψ, i.e.

$$(\forall x)(\phi x \supset \psi x)$$

In the discussion in this chapter, protolaws are to be considered as statements in isolation from the general body of knowledge, and the evidence for them confined to instances of what they assert, and evidence against them confined to cases where the expected association of properties, or sequence of kinds of events, does not occur, as for instance when one of the properties expected occurs without the other. So the form of unfavourable evidence is $(\exists x)(\phi x \,\&\, \sim\psi x)$.

It would be unwise to assume that, because these modern forms of the Aristotelian proposition are widely known and widely used, they are beyond question as representative of the structure of protolaws and the evidence for or against them. Indeed rather the contrary is the case. Their widespread use has perhaps led to them being taken for granted, and left their form in the status of an unexamined elementary assumption in terms of which all other discussion takes place. However, since they exert a powerful influence on what it is *possible* to conclude, they must not escape examination. As we shall see, the arguments against them are actually overwhelming.

The Induction Dilemma

However many objects can be found which satisfy the requirements for both classes α and β, or present both properties or qualities ϕ and ψ together, provide no grounds of a *logical kind* for the assertion that all other cases will be similar. There cannot be really favourable evidence for an Aristotelian generalization. 'Some *a* is *b*' does not imply 'All *a* is *b*.' Of course, a counter-instance would constitute really unfavourable evidence, but, confining evidence to counter-instances, would have the absurd conclusion that scientists could only falsify hypotheses by their experimental discoveries. If only instances of what is asserted in proto-laws are considered as evidence, and since that is all there can be by way of evidence for protolaws, they do seem, in the Aristotelian forms, to be subject to this dilemma:

Evidence is either favourable instances or counter-instances.
If favourable instances, it is not really evidence for the hypothesis.
If counter-instances, they can only falsify the relevant hypothesis.
This leads to the absurd conclusion that hypothetical protolaws can only be falsified by experiment. Absurd because an *instance* falsifying a hypothesis depends upon the Uniformity of Nature equally with the confirmation of a hypothesis by an instance.

Clearly both the modern and the ancient Aristotelian forms, and their logic, must be rejected as exhaustive of the representation of the logic of the relations between protolaws and the description of favourable and counter-instances to them; and indeed as the proper formulation of the structure of protolaws. If any further reason were required Hempel's Paradox of confirmation would supply it (see Chapter 4). It shows that under the assumptions of Aristotelian form, and instance-confirmation, mere arbitrary change of logical form, as sanctioned by the Aristotelian system, can change the degree to which instances confirm a protolaw, which is absurd. In discussing the application of this paradox to laws of nature, we saw that both the assumption about the form and content of laws of nature, and the instance-confirmation theory had to be rejected. Laws of nature do not exist in isolation from a picture of the actual or possible workings of nature, and their confirmation, therefore, is a product both of favourable instances, and of the plausibility of the ideas about natural mechanisms which, together with the statement of the observed regularity, forms the total content of the law of nature. In the present field, there are only the instances by way of evidence, so the only culprit left must be the Aristotelian form and its associated logic. It will emerge that the Aristotelian form and its logic does have a use, but not for expressing protolaws and their relation to evidence. Its use lies in taxonomy.

In our search for the right way to handle protolaws the next step will be to distinguish taxonomic from non-taxonomic general statements.

The general form of many scientific law-statements is;

> For all substances; and for all experimental conditions; and for all
> space-time locations $F(x, y \ldots.)$ (I)

Each quantifactory phrase I shall say expresses a 'dimension' of generality; so that the three dimensions of generality of many law-statements are 'substance-generality', 'experimental-generality' and 'spatio-temporal-generality'.

In science very few law-statements are not subject to qualification of some sort. In practice, I becomes

> For all substances of a certain kind; and for all experimental
> conditions within a certain range; and for all spatio-temporal
> locations within a certain region $F(x, y, \ldots.)$ (II)

I shall call such qualifications 'substance-restrictions', 'experimental-restrictions', and 'spatio-temporal-restrictions' respectively. In general, restrictions are formulated on the basis of the discovery of counter-instances. I propose to argue for the view that the effect of a 'counter-instance' is different in each dimension of generality. Further, I want to argue that the important features of the general form of the inference from law-statements of Type I to law-statements of Type II, mediated by a counter-instance to the law $F(x, y, \ldots.)$ are concealed by the expression of the inference in the Aristotelian form 'From the truth of 'Some S is not P' we infer to the falsity of 'All S is P'. The use of the Aristotelian form suggests the following picture of scientific advance.

> All S is P (hypothesis)
> Some S is not P (experimentally discovered counter instance)
>   *therefore*

All S is P is false
Try All S is Q (hypothesis)
  Some S is not Q (experimentally discovered counter instance)
    therefore
All S is Q is false
Try All S is R

As an outline sketch of scientific method, this picture has a certain degree of plausibility. But it fails to give any rationale to the 'Try-' moves; for in real science 'All S is Q' is derived from 'All S is P,' and 'All S is R' is derived from 'All S is Q.' A new hypotheses is not independent of the one it succeeds. The origins of new hypothesis are not to be dismissed, I shall argue, as intuitive, a-logical or irrational.

The relations which, in fact, obtain between rejected hypotheses and those proposed to replace them, are, on the surface, very complicated; but by attending to the differences between the three dimensions of generality we do find, in each dimension of generality, the beginnings of a systematic logic in which these relations can be rationalized, though it will emerge that in each dimension of generality the logic is different. To find these systematic logics we need to notice:

(1) that both before and after restrictions ! ve been placed on each dimension of generality there is still a poten..al infinity of cases within the restriction. For instance, if we restrict the application of a law to the surface of the earth there are still infinitely many places for which an hypothesis claims the law to hold; similarly restrictions on substance-generality or experimental-generality still leave infinitely many instances of the substance, and infinitely many experimental conditions empirically unexplored. All this is, of course, under the assumption of various continuity hypotheses. The distinction between restricted and unrestricted hypotheses cannot then be expressed in the symbolism of Russellian quantifiers alone, since both are universal.

(2) that the effect of a counter-instance is determined by which dimension of generality it restricts. The different effects can be brought out through a new classification of generalizations.

$$\text{Taxonomic}\begin{cases} \text{Potentially necessary} \\ \text{Non-potentially necessary} \end{cases}$$

$$\text{Non-Taxonomic}\begin{cases} \text{Disguised identities} \\ \text{Non-identities} \end{cases}$$

I call a generalization *taxonomic* when the order of the members of the extensions of the general terms in the generalization is not relevant to the import of the generalization, and non-taxonomic when the order or orders are relevant.

Taxonomic generalizations are characteristically found in classifications. The classification of organisms depends upon general statements about the characteristics of species, genera, etc.; the classification of stars depends upon general statements about the characteristics of different kinds of stars; the classification of the chemical elements (as distinct from the physical elements) depends upon general statements about characteristic reactions which each element exhibits; the classification of the physical elements (which providentially is nearly identical to the chemical classifications) depends upon general statements about the micro-electrical character of characteristic atoms of the elements; and so on. A necessary requirement for the use of such general statements as the bases of classificatory schemes is that the extension of their terms should not be ordered. If we depend upon the general statements 'All mammals suckle their young', 'All mammals possess hair' etc. to demarcate a family of vertebrates, the properties or characteristic behaviour referred to by the predicates of the general statements must be attributable to particular mammals, irrespective of the order in which these instances are examined. Only thus would we have a classificatory *system*, i.e. a scheme which can be used to place individuals, wherever and whenever found, into their appropriate subclasses.

Within the class of taxonomic generalizations, I suggest we distinguish those which are potentially necessary from those which are not. The potentially necessary are those generalizations which are, have, or might be employed in a formal definition of a class in a classification. For instance, the generalization 'All mammals suckle their young' is potentially necessary for we could use it as a definition of the class 'mammal'. However, 'All mammals are warm blooded' is not potentially necessary, for it could not be used definitively since we already know that all birds are also warm blooded. If we find that a certain organism maintains a constant temperature irrespective of changes in the temperature of the environment (within certain limits), then we can say 'This is either a bird or a mammal'. Of course, 'potential necessity' is relative to the classification scheme which is actually in use, for we could choose to have a quite different classification of organisms in which families like 'mammals' figure, defined by the maintenance of a constant temperature. The classification 'potentially necessary': 'not potentially necessary' is not an absolute logical categorization which could be determined by analysis and inspection of isolated general statements, in particular it is not coextensive with the formal distinction between tautologies and their substitution instances, and non-tautologous statements. Nevertheless, at any given time, for any given classification system, those general statements which are functioning as the sources of criteria of classification, and are being treated as necessary, can be picked out.

Not only can we talk of taxonomic generalizations, but we can also use the notion of general terms having taxonomic generality. I shall say that a general term has taxonomic generality when it is employed in such a way that no order of the members of its extension is presupposed; in Russellian terms I shall say that a variable is *taxonomically quantified* when no order of the range of individuals quantified over is presupposed. The ordinary language quantifiers 'all' and 'each' are ambiguous, since they could be used for either taxonomic or non-taxonomic quantification, but the quantifier 'any' is unambiguously taxonomic, since it is used explicitly to rule out a presupposition of order in the members of the range of individuals quantified.

A non-taxonomic generalization is one in which order in the extension of the general terms is presupposed, since the generalization is arrived at through the examination of instances which are members of an ordered set. Non-taxonomic generalizations are characteristically found in those branches of science where functional dependence, for instance mutual variation, determines the characteristic form of a law of nature. Wherever the members of the extension of terms are numerical we have a built-in ordering principle, since the members of the extension of the independent variable can always be ordered numerically, and once they are ordered, the members of the extension of the dependent variable or variables are ordered by the functional relation between the variables.

In much of traditional logic, we recognize two propositional appraisals which are, in elementary logic, exclusive and exhaustive. A proposition can be either true or false, and if it is true then it is not false, and if false then it is not true. Our problem is to decide whether the elementary logic of truth and falsity is adequate for the logic of those scientific statements which I have called 'protolaws' and their evidence. To decide about this, we need to examine the effect of the counter-instances upon the different kinds of generalizations we have recognized, and to compare the actual effect of a contradictory instance with the sort of effect it would have if we were confined to the principles of traditional formal logic. The situation in traditional formal logic is summed up in the square of opposition, in which, disregarding the complications of existential presuppositions, contraries are defined as propositions which cannot be true together but can be false together; and contradictories are propositions which cannot be true together or false together; so that if one member of a contradictory pair is true (or false) the other member must be false (or true). In this system, if we have the hypothesis that 'All X are Y' and we discover that 'Some

X is not Y' is true; 'All X are Y' is false, and that is all there is to it. We have no means of distinguishing different effects of the contradiction, which may depend upon *which X is not Y*. Let us then look at the actual effect of contradictions and compare them with this schema.

(A) *The Taxonomic Generalization*

Suppose we enunciate the hypothesis

'All birds are feathered.'

Case 1. 'Having feathers' is not part of the differentia of the family *birds*. We find a creature, or creatures having the differentia of *birds*, but lacking those epidermal structures we call feathers. We express this discovery as 'Some birds are not feathered,' and conclude

(*a*) 'All birds are feathered' is false.

(*b*) 'Most birds are feathered' is true.

From this, some further moves are mandatory for scientists.

From (*a*) *Either* (i) 'Being feathered' should not be included in the differentia of *birds*; that is, it should not be used as a characteristic for classifying creatures into families.

Or (ii) 'Being feathered' could be included in the differentia of *birds*, so that we conclude

'Some creatures apparently birds are not true birds' and hence we revive the proposition

'All birds are feathered'

in the form

'All true birds are feathered.'

From (*b*) If the *a* (i) move is chosen we can use 'feathered' and 'non-feathered' as the differentium of a new sub-classification.

If the *a* (ii) move is chosen we reject (*b*), 'Most birds are feathered' since it implies, though it does not entail, that

'Some birds are not feathered.'

Faced with an instance contradictory of our original hypothesis we can react to the situation either by rejecting the hypothesis as false or by so modifying the hypothesis that it remains true. Clearly, what move we choose to make depends upon whether we are treating the taxonomic generalization as potentially necessary or not potentially necessary. If not potentially necessary, then we can conclude that the hypothesis is false, but if potentially necessary, then what has to change is not the truth of the hypothesis but the sense of the word 'bird'.

Is the modified hypothesis the same hypothesis or a different hypothesis? In one sense, it is the same hypothesis for it could still be expressed in the sentence 'All birds are feathered', but in another sense it is not the same hypothesis, since the word 'birds' now has a different sense, i.e. there are changes both in its intension and extension.

Case 2. 'Having feathers' is already part of the differentia of the family *birds*.

(i) We find a creature having all the differentia of *birds* except feathers. We conclude

'This creature is not a bird'

leaving the truth of 'All birds are feathered' unimpaired.

(ii) We find a great many creatures having all the differentia of *birds* except feathers. Two moves are open, either

(1) We conclude that 'Having feathers' should not have been included in the differentia of *birds*, and the hypothesis loses its privileged status as a necessary truth; or

(2) We make 'Not having feathers' the prime differentium of a new sub-family, say *dribs*, which are the featherless birds. In both cases, (1) and (2), the superficial conclusion is that

'All birds are feathered' is false.'

Formal logic recognizes only the superficial conclusions, in the following schemata:

I. From ' "Some S are not P" is true'
we conclude
' "All S are P" is false'
II. From $(\exists x)(Gx \ \& \sim Fx)$
we conclude
$\sim (\forall x)(\sim Gx \ \text{v} \ Fx)$

Clearly, *formal* logic will not do to express all the possibilities of inference, since the radically different moves (ii, 1), and (ii, 2) are not differentiated under schemata I and II. The problem is to decide what is missing from the formal schemata which is present in the actual reasoning. The problem is not hard to solve. What is missing is what, as a matter of fact, the expression 'bird' *means*. Scientific reasoning depends for its cogency upon two necessary conditions:

(*a*) That the *forms* of premises and conclusions are related according to the rules of formal logic: that is, that the acceptable formal relations hold between premises and conclusion.

(*b*) That the expressions used in premises and conclusions are related according to the appropriate rules of linguistic usage: that is, that acceptable intensional relations hold between the non-logical components of premises and conclusion.

Where are we to put these intensional relations? Are we to say that they appear as part of the postulates and definitions of a formalized science whose *logic* is still that of Aristotle and Frege, or are we to say that two sorts of logic are deployed in taxonomic reasoning, the one taking over where the other has left off? If we define 'logic' to be the study of the relations between premises and conclusions in virtue of which the conclusion is said to follow from the premises, then I think we are obliged to choose the second alternative; just because, though the definitions we use in the intensional reasoning are part of the postulates or premises for any piece of reasoning, the use to be made of them in the reasoning is not expressible in the formalism of extensional logic. In particular we do not succeed in expressing the sense of a definition or a casual relation if we express these as

$$(\forall x)(Fx \supset Gx).$$

This is because what we want of any piece of reasoning is that it not only be valid, that is be constructed according to the rules of traditional formal logic, but that it be cogent as well. This depends upon other factors. In particular neither our attitude to a proposition, upon which its necessity and definitional status depends, nor our ideas of the causal mechanism whose behaviour it describes, and upon which its status as a causal proposition depends, are captured in the traditional formal structure.

I conclude that for taxonomic generalizations formal logic provides an outline or framework of inference, but that what happens in any particular case is determined by the intensional relations holding between the actual terms employed, and these, in turn, depend upon considerations external to traditional formal logic. I shall return to this question again in discussing natural kinds.

(B) *The Non-Taxonomic Generalization*

It will be remembered that non-taxonomic generalizations are those in which the nature of the terms related in the generalization is such that the members of the extension of the terms are, or could be, ordered. It is specially characteristic of the physical sciences that particular observations and experimental results are orderable, while in the biological sciences, though classification is an important procedure for these sciences, it is the *classes* of organisms which are ordered, not the particular organisms which are the extensions of those classes. This is partly due to the fact that quantified, and hence numerical, results are characteristic of an experimental investigation in the physical sciences and where there are numbers there is the possibility of ordering, and partly due to the connected fact that the subject matter of experimental investigations in the physical sciences is mutual variations. This can all be expressed by saying that in the physical sciences we are often concerned with ranges of particular cases; within certain limits, these ranges are indefinitely extensible from the part of the range where the original observations of a mutual variation were carried out. It is characteristic of a taxonomic generalization that since *any* member of the extensions of its terms is as good as any other for the exemplification of the type, a counter-instance is as much falsificatory of the generalization, wherever it is drawn from, and it is upon this fact that the simplicity of the formal schemata above depends.

However, I propose to argue that it is characteristic of non-taxonomic generalizations that it does matter from what part of an ordered range of cases an instance is taken, and that the position of an instance, favourable or unfavourable, in the ordering, is an essential ingredient in the reasoning about truth and falsity which follows the discovery of such a case, and in the derivation of new hypotheses.

Case 1. Counter-Instances. Suppose we design an apparatus to determine the law of mutual variation between thermal and electrical conductivity. Provided that we experiment with temperatures around those commonly encountered in the laboratory we find that the following relation holds:

$$\theta/k = T \times \text{const.}$$

where θ is thermal conductivity, k is electrical conductivity and T is the absolute temperature. This law, called the Wiedemann-Franz law, is in fact derivable from Drude's theory of metallic conductivity, a theory constructed by treating the 'free' electrons in a metal as behaving analogously to the molecules of a gas. For the purpose of the illustration, I shall be considering this hypothesis as a protolaw, that is before Drude's theory appeared with its sketch of a hypothetical mechanism, transforming the status of the hypothesis to that of a law. The extension of the general term T, is the ordered range of temperatures from absolute zero.

Experiment discloses that there are the following counter-instances:

1. At temperatures close to $0°K$, $k \to 0$, and the Wiedemann-Franz relation fails to hold;

2. At temperatures high with respect to temperature conditions commonly encountered in the laboratory, k falls off more rapidly than

the law predicts. So that at each end of the temperature range experiment discloses counter-instances to the law. Suppose we represent the Wiedemann-Franz relation as

$$L$$

then we have discovered experimentally that there are cases in which

$$\text{not-}L$$

Received formal logic would tempt us to represent the situation as follows:

'For all temperatures L'   (Hypothesis)
'For some temperatures not-L' is true

and using received Logic we conclude either

A. 'For all temperatures L' is false

or we conclude

Q. 'Not for all temperatures L' is true.

But neither move, to (A) nor to (Q), formally valid though each may be, represents the situation in science. (A) does not, because it obscures the fact that we would still wish to claim that for some temperatures L is true; (Q) does not, for it fails to record the most crucial scientific fact, namely for which temperatures L is true.

This situation arises quite generally in the testing of non-taxonomic generalizations by instances. For instance if we consider the effect of counter-instances upon say

$$\frac{\text{Pressure} \times \text{Velocity}}{\text{Area of flow}} = \text{const.}$$

we find, experimentally, that provided the values of the variables fall within certain regions of the ordered range of cases, the relation holds, but that should any or all of the variables be given values outside these regions the relation holds no longer. Instances of this kind of thing could be multiplied indefinitely. It is the typical relation between the non-taxonomic generalization and the values of its variables.

In practice we can recognize three different responses of scientists to this kind of counter-instance; three different kinds of inference-pattern employed consequent upon the discovery of a counter-instance.

(a) *Limitation*

The counter-instances are used, not to lead us to a rejection of the relation expressed in the law, but to lay down the limits in which it holds. And this is only possible with non-taxonomic generalizations where the ranges of the variables are ordered. Formally we can express limitation thus:

(i) *Upper Bound*

$$L \text{ under all conditions } c_0 \dots \qquad \text{(Hypothesis)}$$
$$\underline{\text{not-}L \text{ under } c_n \dots c_r}$$
$$L \text{ under conditions } c_0 \dots \dots c_{n-1}$$

(ii) *Lower Bound*

$$L \text{ under conditions } c_0 \dots c_{n-1}$$
$$\underline{\text{not-}L \text{ under conditions } c_0 \dots c_k}$$
$$L \text{ under conditions } c_{k+1} \dots c_{n-1}$$

It should be noticed that there is an assumption involved in the upper bound cases, not involved in the lower bound cases. The assumption is that if the relation L is found not to hold for some set of cases in the upper regions of an ordered set of conditions, it will not be found to hold in conditions still further removed from those conditions in which we know experimentally, that it does hold.

(b) *Modification*

Having discovered bounds to the ordered set of conditions in which a relation holds between observables, a new relation is formulated by combining the old relation with the bounding conditions. *A fortiori* the new relation holds for the conditions in which the old relation held, and for some part, as yet undetermined, of the conditions for which it did not hold. For example: the incorporation of the high pressure and low temperature bounding conditions on the general gas equation, into that equation to formulate Van der Waal's equation; the incorporation of the bounding conditions on Newton's Laws of Motion, into those laws to yield Relativity; the incorporation of the bounding conditions on the laws of the viscous flow of drops, in the form of the effect of internal circulations, to yield Mathur's relation between phase viscosities, drop velocity, and the drop deformation constant. From the most commonly known laws to the most obscure we can recognize the modification technique being applied.

Formally we represent the process as follows:

$$L \text{ under conditions } c_0 \dots \qquad \text{(Hypothesis)}$$
$$\text{not-}L \text{ under conditions } c_n \dots c_r$$
$$\underline{\text{not-}L \text{ under conditions } c_0 \dots c_k}$$
$$(L \ \& \ c_0 \dots c_k \ \& \ c_n \dots c_r) \text{ under conditions } c_0 \dots$$

For this law we renew the search for bounding conditions.

(c) Idealization

When we idealize, we use the bounding conditions to define, by negation, an ideal universe: ideal with respect to the original relation L. The bounding conditions are interpreted, in the appropriate theory, and a universe is imagined in which this interpretation of the bounding conditions is denied. Ideal gases, perfectly rigid bodies, rational men, etc., are arrived at in this way. There can be no formal logic of idealization, since this moves works, not through the deployment of some calculus, but through the interplay between the phenomena to be idealized and the teleiomorphic models which embody the idealization.

Case 2. Favourable Instances. To find a favourable instance within that part of the ordered range of cases where we already know a functional relationship to hold is useless for science, since it adds nothing to our knowledge, in that dimension of generality. It may add something to our knowledge of spatio-temporal generality but, for the moment, we are considering only the dimension of experimental-condition generality. What would add to our knowledge in this dimension of generality is a favourable instance in a part of the range of ordered experimental conditions where we have not, as yet, experimented. To accommodate this sort of case we need to add another feature to our formal schemata developed for the counter-instance. We have to distinguish explicitly between known and hypothetical cases. Suppose our functional relationship to be again 'L', and suppose that we know, by experiment, that within the range of experimental conditions $c_0 \ldots$, L holds for the conditions $c_1 \ldots c_m$. We suppose that no experiments have been done in conditions $c_0 \ldots c_{l-1}$, and $c_{m+1} \ldots$. The situation can be expressed as follows (where (def) means 'definite', and (hyp) means 'hypothetical'):

$$L[c_0 \ldots c_{l-1} \text{ (hyp)}; c_1 \ldots c_m \text{ (def)}; c_{m+1} \ldots \text{ (hyp)}].$$

The effect of a favourable instance in the range $c_1 \ldots c_m$ is nil. (Repetition of classical experiments, except for sharpening up the values of numerical constants, is of educational value only.)

The effect of a favourable instance, in either $c_0 \ldots c_{l-1}$, or $c_{m+1} \ldots$ is to extend the range of definite cases into the range of hypothetical cases. This can be expressed as follows:

$$
\begin{array}{ll}
L[c_0 \ldots c_{l-1}(\text{hyp}); c_1 \ldots c_m(\text{def}); c_{m+1} \ldots (\text{hyp})] & \\
L[c_k(\text{def})] & k < l \\
\underline{L[c_r(\text{def})]} & r > m \\
L[c_0 \ldots c_{k-1}(\text{hyp}); c_k \ldots c_r(\text{def}); c_{r+1} \ldots (\text{hyp})] &
\end{array}
$$

Favourable instances alter the boundaries between definite and hypothetical applications of a functional relationship; counter-instances put boundaries to the range of hypothetical plus definite cases.

Limiting cases of these schemata give us Aristotelian Logic. From the hypothesis $L[c_0 \ldots c_{l-1}(\text{hyp}); c_1 \ldots c_m(\text{def}); c_{m+1} \ldots (\text{hyp})]$ we can deduce $Lc_l, Lc_{l+1}, \ldots, Lc_m$; any of which could be an S of which it is true that some S are P. Similarly should we find that a random sample of cases $c_0 \ldots$ are all cases in which L does not hold (and here L must be an hypothesis with no evidence at all to support it), we can infer that $L[c_0 \ldots]$ is false. So there is a limiting case in which, from the truth of 'Some S are not P' we can infer the falsity of 'All S is P'.

To summarize our conclusions:

(1) Taxonomic generalization is correctly represented by the Aristotelian Logic, as far as the formal relations between particular and general propositions are concerned. But, when one comes upon a counter-instance to a generalization, one can save the generalization, if need be, by treating it as potentially necessary and claiming it as a definition, i.e. as setting the criteria for class membership. This salvation cannot be expressed in terms of the traditional formal relations between the propositions expressing the generalization and counter-instances, but requires that we consider the meaning of substantive terms, and the consequent intensional relations between generalization and counter instances.

(2) The effect of counter or favourable instances on generalizations over ordered sets of conditions is not even formally expressible in the Aristotelian or Quantifier Logic, since the essential true/false relations of those logics are not used.

These conclusions can now be used to discuss the general form of a protolaw. It was:

> For all substances; and for all experimental conditions; and for all space-time locations $F(x,y, \ldots)$

Each quantifier was said to define a 'dimension of generality,' and it was recognized that, in practice, protolaws are often subject to substance-restrictions, experimental-restrictions (e.g. the Wiedemann-Franz law applies only if the temperature is above $4°K$) and perhaps spatio-temporal restrictions (e.g. that the gravitational acceleration is 980 cm per sec, per sec). From the above conclusions about the logic of counter-instances it follows that *in each dimension of generality a different 'logic' is deployed*, i.e. different inference patterns are used.

If the above reasons are not sufficient to command assent, the point can be further established by considering how generality is saved in each dimension.

(i) *The Substance-dimension*

Generality in the substance-dimension is taxonomic generality, for by hypothesis, each specimen of a substance, is, with respect to the properties of that substance, identical with any other. Hence substance-generality is subject to the logic of taxonomic generalizations, and hence substance-generality can be saved in the face of counter-instances by treating the law as defining or partly defining the substance in question, i.e. as settling the criteria for identifying a sample or an instance. For example if the laws of magnetism fail to hold for phosphorus, we do not drop the laws of magnetism, but deduce that phosphorous is not a magnetic substance.

(ii) *The Experimental-Conditions Dimension*

Since the conditions under which a law holds are orderable, we can save the law in the face of counter-instances by treating the counter-instance as indicative of the boundaries to the range of experimental conditions for which the law holds, or might hold. The appropriate logic, then, for this dimension of generality is that of the non-taxonomic generalization, and we can proceed by limitation, modification or idealization, the principle of which will be further extended below.

I shall return to the consideration of spatio-temporal generality in Chapter 10.

To develop still further the theory of confirmation for ordered ranges of cases of protolaws or non-taxonomic generalizations, a new set of symbols expressing the novel logical concepts introduced above must be devised. The main conceptual innovation that requires expression is the range of ordered cases over which a relation holds. As above, I shall use the ordinary function-argument symbolism for the expression of the relation that is to be said to hold. To this I add a new symbol, the bindor. A bindor has the following powers:

1. It conjoins cases of the instantiation of the relation, in the manner of the universal quantifier.
2. It conjoins the cases in an order, which is the order of the range in which the cases are ordered.
3. It conjoins cases in a given interval.

In short it indicates in what part of a range of ordered cases a relation *holds*; the 'range of holding'. In the theory now to be developed, the principles will be concerned mostly with the use of instances to limit or extend a range of ordered cases in which a relation holds. A protolaw will be expressed by means of a bindor and a propositional function. I choose the symbol π for the bindor to indicate logical product, and super and subscripts to indicate the part of the range in which the relation holds. Thus

$$\overset{r}{\underset{s}{\pi}} F(x_1 \ldots x_n)$$

expresses the hypothesis that the relation F holds between n individuals in that part of the range of ordered cases, as previously explained, that lies between the 'r' and 's' case inclusive. I call a bindor and a function a 'conjecture', but it is to be understood that in this system conjectures do not and cannot become true or false, they simply survive at the end of runs of trials over various parts of their range, and since most ranges are, at least in one direction, potentially infinite, our knowledge of the part of the range in which a relation holds can always be enlarged. The only statements that are true or false are statements of instances of the occurrence of whatever relations among phenomena are expressed in the law. These are written as follows:

$$F(x_{1l} \ldots x_{nl})$$

where x_{1l} is the 'l'th individual of kind x_1 and x_{nl} is the 'l'th individual of kind x_n. In writing out what corresponds to a deduction in this theory, conjectures are written one above the other in the left hand column, and instances one above the other in the right hand column. A column of conjectures is called an 'investigation'. The mark '*' is used to indicate the point in an investigation when we finally decide we know nothing about the relations in the range of cases. '*' is called an 'encumberance'.

In setting out the principles, I assume an experimental setup in which only one parameter, x_1, is being varied, and the other parameters, $x_2 \ldots x_n$, maintained constant. So the bindor appearing at the head of the formula refers only to the range of x_1. In such a course of experiments $x_2 \ldots x_n$, having the values of constants, are not being supposed to hold over a range. The principles can easily be supplemented to express more complex conjectures in which more than one parameter is varied. For instance, the conjecture that $PV = RT$ holds for all values of its parameters above the critical point can be expressed as follows:

$$\overset{p=\infty}{\underset{p=pc}{\pi}} \overset{v=\infty}{\underset{v=vc}{\pi}} \overset{t=\infty}{\underset{t=Tc}{\pi}} F(p,v,t)$$

where pc, vc and Tc are critical values.

P1   The Principle of Truncation

$$\frac{\overset{r}{\underset{J}{\pi}}\,F(x_1\ldots x_n)}{\overset{m}{\underset{k}{\pi}}\,F(x_1\ldots x_n)} \qquad \begin{array}{l} \text{not-}F(x_{1k}\ldots x_{nk}) \\[6pt] \text{not-}F(x_{1m}\ldots x_{nm}) \end{array} \qquad r>m>k>j$$

P2   The Principle of Minimum Confirmation

$$\frac{\overset{m}{\underset{k}{\pi}}\,F(x_1,\ldots x_n)}{\overset{r}{\underset{j}{\pi}}\,F(x_1\ldots x_n)} \qquad \begin{array}{l} F(x_{1j}\ldots x_{nj}) \\[6pt] F(x_{1r}\ldots x_{nr}) \end{array} \qquad j<k<m<r$$

P3   The Principle of Trivial Confirmation

$$\frac{\overset{m}{\underset{k}{\pi}}\,F(x_1\ldots x_n)}{\overset{m}{\underset{k}{\pi}}\,F(x_1\ldots x_n)} \qquad F(x_{1l}\ldots x_{nl}) \quad m<l<k$$

P4 The Principle of Trivial Falsification

$$\frac{\overset{m}{\underset{k}{\pi}}\,F(x_1\ldots x_n)}{\overset{m}{\underset{k}{\pi}}\,F(x_1\ldots x_n)} \qquad \begin{array}{l} \text{not-}F(x_{1r}\ldots x_{nr}) \\[6pt] \text{not-}F(x_{1j}\ldots x_{nj}) \quad j<k<m<r \end{array}$$

P5   The Principle of Rejection

$$\frac{\overset{m}{\underset{k}{\pi}}\,F(x_1\ldots x_n)}{\overset{m}{\underset{k}{*}}\,F(x_1\ldots x_n)} \qquad \begin{array}{l} \text{not-}F(x_{1k}\ldots x_{nk}) \\ \cdots\cdots\cdots\cdots\cdots \\ \text{not-}F(x_{1m}\ldots x_{nm}) \end{array}$$

P6   The Principle of Hypothesis

$$\frac{\overset{m}{\underset{k}{*}}\,F(x_1\ldots x_n)}{\overset{m+1.\infty}{\underset{k-1.\infty}{\pi}}\,F(x_1\ldots x_n)}$$

With the help of these six principles, an investigation, starting from some hypothesis as to the range of holding of the protolaw, can be rationally organized. It should be noticed that the effect of single counter-instances is to reduce the range of cases in which a law holds, in accordance with intuition. Rejection is only allowed when the range of holding is completely blanketed by counter instances.

Since it is characteristic of protolaws to exist in a theoretical vacuum, they must be of the nature of conjectures while they retain their protolegal status. The six principles here set out are a logic of conjectures, and while the items in the right hand column are either true or false, the items in the left hand column are successive conjectures. Each conjectural step can be expressed by the following form of words:

'It is conjectured that the relationship F holds between the values of the parameters in the interval stated.'

An investigation does not start with particular facts, but with a conjecture. This is emphasized in the formal structure by separating the conjectures in the left-hand column from the particular facts in the right-hand column. Thus every investigation must start with a conjecture.

1. There seems to be no easy way to develop a numerical measure of the degree of belief we should have in a hypothesis. If this degree of belief is related to our expectations as to whether the relation which holds under some conditions will hold at extremes of the range of conditions, then if practice is to be our guide, the hypothesis that a relation will hold at a certain part of the range is the less worthy of belief, as the part of the range departs further from the cases in which we know the relation does hold. It follows from these considerations that for protolaws there is only the simple relationship

$$\frac{C(H_1)}{C(H_2)} = \frac{e^{t-m}}{e^{r-m}}$$

where H_1 is the hypothesis that a relation which holds for case m in an ordered range of cases, also holds at case r, and H_2 is the hypothesis that a relation which holds at m holds also at case t, and $C(H_1)$ is the degree of belief we should have in H_1 and $C(H_2)$ that we should have in H_2. The exponential function is chosen to indicate that when the hypothesis is about regions remote from those cases we have studied, great differences in very remote regions make little difference to degree of belief. For example, suppose we are considering whether we would expect the gas laws to hold at 1,000,000°K: Our expectations about whether they would hold for that temperature can differ little, if anything, from our expectations as to whether they would hold at 2,000,000°K.

Yet there will be a great difference in our expectations as to whether they will hold at 500°K and 1,000°K, though there is a factor of two in each case. And, in any event, the calculation of such an index seems quite pointless, since it is so simply and directly and dependently related to a very simple intuition. However, some further rules can be added, though their utility in practice seems remote.

2. The next stage must be to compare our degree of confidence that H, which we believe holds at m, will hold at r, given the discovery that it holds at various points in the ordered range of cases between m and r. Thus if, letting hp and hq represent 'H holds at p' and 'H holds at q', and 'C_mH^r,hp' our degree of confidence that H will hold at r then

$$\frac{C_mH^r,hp}{C_mH^r,hq} < 1, \text{ if } m < p < q < r$$

Or, expressing this in another way

$$C_mH^r,h\mathscr{P} \to 1 \text{ as } \mathscr{P} \to r$$

where '$\mathscr{P}$' is a variable over the range of the parameter.

3. Suppose, however, that H does not hold at $p(r > p > m)$. Then clearly

$$C_mH^r,h\mathscr{P} \to 0 \text{ as } \mathscr{P} \to m$$

4. Next, a set of neighbourhood principles is required. If it is known that H holds at m, and it is discovered that H holds at r, then we should have a high degree of confidence that H holds at n, $m < n < r$.

We should also have a high degree of confidence that H holds at $r+k$, where k is small. Expressed formally these principles become

(4a.) $w < C_mH^r$, hm & $hr < 1$ where $0 \leqslant 1 - w \leqslant j$ where j is small, if $m < n < r$.

(4b.) C_mH^{r+k}, $hr \to C_mH^r$, hr, as $k - r \to 0$

The value of C_mH^{r+k} is to be a function of the remoteness of r from m, and an inverse function of the remoteness of k from r. Thus

$$C_mH^{r+k} = \phi\left((r - m, \frac{1}{k-r}\right)$$

and this function is such that when $r - m$ is small and $k - r$ is large $\phi(r - m, 1/k - r) \to 0$. And when $r - m$ is large and $k - r$ is small $\phi(r - m, 1/k - r) \to 1$.

A similar function could be defined for our loss of confidence in the neighbourhood of a counter-instance.

Accidents. When a regular sequence of kinds of events, or states or relations of things, or some other form of orderliness is observed, and nothing much in the way of theory has been formulated for such

F

phenomena, a protolaw is used to describe that discovery. But the orderliness has to be treated not only as possibly the result of causal mechanisms operating but also as possibly accidental, since the kind of reasons which justify any confidence we may have that any orderliness we think we have already observed will continue to be manifested, are properly derived from theory and not from instances. In short, in the situation I have described, a scientist does not know whether there is truly orderliness: that is, whether there is an enduring natural mechanism at work generating a sequence of events or states, or only the appearance of orderliness, that is, a sequence of events, each produced by some different, and relatively independent natural process, but which taken together seem to constitute some orderly process or non-random structure. This might be because the constituents of the sequence happen near one another in space, and sequential in time, or because they seem to show similarities repetitious enough to warrant the guess that they might have been produced by the same process; that is that their continuing orderliness is due to the endurance of a common generating mechanism. In a state of ignorance as to the productive processes and powers of nature in a certain field, a scientist has to operate with general statements which might be merely accidental, but by treating them as protolaws, while searching for the deeper knowledge of nature which would show the protolegal regularities to be the outcome of the operation of enduring mechanisms, he reserves the possibility that they might come to be treated as necessary, in the sense defined in Chapter 4.

There are, though, in a sense, accidental phenomena in nature. For instance when two determinate properties or qualities are present in an object, each the product of separate, relatively non-interacting chains of causation produced by mechanisms acting in relative independence of each other, we may call their copresence properly accidental. We shall see later that once two causal chains intersect in an object, they determine the inner constitution of that object, and fix it in its natural kind, so that by reference to its total inner constitution and external relations its bouquet of observable qualities is not accidental. The elementary qualities are only accidental considered with respect to each other, they are not accidental considered with respect to the mechanisms which generate them and which form the constitution of the object of which they are the qualities. Considered with respect to the chains of causation by which, say, blue eyes are found with black hair in a human being, their concomitance is accidental, but considered with respect to the essence of that person, the inner (genetic) constitution which is productive of his external appearance and other heritable characteristics, that is his genetic code, the joint appearance of blue eyes and black hair in *him*, is necessary. That is to say, given his genetic code and that he has survived at all, there are no alternative eye and hair colours that he could have, and still be biologically him: that is, still have that genetic constitution.

It is an essential part of the deductivist view, in both its inductivist and non-inductivist or Popperian forms, that all generalizations are in the last analysis accidental. This is because all non-logical relations between predicates are supposed ultimately to be accidental. Thus all predicate relations ought to be capable of being expressed extensionally, that is, in terms of the sets of objects exemplifying them, and their relations, such as inclusion and exclusion. Given this view about predicates, it seems natural to hold that the properties and qualities which they are used to ascribe should also be held to be only accidentally associated in things, related only by the statistical fact that they are frequently found together or successively, in the same set of objects. The idea that one state of a body may be productive of another, or may generate some state, is scouted. Predicates follow properties in their alleged total logical independence on this view, so that it is assumed that any combination of properties *might* be found in nature, however much we know about the world, and any combination of predicates could figure in a law of nature, provided that they are not formally contradictory. Any group of non-contradictory predicates are supposed to be definitive of a possible class of things. And given the further idea that the only predicates which could be mutually contradictory are determinates under the same determinable, such as specific colours, it would seem to follow that provided predicates were chosen each under a different determinable, any random combination is possible. It is also part of this view that, for all we can know, any set of properties can *succeed* any other as attributes of a thing, and we must suppose, might. So it is alleged, from an epistemological point of view, the future must be considered entirely open. Laws are reduced to protolaws, causality becomes nothing but statistical regularity and prediction a mere alogical inductive guess.

In fact the future is not open, causality is generation, prediction can be rational, and our confidence in our knowledge of the future well-founded. Where does the disparity between the philosophers' theories and the practice of the scientific community lie? It lies, as we have seen over and over again, in the logical theories that are the source of the epistemological attitude, and among these logical theories is the Myth of the Independence of Predicates.

The myth comes out in many ways. For instance it can be found in the background of 'Old Moore's Proof', that goodness is a non-natural property. At the heart of that proof is the assumption that all natural qualities can change their determinate character under their appro-priate determinable without entailing or causing changes in the deter-minate qualities under any other determinable.* That, for instance, a heavy red thing can be made light, without changing its colour. But when the determinate qualities of a thing are referred to the internal constitution of the body in which they appear, and which is responsible for them, it is clear that the kind of changes envisaged by Moore, and assumed as possible by deductivists, just cannot be achieved. For example, if colour is referred to the electronic constitution of the molecules of a substance, then any change in colour, other than that caused by a change in the incident light, or in the nervous system of the percipient, that is any 'real' change of colour, will involve a change in the colour-producing power of the substance. That in turn will involve a change, or, in some cases, simply be a change in the chemical constitution of the substance of which the coloured thing is made. This change will lead to further consequential changes in determinates under other determinables, because in the inner constitution or real essence of bodies, i.e. in the electronic structure of the atoms, the powers of things and materials to produce manifested qualities find a common source.

When a predicate figures in a law of nature about certain things or materials, then, it is not independent of the other predicates that could be asserted of the same subject matter, and it may not be negated without leading to changes in the other predicates of the things or materials. In a law of nature, the properties whose behaviour or regularity is described, are tacitly referred to the inner constitution of things and substances, to the generative mechanisms responsible for their appearing under the circumstances that they do. So the logic of predicates in laws of nature is not simply extensional. But since acci-dental general statements are marked off from laws of nature by the absence of tacit or explicit reference to theory, to pictures or models of the generative mechanisms of nature, to the inner constitutions of things, the predicates figuring in accident statements can be treated as independent. But the moment an accidental generalization is suspected of being a protolaw, the predicates must no longer be supposed independent, though until full law status is reached by the clear formulation of ideas of a generative mechanism, we cannot be said to know their connection.

Now a neat account of essences and natural kinds follows very directly. The essence of a type is the totality of kinds of generating mechanisms it contains. What generating mechanisms a thing or material contains determines its powers, that is what it can do by way of reaction

* G. E. Moore, *Principia Ethica*, Cambridge University Press, 1903, pp. 40–41.

to affect, and by way of action upon and interaction with other bodies. The properties which things actually manifest to perceivers or instruments are those determinate qualities and properties of a thing, or substance or region, which are actually developed under the stimuli by which things etc. are actually affected. In this way one's intuition suggests that the initial conditions of the universe are, as one might say, the ultimate accidents.

On this account the gene pool is the essence of a species, since it is the genetic mechanism in each individual, and in each cell, which determines the final structure that appears. The essences of substances are their electronic, atomic, and molecular structure. The natural kinds are sets of possible generating mechanisms known already, or conceived, in the current scientific genre, which can coexist in individual things and in pieces of material.

In the end true generality must be confined to protolaws. If a sequence of like occurrences or a repetition of a like structure seems to be *truly* accidental, that is, produced on each occasion by different kinds of natural causes, then, even if it seems idiomatic to describe what happens, in a tone of surprise, in general terms, the content of such a seeming general statement can be no more than a finite conjunction of particular statements describing each case. And such a statement is true, and the question of our degree of belief in it does not arise. So in the end the situation seems to be this: the rough form 'All *a* are *b*' may express either:

1. An accidental general statement, really a finite conjunction of particular statements, that is, the statement means 'Each *a* is *b*'. Or:

2. A taxonomic general statement, in which case it asserts a putative attribute of a kind of thing. Such statements can be given a privileged status and treated as necessary. But circumstances can arise in which they no longer are accorded that status, and may be declared false. Such statements can be expressed as 'Any *a* is *b*'. Or:

3. A non-taxonomic generalization or protolaw, whose logic has been outlined above, that is, the statement means 'Every *a* is *b*'.

Instances operate differently in each case. A further instance of an accidental general statement simply joins the conjunction and there cannot be counter-instances. A further instance of a case under a taxonomic generalization does nothing in the period of the principle's necessity, but a counter-instance, once admitted as such, can generate a crisis, and even conceptual revision. This is because to admit it at all as a counter-instance is to begin a period of, at the least, conceptual discontent. Finally the complex relationships between non-taxonomic generalizations and their instances, pro and counter, have already been sketched.

The important idea of a statement being potentially necessary was introduced in the course of this chapter. This was the idea that a statement which is not tautologous can be treated as immune from falsification by instances. I shall return to a more detailed discussion and examination of this idea in Chapter 8 where I shall look, in detail, into the manner in which the use of concepts is controlled, and criteria for their application are generated, by our treating certain statements as immune from falsification by instances.

I have referred on several occasions in this chapter to the confirmation of a hypothesis by evidence, and to the degree of belief we might have in a hypothesis. The positivist and Humean epistemology and logic leads to the analysis of these notions in terms of the concept of probability, and particularly of probability conceived of as the relative frequency of cases in a class of similar cases. We must now turn to an examination of the concept of probability, and to the attempts that have been made to construe degree of confirmation in terms of it.

SUMMARY AND BIBLIOGRAPHY

A class of statements is identified, the members of which could, in some future state of knowledge, become laws of nature. Their logic and epistemology is discussed with respect to the evidence, restricted to favourable instances, upon which they are based. A detailed examination of the logic of systems of protolaws can be found in

151. S. Körner, *Experience and Theory*, Routledge and Kegan Paul, London, 1966, Chs. I–V, VII and X.

The fact that acceptance of the simple universal form for the analysis of laws leads to the dilemma between induction and falsification can be seen in

152. J. W. N. Watkins, 'When are statements empirical?' *BJPS*, **10**, 287–308.

The treatment of certain universal statements as definitions, and hence as for the time being immune from falsification raises the problems of the logical status of definitions, which is barely touched upon in the Chapter.
For extensive discussions of this problem see

153. F. Waismann, *How I see Philosophy*, edited R. Harré, Macmillan, London, 1969, Chs. IV and V.
W. V. Quine (*see* 90), Ch. II.
P. Achinstein (*see* 73), Chs. I and II.

Reference should also be made to the bibliography for Chapter 8, where the same problem arises through the study of the logical and epistemological status of principles. The problem of the independence of predicates, widely assumed in logical positivist writings, is also raised in this chapter.

154. J. M. Keynes, *A Treatise on Probability*, Macmillan, London, 1921 (new edition, 1963), Ch. XXII.
155. A. N. Prior, 'Determinables, Determinates and Determinants', *Mind*, **58**, 1–20, 178–94.
156. D. H. Mellor, 'Connectivity, Chance and Ignorance', *BJPS*, **16**, 209–25.

6. Probability and Confirmation

The Argument

PROBABILITY

1. *The Problem.* Can the notion of the probability of an event be extended, or developed by analogy to provide an account of the degree of confirmation of a hypothesis?

2. *Probability*

   (a) p = likelihood of the occurrence of an event or of the coming to be of a state;

   (b) probability statements essentially have *future* reference;

   (c) the distinction between meaning of probability statement, and grounds for probability assessment:

   (i) is confused by verificationists, e.g. statistics are grounds for, not meanings of probability statements;

   (ii) there are various grounds for probability statements, but their meaning is degree of belief, commitment, hope, etc.;

   (iii) it is a fallacy to use grounds as definitions.

3. *Conditions under which Probabilities are assigned*

   (a) things in state A *can* become either B or C, but not both;

   (b) causes of the differential manifestation of B and C are not known.

4. *Probability Situations*

   Degree 0. Not even the grounds for equipossibility assumption are known, so there is no rational assignment of probability to B or C etc.

   Degree 1. Conditions for the Assignment of Probabilities

   Condition 1; B, C etc. are assumed to be the only determinates; Condition 2; B, C etc. are assumed to be equipossible.

   (a) the more that is known about A and the differential origins of B, C etc., the more Conditions 1 and 2 are justified;

   (b) the more knowledge of the ratios of occurrences of B, C . . . N we have the better are we able to assess Condition 2.

   Degree 2. Knowledge of, or best possible estimate of, actual statistics of B, C . . . events, grounds Condition 2 (Frequency theory of probability).

   Degree 3. Knowledge of the nature of A, and the causal mechanisms involved in the production of B, C . . ., grounds Conditions 1 and 2 (Range theory of probability).

5. *Summary*

   The grounds are objective, but assessment being an expression of degree of confidence, expresses a 'subjective' feature of discourse.

Degrees *0–3* are degrees of knowledge associated with joint growth of knowledge of causes and knowledge of statistics.

II
CONFIRMATION

1. Objects of confirmation-statements are propositional, e.g. beliefs, conjectures, etc.

2. Relation of confirmation assessment of hypotheses to grounds has something to do with logical relations between hypotheses and statements of grounds;

3. To say a hypothesis is probable is to say it is likely to turn out to be **true**; Can confirmation be thus linked to probability?

4. Examination of metaphor, 'turning out true';

5. Examination of the true/false dichotomy;

6. ['Probable' means 'likely to turn out to be true'; ['Confirmed' means 'has (sometimes) turned out to be true'. *Solution.* 'Probability and Degree of Confirmation' are linked by common evidence, but syntactically different;

7. Examination of transfer of numerical probability of events to the numerical probability of a statement describing the event;

8. Examination of the use of relative frequency of times true to times put forward, as a measure of numerical probability of hypothesis.

III
EXAMINATION OF ALLEGED SYNTACTIC ISOMORPHISM
BETWEEN 'PROBABILITY' AND 'CONFIRMATION'

(A) *Carnap's Desiderata*
   (i) Preservation of confirmation through logical transformation of evidence is acceptable;
   (ii) Preservation of confirmation through logical transformation of hypotheses is objectionable, because of Hempel's Paradox, unless such transformation is severely restricted on extra-logical grounds;
   (iii) That the confirmation of a conjunction of hypotheses is less than the confirmation of the worst conjunct is counter-intuitive;
   (iv) That the confirmation of a disjunction of hypotheses is greater than the confirmation of the best disjunct is counter-intuitive;
   (v) That the confirmation of the negation of a hypothesis is 1 – the confirmation of the hypothesis is counter-intuitive (*a*) for empirical hypotheses, (*b*) for principles which are only falsifiable in the long run;
   (vi) Remarks on the negation of theories with respect to confirmation.
   (vii) That extremal confirmation values are also assigned to logically true and logically false propositions is counter-intuitive. Disconfirmation function required.

(B) *Popper's Desiderata.* These have certain merits but:
    (i) Fail to distinguish extremal values of corroboration, from values for logically true and logically false statements;
    (ii) Fail to distinguish failure to falsify, which entails suspension of judgement, from corroboration.

Ever since Laplace's *Philosophical Essay* there have been attempts to develop extensions of the gamblers' notion of likelihood of events so as to supply mathematically sharp concepts for many other jobs of assessment. I am concerned in this chapter with the extension of probability concepts to the assessment of the satisfactoriness and correctness of hypotheses, principles and theories in the sciences, in short the attempt to define the epistemological concept of confirmation and related notions, in terms of the concept of probability: the rival procedure to the naturalistic attempts of the last chapter. The most interesting case is that in which the concept of probability is made mathematically sharp by requiring it to be at least isomorphic in its syntax and usage, with the concept exemplified in the classical 'theory of probability'. First I shall briefly examine the classical concept of probability and then outline the notions of confirmation that are relevant to science, with the intention of bringing out the problems that would have to be solved if these concepts were to be connected logically. I want to address myself finally to the problem of whether a function $C(h, e)$ having as values the degree of confirmation of hypothesis h on evidence e, behaves analogously to the already well-established function $P(h, e)$, where h is a happening or a description of a happening, e the evidence we have for its occurring, and $P(h, e)$ the probability of its occurring, or the probability of the sentence h being used to make a true statement.

Typically, primitively, primarily and basically, probability has to do with events. It is used in the prior specification of particular events. It is the winning of *this* race by *this* horse, the coming up heads on *this* occasion by *this* penny, the maleness of *this* birth, that are assessed as more or less probable. If the event in question initiates a subsequent enduring state, we also make a probability assessment of the likelihood of the 'coming to be' of the state so initiated. So we can speak of the likelihood of his finishing up in gaol, the probability of getting a certain disease, the chances of being cured, and so on.

It seems odd to say of some particular event that it will be probable, though not so odd to say that it will become probable. I think this has some significance, which derives from the fact that probabilities are assigned to future events now. If we suppose at some future time our changing an assessment of probability, with differing evidence and differing external circumstances, the assessments of probability will be

F 2

of some, yet more future, event or initiated state. One *might* say that probability statements have essentially future reference, whatever their tense may be. To say that a certain event is probable, is to say, with an escape clause, that it might or will occur. To say that some event was probable is to say that it was likely that it would have occurred. A pretty intractable philosophical puzzle is generated by this. It is to ask for the status of the alleged future events to which probabilities have been assigned in the present, and yet which do not occur. To what does a probability statement refer? A possibility, perhaps? The reaction of some to this problem has been to propose that probabilities never be assigned to particular events at all. They argue that probability assessments of particular events should be understood as short for statements about the statistics of classes of which the particular event is a typical member.

To this, as to some of the other difficulties with which the traditional notion of probability is surrounded, an elegant dismissal can be given by being very clear about the distinction between the meaning of an assessment of probability, and the grounds for that assessment. It is yet another instance of the old naïve verificationist fallacy to confuse, and so to identify, the content of a statement with the grounds for making the statement, or worse still with the grounds for deciding whether it should be assessed as true or false. In short it is to confuse the meaning of a statement with its method of verification. It is, no doubt, pretty often the case that the grounds for the assessment of the probability (likelihood) of a particular event are the statistics of like events. That fact provides no justification for claiming that an acceptable paraphrase of the assessment of the probability of a particular event is given by citing the statistics of that kind of event, of its relative frequency in a class. The grounds for the assessment of probabilities are very various, some *a priori*, some empirical, and of the empirical some are statistical and some are not. There is no problem to be resolved because of this diversity, since each may be quite proper as one of a variety of grounds for making the same assessment. That common assessment, to take one eminently sensible proposal, could be an expression of the most rational betting odds to take on a certain particular event occurring, expressing a degree of belief in, or degree of commitment to, or of hope of the event occurring.

An instance of the muddle of the meaning of a probability statement with the grounds for the assessment, is the misguided attempt to *define* probability in terms of relative frequency. So let us not get bogged down in pseudo-problems about whether probability really is relative frequency, or proportion of alternatives, or whatever. At best such disputes could concern the propriety of making use of empirically found frequencies of states of a system, or logically possible alternative states of a system, as the grounds for assessing probabilities.

The conditions under which probabilities tend to be assigned to events are fairly easily identifiable. There must be some set of events, or states say A, some members of which are associated with some characteristic B, and others with a different characteristic C. But being associated with C excludes the possibility of being associated with B. There may also be some As which are D, and being D they cannot be either C or B, and so on. The exclusion condition can often be satisfied by requiring that the characteristics B, C, D etc. be determinates under the same determinable, e.g. specific hues under the determinable colour. Then we ask of some possible event or state, of kind A, which we have not yet observed, say because it is an event or state that we expect to occur at some future time, whether it will be B or C or what. If it is not known what factors determine which of the states B, C etc. will be manifested, then it cannot be said with certainty what characteristics will accompany A. We may know that some red caterpillars are poisonous to birds and some not, without knowing what factors are responsible for a caterpillar being red and poisonous, and what factors are responsible for another being red and nourishing. Simply given that a caterpillar is red, we will be unable to say with certainty what will happen to a bird which consumes it. All that can be said is that in the present state of knowledge we can say that it will be either poisoned or nourished. This is the point of entry of probability notions. Under various states of knowledge additional to this primitive situation of barely knowing that the two states are both possible, probabilities can be assigned.

In the primitive situation, which I shall call 'Degree *0*', no probability can be assigned to any of the possible states. This, we say, is pre-probabilistic. All that is known is that events or states or generated things of kind A are sometimes B, sometimes C . . . sometimes N. And it is also supposed that B, C, . . . N are mutually exclusive characteristics, such as determinates under the same determinable. By adding the conditions

1. that B . . . N are the only possible determinates for things of kind A, under a given determinable,

2. that B . . . N are equipossible,

probabilities can be assigned. The addition of these assumptions I shall say creates an epistemic situation of Degree *1*. Under these circumstances it is proper to say that if an event, state or generated thing is of kind A, then the probability that it is of kind B is $1/k$, of kind C also $1/k$, and so on, if the number of possible states B . . . N is k.

What of these assumptions? What grounds are there for accepting them? Clearly the more we know about the nature and origin of the characteristic A, the more we are justified in accepting the assumptions which make an assignment of probabilities of Degree 1 possible. Suppose it is discovered that the caterpillar's colour is produced by an organ, which produces, as a byproduct of this process, a substance poisonous to birds when the caterpillar feeds on mulberries, but that the byproduct is harmless when the caterpillar feeds on other leaves. Suppose it also turns out that mulberries make up roughly half the forest, and that caterpillars are evenly distributed through the trees. Taking these facts into account, the assumption of these conditions is clearly justified and probabilities can be assigned with confidence. Subsequent degrees of knowledge with respect to probability assignments arise by the further incorporation of the two kinds of knowledge mentioned in this paragraph, namely, theoretical, explanatory knowledge, and statistical knowledge. It is a central tenet of the doctrine of this book that statistical knowledge is primitive, and can at best be used as evidence in support of protolaws, and that it is essentially preliminary to complete understanding, which can come only with a knowledge of the mechanisms of nature which are responsible for the events and states manifested in the world, knowledge of which explains the statistics of those events or states. So statistics will be added next to give a state of knowledge of 'Degree 2'.

Statistical information is essentially information about the numerical proportions of members of segments of subclasses to the number of members in comparable segments of whole classes. Apart from the difficulties of formulating satisfactory limit concepts for proportions of classes which are potentially infinite, it is not clear just how statistical information functions in the making of assessments of probability. The frequency theory of probability holds that such information just *is* what an assessment of probability actually amounts to. On the view being expounded in this chapter such information takes a place as part of the grounds for the assessment of probability. It bears upon Condition 2 above, which, in the situation of knowledge of Degree 1, is an ungrounded assumption, an article of faith. In the situation of knowledge of Degree 2, Condition 2 is modified and grounded, since the statistics of events, states and generated things affect our view of the relative *possibilities* of events, states or generated individuals of particular kinds. It is rational to believe that if a B-kind of event occurs with greater frequency than a C-kind, then a B-kind of event is a greater possibility. Hence, there is a greater probability that some particular event of kind A will also be of kind B rather than of kind C. In short, the concept of possibility applies to kinds, and that of probability to particular instances of kinds. However, though statistical knowledge serves to give grounds for various particularizations of Condition 2, Condition 1 remains an ungrounded assumption at Degree *2*. The frequency theory of probability halts at Degree *2*. How it is associated with positivism ought now to be clear. Since on that view the content of science is nothing but the statistics of phenomena, there can be no further degree of knowledge. Couple this with the associated principle that the meaning of a statement is its method of verification and we get the frequency theory of the meaning of probability statements, in which the statistical grounds become identified with the total meaning of the probability statement. The concepts of possibility and probability are identified. It follows on this view that a statement of the probability of a single event has to be treated as an elliptical statement of the proportion of events of that kind in the set of events of which it is one.

To add knowledge of the origins of the forms and qualities of events, states and generated things is to move to Degree *3*. Here is added truly scientific knowledge, that is, information about the generative mechanisms in nature which produce the kinds of events, states and generated things, knowledge of whose proportions in their kinds, i.e. their relative possibilities, made up the statistical information of Degree *2*. This kind of knowledge bears upon both Condition 1 and Condition 2. Knowing what kind of generative mechanism is at work closes down the possibilities for the future. For instance, knowing what the generative mechanism is for human beings, in particular knowing about the role of the X and Y chromosomes in the determination of sex, it becomes clear why there are only two possibilities for newly generated humans, to be boys or to be girls. It also explains why, at first sight, these possibilities are equally possible. Adding the statistical information that 53 out of every 100 births are births of boys, we arrive at the grounded statement that the probability of some specific birth, say the very next, being that of a boy, is 0·53. Knowledge of the generative mechanism at work precludes the possibility of, say, a kitten being born at the next human birth. Finally, further knowledge of the generative mechanisms can be expected to supply an account of the reason why there is a slight preponderance of boys born. Ultimately, the statistics are depreciated in importance as the knowledge of the generative mechanisms at work becomes more complete. This is the situation of Degree *3*. The range theory of probability can be associated with this degree of knowledge.

The range theory treats an assessment of probability as meaning the number of ways in which a thing or event or state which is A, can acquire determinates under a determinable. If a thing is an apple, in how many ways can it be coloured? If the answer is two ways, then the probability of its being one of those determinate colours is 0·5. The restriction of determinates under a determinable, for some kind of thing, is achieved by reference to scientific knowledge, that is, to laws or theories. If it is in the nature of ripe apples to be either red or yellow, then the probability of *an* apple being yellow if ripe is 0·5. Knowledge of the ripening process, and the chemical changes associated with it, enters into the assessment through increasing our confidence that the only ways in which a ripe apple can be coloured are yellow and red. Difficulties for the range theory arise in incorporating statistical information into the assessment, unless it is through the bearing statistics have on our knowledge of the ways things can be. This difficulty disappears when it is seen that there are two conditions which have to be satisfied to allow us to make rational probability assessments, and that scientific knowledge bears upon both Condition 1 and Condition 2, while statistics bears exclusively upon Condition 1. At Degree *3*, both the conditions upon which rational assessment of probability depend are grounded.

Thus, the grounds for making probability assessments are objective, and do not involve any consultations of our subjective feelings of certainty or doubt, or degree of belief in an assertion. But the objectivity of the grounds does not make what is expressed by the probability assessment objective, i.e. a property of the event or state mentioned. It does make such assessments rational and properly grounded. Probability assessments, I believe, express our degree of confidence that an event or state will occur, or an object of certain kind be generated. This is publicly shown by, for example, the betting odds a man may take. So we resolve the subjectivist-objectivist controversy. A rational man adjusts his subjective degree of confidence to the objective facts as far as he knows them, that is, he grounds his expectations in his statistical and scientific knowledge. And his knowledge passes through various degrees as different kinds of objective information are incorporated in the grounds of his judgement. But his assessment of probabilities expresses throughout just the same thing – namely, his degree of confidence that the event, state or generated thing which he is expecting will be the way he expects it to be.

A considerable argument would be needed to turn this sketch into a doctrine, particularly at the point where I simply assume that the acceptance by a rational man of certain betting odds shows his degree of belief.* I also assume that the betting odds a rational man would take follow from his state of knowledge, and change with it. These assumptions would all need arguing for if the position briefly outlined here were to be established.

* Only at this point do I part company with Lucas; cf. J. R. Lucas, *The Concept of Probability*, O.U.P., 1970, Ch. I & IV.

I now turn from the notion of the probability of an event to the notion of the confirmation, in various degrees, of hypotheses. The typical objects of confirmation are propositional in character, and provisional in logical status. Typically it is beliefs, conjectures, guesses or hypotheses that are confirmed or disconfirmed. It is sometimes said that one is confirmed in one's beliefs; confirms a guess, a hypothesis, even a theory. To say that a hypothesis is confirmed is to make an assessment of the hypothesis, and the grounds for the assessment are the pieces of evidence that are taken to be relevant to the confirmation of that particular hypothesis. Frequently, in very simple cases, if a hypothesis implies certain consequences and these are found to obtain, their being found thus confirms the hypothesis. The relation of evidence to assessments of confirmation, in some cases at least, looks quasilogical.

One sometimes says that one's hypotheses are rendered more probable, or that a certain theory seemed much less probable after certain facts became known. Asked what these expressions might mean, it seems natural to say they mean that it is more or less likely the hypothesis or theory will turn out to be true. 'Turning out true' seems something like an event, as if the turning out to be true of a hypothesis is like the turning out well of a cake. One might be inclined to see the probability of a hypothesis as an expression of the odds on the event of its turning out true. One might formulate the hypothesis that the concept of 'confirmation of a hypothesis, theory, etc.' can be understood in terms of the concept of probability.

There are some fairly obvious objections to this way of treating the confirmation of hypotheses. We started with one concept *typically* applied to statements, and another *typically* applied to events, entities of as distinct categories as one could very well find. The idea that one can be explained in terms of the other depended upon identifying a kind of event in which hypotheses could figure. That is the use of 'probability' and cognates which I have just described seems to be confined to statements and statement-like entities. This way of thinking would lead one to say that there was no special 'logical' concept of probability. The probability of hypotheses is the same concept as the probability of events, but in the confirmation context it is used metaphorically, the metaphor operating through the analogy of 'turning out true' to, say, 'turning out edible', the latter being a genuine event-like phenomenon of which a probability assessment can properly be made; but in the case of edibility, we can distinguish between the evidence for the edibility of a strange fruit and the independent fact of its edibility, i.e. between the fact that chimpanzees eat it without coming to harm, and the fact that it nourishes human beings. But there is no independent fact of the truth of a hypothesis. A hypothesis is true just when the evidence verifies it. This might make one suspect the analogy. It certainly is not a simple parallel. Let us look more closely at 'turning out to be true', to see just how event-like it is. To start with, it is the case that theories do turn out to be true, and sometimes the date of this event can be given, though it is pretty difficult, if not impossible to show a theory to be false in any crisp way or to date its turning out false. One would have to hold a minimal deductivist notion of theory, to hold that view. Falsification of theories is a gradual process. One can date only in the sense that over an interval of time there is decline in the number of adherents to a theory. But for a long time there will be people, not all of them elderly, who refuse to abandon an old theory for a new one. Yet there does come a time when we can quite properly look around and say, 'Well, that theory, after all, has turned out to be false.' 'Turning out true' and 'turning out false' are more processes than events. Even so, such processes do reach some sort of termination, and our rational bets may be conceived of as laid on the nature of these.

Truth and falsity do not seem to be the only possibilities for how a theory may turn out. Take the hypothesis that the earth is a sphere. It hardly seems to express the matter rightly if one says belatedly that the discovery that it is an oblate spheroid falsified the hypothesis that it is a sphere. For one thing, for many practical purposes the earth can be treated as a sphere. It would have been quite a different matter had it been shown that the earth was not a spherical shape at all, but, say, a disc or a cube. A spheroid and a sphere are siblings, but a cube something quite different, i.e. we can argue that a sphere is a special kind of spheroid. Many theories are much more complicated in their relations with evidence, as we have already seen. The fact that theories are concerned with explanations and postulate mechanisms, some of which are not known to exist for sure, leads to two radically different kinds of evidence affecting our judgement of them. There is the evidence which derives from successful and unsuccessful attempts to make accurate predictions, and the evidence which derives from the plausibility of the mechanisms postulated in the theory. The hypothesis that the earth is a sphere stands somewhere between a simple empirical hypothesis like 'Bordeaux mixture will prevent the appearance of *any* white butterfly" which is vulnerable to the appearance of one member of the species after the crops have been dusted, and a theory like 'The malaria syndrome is due to the inhalation of swamp miasma', where the existence or non-existence or even the plausibility of such a substance as 'miasma' may turn out to be crucial in our assessment of the theory.

Even though the turning out true or false of hypotheses and theories is not exactly event-like it is something which occurs in time, and, despite the reservations I have been considering, it does not seem too implausible to treat relative assessments of probability as rational betting odds on the hypothesis or theory 'turning out true'. But 'probability' and 'confirmation' are not quite the same concept. The first point to notice is that, using the phrase 'turn out to be true', we get the following relationship:

'probable' means 'likely to turn out to be true';
'confirmed' means 'has turned out to be true on many or all occasions'.

It would seem to follow from this that, if a proposition is highly confirmed, that is, has turned out true on many important occasions, it will be proper to say that it is probable, that is, *likely* to turn out true on other important occasions. So the grounds for saying that a statement is highly confirmed are, via hypothetical syllogism, capable of functioning as the grounds for saying that it is probable. But clearly, the two judgements, 'confirmed' and 'probable' are not the same in meaning, and the one cannot be substituted for the other. They differ, as it were, in tense. To say of a proposition that it is probable, is to say that it *will*, or is likely to turn out true; to say that it is confirmed is to say that it has turned out to be true. Only via the discredited verification principle, identifying grounds and content of propositions, could we come to say that 'highly confirmed' and 'probable' are synonyms. These last expressions are assessments made, quite frequently, on the basis of the same evidence. The same evidence that convicts a person of the crime of stealing may be the basis for a diagnosis of kleptomania. But to say of someone that he is a kleptomaniac is not to say the same thing of him as to say that he is a thief.

The next step in our analysis will be to examine the relationship between the probability of events and the probability of hypotheses about those events. The identity of the grounds for both probability and confirmation assessments that have been mentioned above might tempt us to transfer the numerical probability of the event whose occurrence would constitute the evidence for the probability of the hypothesis turning out true, to the hypotheses, as its degree of confirmation. If I say that the probability of the occurrence of a certain event is 0·75, I can easily slip into saying that the probability that the event will occur is 0·75, and then go on to say that the hypothesis 'This event will occur' has a probability of 0·75. According to the view sketched above, this should be glossed as asserting that the likelihood of its turning out true is 0·75, which is a confirmation-like concept, since it is attributed to a proposition.

Another move might be to try to develop statistics for statements.

We would assess the likelihood of a statement turning out to be true on yet another occasion, by reference to the number of times it has turned out true with respect to the number of tests to which it has been subjected.

Both suggestions are well known and open to objections. To the former it can be objected that the probability concept changes in the course of the derivation of the degree of confirmation from event-likelihood. It might be argued that this is to change from a relative frequency sense of probability to the sense of a logical relation between a hypothesis and the evidence for it. My reply to this would be that, while the two statements of probability do not mean the same thing, the concept of probability is the same in each, and the first, the event-probability statement, constitutes the best possible evidence for the estimate of the likelihood of the hypothesis that the event *will* occur turning out to be true. To the latter it may be objected that it runs counter to strong intuitions about the way degree of confirmation is modified with successive testing. If a hypothesis turns out true on only 75% of occasions of its testing, then far from it having a 0·75 probability, the 25% of failures show it to be false, and to have a probability, in the sense we are canvassing, of 0.

To complete this rather brief discussion of these concepts, I will examine the view that a confirmation functor $C(h, e)$ can be defined, the logical syntax of which is isomorphic with the logical syntax of a functor $P(h, e)$, which obeys the principles of classical probability theory. In practice this amounts to the idea that the usage of the concept of confirmation can be formalized with the help of the already existing mathematical theory of probability. I now turn to a more detailed comparison of the structure of the laws of the classical probability calculus with some intuitively given principles governing the syntax of the notion of confirmation of hypotheses by evidence. I shall test the correlation in various cases. At the very best, this idea can yield only the conclusion that for parts of confirmation theory there is a model in the probability calculus.

The principles of R. Carnap's *Logical Foundations of Probability* make a useful starting point for these discussions. After distinguishing, in what seems to me to be entirely the right spirit, between statistics and confirmation, in his two senses of probability, he then lays down some conditions on adequacy for a confirmation functor. As we shall see below, they are such as to prejudge the issue of the shape, as it were, of the syntax of confirmation. If they are accepted, they ensure that the shape is probabilistic, and specifically that confirmation theory will be modelled on the classical probability calculus. Of the four conditions Carnap adduces, three seem to me to be clearly mistaken. I shall discuss each in turn, and after being forced to reject three of them, we shall then be in a position to see that the logic of confirmation is not, after all, to be modelled on classical probability theory.

Carnap's desiderata: '*h*', '*h*'', etc. are hypotheses, '*e*', '*e*'', etc. are evidence for hypotheses, and $C(h, e)$ is the confirmation of hypothesis *h* on evidence *e*.

1. If *e* and *e*' are *L*-equivalent, i.e. logically equivalent, then $C(h, e) = C(h, e')$. This desideratum seems to be just barely tenable. A good deal hangs on what 'logical equivalence' is supposed to cover, but if it remains restricted to simple transformations of form, such as occur in immediate inference, and provided no substitution of alleged synonyms of content-bearing terms is permitted, it seems unobjectionable. For instance, if the hypothesis is 'All trans-specific hybrids are sterile', and the evidence adduced is that some trans-specific hybrids are sterile, then it seems reasonable to say that the hypothesis is confirmed to the same degree by the true statement 'some trans-specific hybrids are sterile' as by the logically equivalent statement that some sterile animals are trans-specific hybrids.

2. If *h* and *h*' are *L*-equivalent, then $C(h, e) = C(h', e)$. This is clearly mistaken. Hempel's confirmation paradoxes show it to be simply wrong. It is simply wrong to say that all crows are black, and all non-black things are non-crows is equally well confirmed by the discovery of a black crow. The reasons behind this are complicated, as we have seen. They amount to the requirement forced on us by an adequate understanding of the distinguishing marks of scientific laws, that laws of nature depend for their strength of confirmation both upon knowledge of regularities among phenomena, and knowledge of persisting generative mechanisms. It follows that certain traditionally sanctioned logical moves, one of which is contraposition, are not allowed in the logic of statements having nomic necessity, i.e. that are laws of nature. In the case of the original law, the mechanism which generates the regularity is a black-generating mechanism in crows, the existence of which helps to confirm the law. The terms of the contrapositive are too weak, i.e. non-black as the complement of black, includes all the rest of the universe, in which case it makes no sense to talk of generative mechanisms. Without the possibility of reference to such mechanisms, it is impossible to describe relations among things and qualities as lawful. Even in a restricted universe in which there are only two kinds of things (shoes and crows) and two qualities (black and white) 'All white things are shoes' would qualify as law-like only if a shoe-generating mechanism in whiteness could be made plausible. So the simple contrapositives are not equivalent for science to the original statements of which they are the contrapositives. Quite generally, in a universe of any

tolerable complexity, if a proposition is a law of nature, then its contra-positive is not a law of nature. The case is more complicated than this brief sketch comprehends, and the reader is reminded of the more ex-tensive discussion in Chapter 4. So L-equivalence in Carnap's sense certainly fails to preserve confirmation. We can then say that either L-equivalence must be understood differently, or that logically equiv-alent hypotheses are not equally confirmed by the same evidence. Whichever of these alternatives we choose, we are obliged to abandon Carnap's desideratum 2.

3. The general multiplication principle states that the confirmation index of the conjunction of two hypotheses h and j, on e, $C(h \& j, e)$ is equal to the product of the confirmation index of h on e, and the con-firmation index of j on e and h.

$$C(h \& j, e) = C(h, e) \times C(j, e \& h).$$

Since, according to Carnap, C-functions always have values between 0 and 1, the confirmation of a conjunction of hypotheses is always less than the confirmation index of each taken singly. This is surely not so in general, and is true only for probabilistic, statistical hypotheses. For example, if h is Galilean mechanics, and j is Kepler's planetary law, then Newton's cosmology, which might be said to be the conjunction of h and j, is much more highly confirmed by the evidence e, the facts about motion of bodies known to Kepler and known to Galileo, than is either taken singly, though this example is actually more complicated, since Galileo's Inertia Law was unacceptable both to Kepler and to the Newtonian Synthesis. Theoretical synthesis, with no addition of new evidence, is classically taken to lend further support (by the very fact of its being successful at all) to the joint theory than to either theory taken separately. As a further example, compare the change in degree of confirmation of Darwin's Theory conjoined with Mendel's Theory, with Lamark's Theory conjoined with Mendel's Theory. It is not clear that this desideratum will work for theories. For further dis-cussion, see the bibliography on 'consilience'.

But what about the synthesis of hypotheses? Black body radiation is a nice example of the synthesis of laws. The evidence provided by experiment lends some confirmation to Rayleigh's Law, but the law tails off, and the same is true for Stefan's Law. Each law covers part of the range of temperature and radiant energy. Planck's Law, which is Rayleigh-like at low temperatures and Stefan-like at high temperatures, can be shown to be something very like a logical conjunction of the two more restricted laws. Further, the Carnap principle requires that the evidence be the total evidence, that is, the evidence which is for Rayleigh and against Stefan in the Rayleigh region, plus that which is for Stefan but against Rayleigh in the Stefan region. It is then surely quite absurd to say that the synthesis for which all the evidence is now favourable should be less highly confirmed on that evidence, than is each of the primitive laws on that very same evidence, some of which is *dis*confirmatory of the restricted laws. And this situation is far from atypical.*

4. Finally, there is the so-called Special Addition Principle, which states that if the conjunction of *e* and *h* and *j* is a contradiction, that is *L*-false, then

$$C(h \vee j, e) = C(h, e) + C(j, e).$$

Since *e* is evidence for *h* and evidence for *j*, the condition requires that *h* and *j* be incompatible hypotheses. This principle would have the effect of making their disjunction more highly confirmed than either. Certainly this is not a syntactical principle, for the notion of the confirmation of theories. If *h* is the oxygen theory of combustion, and *j* is the phlogiston theory of combustion, then the disjunction 'Either oxygen is absorbed or phlogiston is evolved' would have to be more highly confirmed by the work of Lavoisier *and Stahl* than the oxygen hypothesis alone. But the point about the oxygen theory is that it makes the evidence which was thought specially favourable to the phlogiston theory as much evidence for oxygen as for phlogiston. In fact, the disjunction is such that the phlogiston theory becomes a mere cypher, and the disjunction is exactly as well confirmed as the better of the two theories is disjoined. It is perfectly clear that no confirmation theory meeting Carnap's conditions could come anywhere near expressing the main confirmation relations for theories. It is all far too crude. It is a very striking fact that Carnap's examples, which do seem to have some plausibility when understood in the light of his desiderata, are taken from psephology, not from any of the more theoretical sciences.

Another important principle of classical probability theory is the law governing negation. In general, the law is that if *p* is the probability of *e*, then *1-p* is the probability of not-*e*. If the probability that a die will lie 6 up is 1/6, then the probability that it will not lie 6 up is 1-1/6, i.e. 5/6. Does this principle extend to confirmation theory? Is it correct to say that the degree of confirmation of the negation of a theory or hypothesis on some given evidence is 1 – the degree of confirmation of its affirmation, *on the same evidence*? We shall see that this, too, will not do. The typical confirmation situation is very different. Consider the hypotheses that smoking causes lung cancer, highly confirmed on the available evidence. What degree of confirmation should we say the

* Feyman's interpretation of the 2 slit experiment is a case where $C(h, e) < C(h \,\&\, h', e)$ and $C(h', e) < C(h \,\&\, h', e)$ where *h* and *h'* are hypotheses about the path of single electrons.

same evidence gives to the hypothesis that smoking does not cause lung cancer? The answer is, surely, 'None!' The typical law for confirmation of negations must surely go something like this:

$$\text{If } C(h, e) = k \text{ and } 0 < k < 1, \text{ then } C(\sim h, e) = 0.$$

The case is even more striking for those scientific propositions we single out and elect as principles. Principles are those general statements that are *treated as* regulative for the time being. They determine in advance our way of treating phenomena, conceptually. They are not analytic propositions, though we do not usually allow them to be falsified. If contrary evidence seems to appear, we make adjustments elsewhere to accommodate it. Sometimes, we come to stop treating such principles as immune from counter-instances and allow them to be falsified. This is usually when, if we continued to follow them, our handling of phenomena becomes too complex, or too full of *ad hoc* corrective assumptions, or both. Suppose we treat the hypothesis of the rectilinear propagation of light as a principle. Then, if a ray (itself a hypothetical entity) is *not* rectilinear, we use the proposition as a principle when we infer from its curvilinearity or abrupt change of direction, that it has been interfered with. If a principle allows us to make sense of a great many phenomena, these can be treated as evidence confirming us in our adoption of the principle, and so metaphorically in confirming the principle, what one might call 'lending approbation' to the principle. But the phenomena which have been made sense of by the use of the principle (by governing the setting-up of a vocabulary for describing the phenomena), and so have tended to confirm us in our approbation of the principle, lend no approbation whatever to the negation of the principle. If the phenomena of reflection and refraction lend approbation, in the way I have described, to the principle of the rectilinear propagation of light, then those phenomena lend no approbation whatever to the principle that light does not travel in straight lines. During the lifetime of a principle, apparent counter-instances perforce become instances of the principle, since they become the occasions for such moves as the introduction of disturbing factors, which explain the counter-instances under the principle. Refraction, say, which is a deviation from rectilinearity, is treated as a special case of rectilinearity, under a specific kind of interference.

A theory can be reduced to the following essential elements, as we have seen:

(*a*) An existential proposition which introduces some entities, M.

(*b*) A general, law-like proposition describing (though this may be in fact laying down in advance) the qualities and behaviour of M.

(*c*) A general proposition stating how the states of M appear, that is, are manifested in, or affect the manifested qualities of, the world that can be observed.

The negation of a theory may be generated by the negation of any, some or all of these elements. The most powerful negation is of (*a*). If we accept that there are no animal spirits, then a complete physiological theory is abandoned with it, and subsequent propositions about the behaviour and visible and tangible effects of animal spirits need not further detain anyone but a historian. This negation, though the most powerful, is, of course, the most difficult to achieve. Or we can retain (*a*) as, say, by retaining a belief in cells as the units of biological structure, and move from, say, Schleiden's cell theory, in which the cell wall is prominent, to one in which the protoplasmic content is prominent, by denying the proposition 'The essential functions of the cell are carried out in the cell wall' and affirming 'The essential functions of the cell are carried out by the cell contents'. Or there might be cases in which it is the proposition of the third sort that is denied. We might continue to believe in the existence of certain entities, and continue to accept the mode of behaviour postulated for them as before, but change the statements of how we believe they affect the things we can observe. We no longer believe that the planets and their relative positions affect the earth and its inhabitants astrologically, but we continue to believe that they do so gravitationally. It was thought they had all sorts of effect, of which only some are now believed to occur.

Consider these three cases with respect to degree of confirmation. Does the evidence for the negation of the theory of animal spirits support in the slightest degree the theory of animal spirits? Of course it does not. But unfortunately other cases are different. It is quite clear that the reduction experiments carried out by Stahl on metallic oxides do support the phlogiston theory, as well the oxygen theory of combustion and calcination that superseded it. In some respects, these cases are alike, in that each involves the postulation of a substance which people later came to believe did not exist. And yet they are different in the syntax of confirmation with respect to the kind of evidence which existed for them. The phenomena which were supposed to be indicative of the animal spirits were as mythical as the spirits and the pneumas themselves. The cases distinguished above are still more equivocal. Some of the evidence for the cell-wall theory supports the cell-contents theory, namely, that evidence bearing on the cellular organization of living things. This is a simple consequence of the fact that particular evidential statements of the form 'Something is a cell and behaves in a certain way' are naturally analyzed so that they have an existential component, namely 'There exists a cell, and it does so and so'. So the evidence for the existence of cells bears in favour of the assertions that some cells perform their vital functions mainly in the cell wall, and that some cells perform their vital functions mainly in the cell contents. In short, there is a part of the evidence that gives the same degree of confirmation to part of each hypothesis. Let that evidence be

$$E \text{ (the total evidence)} - ew \text{ (the cell wall evidence)}$$

but if W is the wall component of the wall theory, and L, the contents component of the contents theory, then it seems clear that whatever the values of $C(W, ew)$ and $C(L, el)$, there is very little chance of being particularly successful in formulating any illuminating functional relationship between them.

Yet another of the dubious features of Carnap's system concerns the assignment of the extreme values 0 and 1. Effectively, Carnap introduced the value 1 as the product of the confirmation of a hypothesis and its complement, so that 1 becomes the confirmation value of a tautology on any or on no evidence.

$$C(T, e) = 1 \text{ whatever } e \text{ may be.}$$

And similarly, we get the principle that an L-false proposition or self-contradiction takes the value 0, so

$$C(\text{not-}T, e) = 0 \text{ whatever } e \text{ may be.}$$

But any system adequate to the kinds of propositions that are tested against evidence in scientific contexts must assign full truth, 1, to confirmed existential propositions, and full falsity, 0, to disconfirmed general propositions. But to no existential proposition (Ex), however unsatisfactory, must 0 be assigned, nor must 1 be assigned to any general proposition (Gen), however well authenticated. So we seem to have the following rules:

$$0 < C(Ex, e) \leqslant 1$$
$$0 \leqslant C(Gen, e) < 1$$

Now this, combined with Carnap's method of introducing 0 and 1, leads to a situation so counter-intuitive as to amount to an absurdity. It is perfectly proper to say of existential statements that they are confirmed, but surely absurd to say of tautologies that they are confirmed. Giving the same value to both confuses the logico-mathematical part of science with the empirical. A tautology does not have a very superior kind of empirical truth. Phrases like 'true in all possible worlds' tend to suggest such a confusion, but it might be better to say, not that tautologies are true, but they are, say, necessarily-true, so that, while the pro-assessment force of 'true' can still be effective, we are under no illusions that any of the usual criteria have been satisfied in some superior way.

Any system adequate to the requirements set by existing standards of judgement ought to contrast what I shall call modal assessments, such as 'necessarily true' for the mathematical and logical propositions common to the sciences, 'possibly true' for empirical propositions, whatever their degree of confirmation, and 'necessarily false' for the self-contradictory. None of these assessments admit of degrees. Only of a proposition already agreed to be possibly true should we ask 'Confirmed?', 'How far?', 'Disconfirmed?'. For these propositions, confirmation for existential statements is

$$1 \geqslant C(Ex, e) \geqslant 0$$

but for general statements

$$1 > C(Gen, e) \geqslant 0$$

while disconfirmation, an equally but *only equally*, important relation between evidence and hypothesis is, for existential statements

$$1 < D(Ex, e) \leqslant 0$$

while for general statements

$$1 \leqslant D(Gen, e) \leqslant 0.$$

We need an independently introduced D-function, for disconfirmation, because, as we have already noticed, to confirm something to ever so slight a degree is not to disconfirm it at all, since the favourable evidence for some hypothesis gives no support whatever to the contrary supposition in many cases.

The conclusion that one is driven to by the general failure of the Carnapian desiderata to provide a plausible syntax for the notion of the confirmation of a hypothesis by evidence, or the confirmation of a theory by evidence in all cases, has been expressed by many logicians, not least by Sir Karl Popper. He says, and I echo the sentiments, in his *Logic of Scientific Discovery* (p. 387), 'The degree of corroboration is not a probability' and 'The calculus of corroboration is not one of the possible interpretations of the probability calculus'. Admirable though these sentiments are, they are not backed up in Popper's case by any more adequate set of desiderata for the syntax of the notion of confirmation than in the efforts of Carnap. Like Carnap, Popper does not clearly disentangle the notion of 'necessary truth' from 'fully confirmed by evidence'. Instead of making sure that truth assessments from evidence are carried out on propositions that are already classified modally, he tries to define his extreme values of confirmation and disconfirmation as follows

$$- 1 = C(\text{not-}h, h) < C(h, e) < C(h, h) = 1.$$

It is clear that this desideratum obliges one to identify the truth of a self-contradiction with the falsity of a disconfirmed general hypothesis, and the truth of a tautology with the confirmation of a confirmed existential hypothesis, both of which identifications are not only question-begging, but absurd.

Again, Popper fails to distinguish confirmation from failure to falsify. Mere failure to falsify, in the absence of positive evidence for a hypothesis, ought to lead to suspension of judgement, and to a value exactly between – 1 and 1, of 0. Furthermore, empirical statements might be either existential or general, and these two classes of statement are differently related to their evidence, as has long been observed. As I noted above, the intuitive scale for existential statements runs from a possible ten out of ten when something of the kind hypothesized does turn up, to lower and lower values as an ever-widening search fails to find anything of the hypothesized kind. But since it can never be conclusively established that something of any conceivable kind does not exist, those values can never reach 0 or – 1, or whatever is chosen as the numerical assessment for the definitely false. So Popper's third desideratum, which requires, *inter alia*, that all empirical hypotheses shall have the same range of confirmation values, must be rejected.

The syntax of confirmation has nothing to do with the logic of probability in the numerical sense, and it seems very doubtful if any single, *general* notion of confirmation can be found, which can be used in all or even most scientific contexts.

I have assumed, in all this, that the only notions which require elucidation are those involved in the assessment of general statements. The truth of particular statements has been assumed. We must now turn to examine our descriptive vocabulary and the problems involved in the truth and falsity of particular descriptive statements of fact.

SUMMARY AND BIBLIOGRAPHY

The complicated and difficult topics of the interpretation and analysis of probability-statements and confirmation-statements are treated briefly in this chapter by sketching a theory without setting out any detailed supporting arguments. The main points in the extensive discussion of the meaning of probability-statements can be found in

157. R. CARNAP, *Logical Foundations of Probability*, University of Chicago Press, 1950, Ch. II.
158. S. TOULMIN, Review of Carnap (157), *Mind*, **62**, 89–99.
159. S. TOULMIN, *The Uses of Argument*, Cambridge, 1958, Ch. II.
160. E. H. HUTTEN, 'Probability Sentences', *Mind*, **61**, 39–56.
161. N. COOPER, 'The Concept of Probability', *BJPS*, **16**, 226–38.

In the text of this chapter, the traditional theories are treated as analyses of the criteria for making estimates of the rational betting quotient in a particular event-possibility. Good accounts of the frequency theory can be found in

162. R. von Mises, *Probability, Statistics and Truth*, Hodge, London, 1939, Ch. I.
163. H. Riechenbach, *The Theory of Probability*, University of California Press, Berkeley, 1949.

For a clear critique of the frequency theory see

164. H. Jeffreys, 'The Present Position in Probability Theory', *BJPS*, **5**, 275–89.
165. W. Kneale, *Probability and Induction*, Clarendon Press, Oxford, 1949, §33.

The range theory is expounded by

A. Pap (*see* 102), Ch. 12.
W. Kneale (*see* 165), 34, 35, 36.

For a critique of the range theory see

166. J. F. Bennett, 'Some Aspects of Probability and Induction', *BJPS*, **7**, 220–30, 361–22.

Subjective treatments of probability are developed in

167. H. F. Kyburg and H. E. Smokler, *Studies in Subjective Probability*, Wiley, New York and London, 1964.
168. R. Jeffrey, *The Logic of Decision*, McGraw-Hill, New York, 1965.

A specific application of decision theory to the scientific context is to be found in

169. I. Levi, *Gambling with Truth*, Alfred A. Kropf, New York, 1967.

The attempt to use probability concepts and calculi to express the meaning and syntax of the confirmation of hypotheses has generated a considerable literature, of which the most important items are

R. Carnap (*see* 157), Ch. IV, §53, where the essentially probabilistic desiderata are set out, and this is challenged by
K. R. Popper (*see* 121), pp. 387–419.
170. L. J. Cohen, 'What has Confirmation to do with Probabilities?', *Mind*, **75**, 463–81.

Popper's own desiderata are set out in

K. R. Popper (*see* 121), pp. 400–1,

and Cohen's non-probabilistic system is sketched in

171. L. J. Cohen, 'A Logic for Evidential Support', *BJPS*, **17**, 21–43, 105–26,

and expounded in detail in

172. L. J. Cohen, *The Implications of Induction*, Methuen, London, 1970.

7. Description and Truth

The Argument

I

DESCRIPTION

1. *The Meaning of Descriptive Terms*

(*a*) A necessary condition for a term to have meaning is that its rôle in some linguistic activity can be described, and this may involve mention of some sample;

(*b*) Learning to use words is to imitate certain paradigms. Successful copying of paradigms ensures regularity of use, which can be described metaphorically as 'obeying rules';

(*c*) This view is challenged by the theory of ostensive definition in which a sample is supposed to be definitive or paradigmatic.

(*d*) *Defects*

(i) Objects are complex;

(ii) A variety of examples teaches a category, and then each example becomes a sample case, rather than a paradigm case.

2. *Attempts to Stabilize the Descriptive Vocabulary*

(*a*) Sensationalism, which is a version of the ostensive theory;

(*b*) operationism.

3. For a scientific vocabulary, observation statements ascribe powers; the quality words become metaphorical, and we escape the subjectivism of sensationalist views, because a power statement directs us to the nature of the material or thing in virtue of which it manifests itself in this or that way.

II

TRUTH

1. *The Concept of Truth*

(*a*) The Logic of Evaluation is applied in a modified form to the evaluation of statements;

(*b*) Location and Individuation: necessary because powers are ascribed to individuals and to actual materials;

(*c*) Uses of the word 'true'; usually as a response to another's statement 'S'.

(i) in confirmation of that statement, i.e. 'I agree' or perhaps that I can prove 'S' in the strongest case;

(ii) in evaluation of 'S'. How do we know how to evaluate 'S'?

(*d*) The notion of evidence; evidence might be a state of affairs, or it might be a proposition. Same category as 'fact'. Why?

(i) fact might be a state of affairs; 'fact that . . .' suggests it is a proposition;

(ii) 'Evidence for.' Does not evidence itself have to be evaluated, and can that be done by reference to more evidence?

2. *Words and Things*

(*a*) Learning to observe things in the world enables us to see that . . ., i.e. observing is like reading the world as expressing propositions; these are the facts;

(*b*) Verbal and pictorial expression of the same fact;

(*c*) Resolution of the difficulties of the classical theories.

    (i) There is no problem about correspondence because of the common proposition which ensures that the statement does correspond to the fact. Correspondence exists in the sense in which translations correspond to each other, by being different expressions of the same fact, or of the same proposition.

    (ii) No problem about coherence because of the world-expressing function of coherent propositions; if the descriptive statement *does* express the same proposition as the facts show then the statement and the fact must entail each other. The identification of this perceptual skill is made through the distinction between sensing and observing.

(*d*) Any further investigations must be of the metaphor of perceiving that . . ., or assessing, as reading, and leads to the psychology of perception, of language and of art.

III

CLASSIFICATION

1. If descriptive language does have a powers analysis in science, then every classification automatically refers to the real essence in virtue of which the things or materials classified have the nominal essences that they do.

2. Taxonomy is a classification that employs differentiation by the nominal essence to classify by the real essence.

The Meaning of Our Descriptive Vocabulary

Theories change, and fashions in explanation succeed each other. It seems that the facts once ascertained should form the permanent part of science. It is a characteristic doctrine of deductivism that *permanent* facts are the basis of the judgement of *ephemeral* theories. If this were to be so it would seem that the terms in which the facts are described ought to be stable in meaning, and that whatever method of definition is used to give them meaning should do so in independence of whatever theories happen to be in the field. We shall see that this idea has influenced the development of theories about the meaning of descriptive terms, but that none of the theories which would ensure independence and stability of meaning for descriptive terms has been satisfactory. I shall offer an account of how descriptive terms have meaning that resolves some of the problems in this area. We have seen how the theoretical terms in a science derive their meanings through the analogies that their subject matter, iconic models and hypothetical mechanisms bear to real and well-known things and processes. What stability of meaning is maintained for terms apparently connected to several theories such as 'gene' or 'atom', derives from the fact that in each employment the source of the model they are used to describe is the same. Differences in meaning from a common source, derive from differences in the analogy between source and model. The matter is made still more complicated by the analogy that a model bears to its subject. In the descriptions of both source and subject, descriptive terms occur. It is to the problem of the stability and modification of meaning of these that we now turn.

The protean notion of meaning includes so much that any general account would be out of place here. Meaning embraces emotions evoked, objects signified, images stimulated, information conveyed, linguistic equivalents, rôle in a socio-linguistic activity, actual use to which words are put, analysis in more familiar terms, definition, intention of speaker, truth conditions of sentences, and no doubt other things as well. Out of all these we can pick one necessary condition for meaningfulness that will be sufficient for my purpose in this chapter. That is that for a word to have a meaning it must play some rôle in some human activity. So we shall be trying to describe the rôle of descriptive words in the scientific enterprise as we have identified it. In practice giving synonyms may be one way of indirectly indicating the rôle, and perhaps this is the reason why traditionally meaning has often been identified with some group of synonyms. I shall aim at explaining meaning by indicating rôle. The idea of rôle has led to the view that since playing a rôle can be achieved by following rules, the using of language for various purposes is a rule-governed activity, and the giving of meaning often best achieved by giving the rules governing the use of a word.

This is compatible with another useful idea, that language using is a paradigm-determined activity. In learning language we are presented with examples to copy rather than rules to follow. It is because of this that grammatical models are of such imporance since they can exert considerable sway over the way we can express ourselves, and to some extent over the kind of theories it is possible to hold. For example the grammatical models, or paradigm sentences, of the languages of the scientific community, use the device of subject and predicate, of noun and adjective in this manner, distinguishing the referent of a statement from what is asserted of the referent. Indeed, Chomsky would go further and argue that there are paradigms common to *all* linguistic communities.* Under this paradigm some of the descriptive content of a statement is carried by the subject term. But compare 'The car is stuck in the mud', where the grammatical subject is both descriptive and referential with 'This is a car stuck in the mud' where the grammatical subject is maximally referential and minimally descriptive. It looks as if the de-

* N. Chomsky, and M. Halle, *Cartesian Linguistics*, New York, 1966.

scriptive model can be transformed into the demonstrative model and all the content of the statement transferred to the predicate, while the referential, attention-drawing function, can be performed purely. (Chomsky would treat this transformation as one of surface structure only.) It is perhaps from the idea that this is a substantial transformation that the theory of ostensive definition ultimately derives. The demonstrative model seems to work by using some words to the accompaniment of a gesture of pointing, an ostensive gesture, by which the words are linked to the world. A pointing finger seems to link words and the world in a particularly unambiguous and striking way.

The theory of ostensive definition can be looked upon as an application of this grammatical model, as a paradigm for a procedure of definition which will provide the stability of meaning aimed at by positivist epistemologists, *and* since the ostension is ideally supposed to be to sensible qualities or even to sensory contents, it will also provide a set of terms with meaning independent of any theory. It supposes that descriptive terms, nouns and adjectives mainly, can be introduced one by one, each by an independent act of ostension, together with the utterance of the word expressing the predicate. So each term, not only has meaning independent of theory, but independent of other descriptive terms. Saying 'square' while pointing to a square is thought of as the paradigm of a definition for a term in a descriptive vocabulary. Since it seems that in this procedure words and the world are brought right up against each other, it seems that it is impossible for a scientific or even a metaphysical theory to be involved, and hence a meaning given this way is permanent and definite. Objections to this theory are not hard to find, mostly deriving from Wittgenstein's treatment of the theory in his *Philosophical Investigations*. In ignorance of the category or kind of the attribute for which the word is being introduced, one cannot tell from the act of ostension alone, how the word is to be taken. Is it a word for a colour, a shape, a number, a relation or what? The simplest object to which our attention can be drawn, say a patch of colour, is complex in a variety of ways: it has a shape, a colour, a texture, a number, a temporal duration and so on. Simple ostension draws our attention to the patch in its primitive complexity, and in the absence of further information as to which aspect of it to attend to we do not have a sufficient background to grasp the sense of the word being introduced. But, it might be argued, a multiplicity of examples will resolve the difficulty. By presenting numerous squares, differently coloured, and of different textures, eventually the learner will have grasped that 'square' is a shape word. But then the definition is not by simple ostension. The variety programme teaches the word correctly and successfully because it also teaches the category in which the word is embedded in our conceptual scheme, as for example one grasps that the word 'mauve' is a determinate of the determinable 'colour'. The act of ostension is not primary to the giving of meaning. It is secondary, since what it actually does is to provide, subsequently to our grasping the category, kind or determinable, a sample case for the use of the word, and so is certainly not independent of metaphysics, for instance the thing/quality distinction.

The search for a system of descriptive terms which will be independent of the scientific climate, that is not affected by shifts in the current theories in the field for which a term is employed, has led to two further theories of meaning and definition. In one, the idea is developed that if we describe only our sensations, by way of observation reports, then, though these have to be categorized and sorted under determinables, if the meaning of the terms can be given in terms of sensations, they must be independent of any theory whatever. This view is particularly associated with Mach* and Pearson†, but it was also an element in the philosophy of science of Berkeley‡, and the theories of Poincaré.§ There are two obvious objections to such a theory. In the first place it has all the defects of the theory of ostensive definition. Secondly, the private character of sensations stands in the way of the formation of a descriptive vocabulary which would have a public rôle, and thus, according to Wittgenstein, have any meaning at all. This suggestion will be discussed again in Chapter 10, where it will emerge that a science based upon the correlation of sensations as its ultimate units of empirical information can be no science at all, since explanation will be impossible. We can safely dismiss the theory in anticipation.

The other theory is that of operationism, in which attention is shifted from what is observed, to the operation of observing, and particularly of measuring. This theory, though vigorously advocated, has never been very thoroughly worked out. Essentially, it involves an identification between the criteria for assessing the numerical value, according to some scale, of some physical characteristic of objects, and the physical characteristic itself. Our confidence in methods of measurement depends upon the theories of current physics: when they change, so do the relations between the methods of measurement and what is allegedly measured. Instead of holding that the characteristic measured remains the same, and that the change in relations is due to changes in the methods of measurement, the operationists, impressed by the immediate observability of the methods of measurement, attempted to give them indepen-

* E. Mach (see 175).

† K. Pearson, *The Grammar of Science*, New Edition, Everyman, London, 1949.

‡ G. Berkeley (see 12).

§ H. Poincaré, *The Value of Science*, New English Edition, Dover, New York, 1958, Ch. X.

dent status. The physical characteristic was eliminated from the subject matter of science by confining the *meaning* of descriptive terms to the description of the operations of measurement. This raised a typical 'problem'. Why are there very diverse sets of operations which all yield approximately the same numerical result? The traditional view explains this by showing, through the application of physical theory, how each set of operations is a method for measuring a common physical characteristic. For the operationist, each set of operations which yields a result, defines an independent physical concept, and there are no further concepts for formulating a theory by which to explain the extraordinary coincidence of their yielding the same or similar numerical values.

It is important to try to identify the exact point of entry of physical theory into the system of physical concepts. I want to argue that at least spatial *interval* and temporal *interval* can be defined independently of any physical theory. It can easily be seen that physical theory, which is, of course, revisable, certainly enters when a metric is required to yield measures of the intervals of space and time and, as I shall show in Chapter 9, it enters even for the empirical application of these concepts, but *not* for *giving them meaning*. The primitive concepts to be employed in our premetrical definitions of spatial and temporal interval are 'thing', 'event', 'coexists' and 'identical with'. These concepts are not, of course, independent of a certain *metaphysics*. The examination of that metaphysics will occupy us in Chapters 10 and 11.

'There is a spatial interval between A and B', means 'A and B coexist, and A and B are not identical.' (1)

But the criterion for asserting that there is a spatial interval between A and B involves criteria for the empirical differentiation of A and B, i.e. independent individuation, and entails that A and B are things or points in a geographical system related to things. They also involve a third thing, C; and an ordering relation, say 'between', for a one-dimensional space C must satisfy the following conditions:

'C coexists with A and B, and is identical with neither, and is between A and B'. The criterion for whether C really is between A and B involves physics, as well as the empirical differentiation of C from both A and B.

If (1) is intelligible, then the meaning of the assertion that there is a spatial interval between two things is independent of the criterion for asserting that there is a spatial interval between two things. We proceed similarly for 'temporal interval'.

'There is a temporal interval between e_1 and e_2' means, 'if e_1 and e_2 are assumed to be events, then e_1 and e_2 do not coexist.' (2)

G

But once again the criterion for asserting that there is a temporal interval involves not only the empirical differentiation of e_1 and e_2, but a third event, e_3, and an ordering relation, say 'before'. E_3 must satisfy the following conditions:

'E_3 does not coexist with either e_1 or e_2, and e_3 is before e_2, and e_3 is not before e_1'. The empirical criterion for whether one event is before another involves the use of causal laws and perhaps other considerations, and hence is not independent of physics.

Spatial and temporal metrics can now be easily devised on the basis of the criteria. They could not be devised on the basis of the above 'definitions', for these do not allow us to make comparisons of degree of separation of things or events, for which at least a third thing or event is required. Such comparisons are the essential basis of metrics. In order actually to devise a metric we need to employ the laws of physics, for these describe how the devices which embody the general criteria will behave, and, in particular, they specify the essential invariance conditions upon which any metric must depend. For instance, the constancy of the velocity of light is required before we can, by identifying the body C with a photon, devise a telescopic, light-year metric for astronomical distances. Similarly, the laws of motion are required before we identify the number of swings of a pendulum with the temporal interval between events. The novel kinematic invariants of special relativity drastically modify the metrical concepts that can be applied to the measures of interval in space and time. If the various specifications of the general criteria into actual methods of measurement are arrived at by a combination of these general criteria with different regularities of empirically known fact, then the operationist case does indeed collapse, since the general criteria which are derived in each case from one and the same 'definitions' of interval, respectively spatial and temporal, are inextricably bound up with each specification of a metrical method. Hence, the methods of measurement are not independent of the characteristic measured, and *a fortiori* by describing them we do not define a set of empirical concepts independent of our concepts as to the nature of the things which are studied. The definitions 1 and 2 give meaning to spatial and temporal concepts which are indeed independent of physics, though *any* empirical explanation they may have is not independent. The special theory of relativity, for example, restricts their empirical application so that they can only be applied *together*, i.e. there is no empirical application of concepts of absolute space and time.

In fact a descriptive vocabulary arises in an exceedingly complicated way, and its meaning is not given by any simple definitional procedure. It involves metaphysical assumptions in the same way that the 'definition' of spatial interval involved the assumption that no two things can be in the same place at the same time. It also involves the workings of a taxonomy, principles of classification, which are determined as we shall see in Chapter 8, by relatively *a priori* considerations. It cannot fail to be influenced by current theory. We shall begin our task of unravelling the problem of how all these factors can be incorporated in the giving of the meaning to descriptive terms by supposing that for scientific purposes our descriptions of things and materials refer to their powers rather than to their qualities. In short, I shall be developing a new metaphysics within which science *and its radical conceptual changes* are possible.

Descriptions as Power Attributions

What do we mean when we say that something or other is yellow? In unreflective and non-scientific contexts we certainly intend to say what hue it is, and we say this on the basis of what hue it manifests to us. But if that were what was meant by the word, then 'yellow', and words like it, could not, it seems, form part of a many-person vocabulary, and per-haps could form no part of any vocabulary, since it is logically possible that the same object may manifest itself differently to each of us, and perhaps differently to any of us on different occasions. 'Yellow' *is* part of a many-person vocabulary, and the apparent contradiction might be resolved by maintaining that the colour word is used to locate the hue manifested to each of us in a system of hues, and that systematic isomor-phism is enough for a publicly usable vocabulary to exist. I know of no objections to this view for ordinary description and identification of things and materials. But it certainly will not do for sophisticated and scientific contexts.

Case 1. A Sophisticated Context. Seeing something red under sodium light, one says to one's naïve companion, 'That may look black, but it's really red'. It is customary to construe 'This is really red' as meaning either 'This would look red under normal conditions' or perhaps 'This would reflect light of 8000 Å only none is now falling on it'. It is important to see that both these statements are attributions of powers and not attributions of qualities. To say of a thing that it is really red, is, in such contexts as these, to say that it has the power to manifest itself as of a certain hue in the appropriate conditions. We usually find out that things have such powers by their manifesting themselves in the appro-priate guise to us. But this is not always true. I might have unwrapped a packet of photo-paper in the darkroom, and, on seeing that it looked red, have known that it was really white, and then burnt it, without ever having seen it look white. A great deal of experience is presupposed in such an epistemological divertissement. It is clear that to use colour words in such contexts in such ways is to treat ascriptions of colours as ascriptions of powers and not as ascriptions of hues, and certainly this meaning of 'red' could not be given by ostension to the sensible quality, though it is not wholly independent of hue.

Case 2. The Scientific Context. What is really going on from a semantic point of view when we start talking about red light, red surfaces and what not, having in mind the wavelength of light? What has happened to the word 'red'? It is certainly no longer being used to ascribe a quality: a hue, which an observer might observe. Red light cannot be seen at all. If the word meant a hue, then red light would surely be a non-sensical mess of words. 'Red light' means 'light of wavelength 8000 Å'. The connection is this: light of this nature has the power to affect us in such a way that we experience a certain hue, qualified as in our discussion of the unsophisticated context. The light has this power whether or not it affects a human being. The colour word can be used metaphorically for wavelength, because it is in virtue of its wavelength that light affects us colour-specifically. Similarly, a body has a temperature even when there is no thermometer leaning up against it, and it is then *hot.* What is it that it then is? It continues to possess the power to affect a thermometer in virtue of the state it is in. To be capable of ascribing the power we do not need to be able to specify the state which is responsible, only to locate the responsibility for raising the level of the mercury as intrinsic to the stuff to which we attribute the temperature. In this case internal and intrinsic states are extensionally equivalent.

All words like 'hot', 'bright', 'green', 'heavy', 'smooth' and so on, when used by scientists within a scientific context are not only power words, they have something of the character of metaphors. In the first stage of preparation for scientific use, the word ceases to be used to ascribe a quality and is used to ascribe a power. This attributes a certain nature to the entity involved as subject, but at this stage the verification of the power statement is through the checking of the disposition to look this or that hue. At this stage there must be some situation in which whatever it is, really does look red. In the second stage the nature is better known and so can be more exactly specified. We may speak of 'redness' in contexts when that nature is engaged, but not in such a way that a hue can be manifested to anyone: for example in making a numerical measure in a spectrometer, and then we have certainly passed from literal to metaphorical language.

We are now two steps away from the ordinary use. The use of many common words for scientific description is metaphorical,* but this is because they have come to be used to ascribe powers, and not to ascribe sensible qualities.

---

* strictly this use should be described as an example of *paronomasia*, which can be given an etymological explanation.

To treat the uses of the descriptive vocabulary, when properly analysed, as ascriptions of powers leads to an important further insight into the use of a scientific vocabulary. If we are to understand 'This is red' as 'This has the power to look red in appropriate conditions', then we must construe the statement according to the normal analysis of a powers statement. A detailed discussion and justification of this analysis is to be found in Chapter 10. It becomes, 'This is of such a nature that if it is in the appropriate conditions it will look red'. This statement is clearly a schematic explanation, since it asserts that the nature of the surface or of the material is such that the conditional component of the statement is true. Scientific description then automatically sets the scene for scientific explanation, and in terms of the constitutions of the things and materials whose behaviour and dispositions is being described, and our conceptions of the natures of things are part of our theories of their behaviour.

All the materials are now at hand for dealing with the general problem of meaning-stability and meaning-change. It is certainly true that the meaning of descriptive terms is not independent of physical and metaphysical theories (*see* 199). It is also certainly true that the meanings of such terms have a measure of independence from theory-change and do have a certain stability in the course of the development of successive theories (*see* 190). Provided descriptive terms are construed as attributing powers, rather than sensible qualities, a solution to the apparently incompatible concessions above can be found. An attribution of a power has the form:

'In conditions C, A will (might) manifest sensible quality φ, in virtue of being of nature N.'

In a *scientific* context the ascription of colour is not the simple ascription of hue, which would be no more than 'A is φ'. Science, we know, cannot be built on such a basis. Nor is it the ascription of surface structure to things in virtue of which they differentially reflect light, by means wholly derivative from theory. Empirical testing of hypotheses would be impossible were that so. Our solution runs like this:

The difference in meaning of a descriptive term which comes from its being used in the context of different theories derives from different specific substitutions for N in the 'in virtue of N' clause, which derive from the theory. The stability of meaning of a descriptive term during embeddings in successive and different theories derives from the necessity for intersubjective identifications of the presence of the sensible quality φ. We have also seen how the sensory element can be attenuated to metaphor.

A residual problem remains. In the use of words to describe things and materials, and in similar uses to describe the reactions of instruments, the powers analysis depends upon a prior understanding of the ordinary qualitative uses of descriptive words. In the end there must be some statements which do state that a person is having an experience of a certain kind, construed non-subjectively, or that an instrument is reacting in a certain way. These statements raise another kind of question. How is it that words can express the state of the world, and how is it that we can know whether they do so correctly or incorrectly? In short, there remains after all this talk of powers a residual problem of truth and correctness, falsity and incorrectness of speech. To this problem we now turn.

Truth

Scientists have not infrequently claimed to be seekers for *truth*. This concept often crops up in popular assessments and explanations of science. Philosophers of science have not devoted any special attention to the concept of truth. Discussion of truth has become a part of general philosophy. But science is specially concerned with truth, and with what seem to be related goals, like accuracy and correctness, objectivity and so on. The reason for the apparent lack of interest in what would seem to be the most central epistemological concept of the subject is to be put down, I believe, to the effect of adherence to the deductivist myth. The logical deducibility of particular statements of alleged fact from principles seems a more prominent desideratum for the acceptance of statements about phenomena than the confrontation of words and things. By considering the way very simple descriptive statements are made, the key to the criteria for using words like 'true' can be found, since we cannot know what are the criteria for a true statement until we know what are the conditions for successful utterance: that is for making an utterance which conveys the information we want it to convey to another speaker of the language.

In making a simple descriptive statement we draw a listener's attention to some more or less specific feature of his environment and convey some information about it. Basically, two kinds of information are conveyed. One may be taken as saying that whatever is referred to is like certain other things in some respect, that is, belongs in a certain class, or one may be saying that it has a certain property, that is, is possessed of the power to manifest a sensible quality, or to affect an instrument.

For the ascription of powers, that part of the world to which a power predicate is to be applied must be circumscribed, that is, the source of a power must be *located*. The simplest device for locating anything is the utterance of 'this' accompanied by an ostensive act, such as pointing. It is the simplest device since no recognition of the subject beyond a very general categorization as an individual is involved, just the following of the pointing finger and the contemporaneous directing of attention to a place at a time. An important feature of most individuals of a thing-like character is that they endure beyond the particular spans of attention of those who are attending to them. For making reference to enduring individuals outside our span of directed attention, we typically use proper names and definite descriptions. Now such linguistic devices both provide a location for the ascription of a power, in so far as they inherit the demonstrative function of 'this', and they individuate, that is, state or imply a distinction between this individual and all others, which is more than its temporary occupation of this or that spatio-temporal location and which enable us to recognize it as the same, that is to reidentify it. The function of referring which logicians have made so much of is a 'vector sum' of two more primitive functions, location and individuation.

In the above account the recognition of an individual as the occupant of a place at a time is supposed to have logical priority over the identification of it as an individual having certain powers, that is, reference is supposed to have priority over ascription of a property or membership of a class: that is over description. Many years ago, J. L. Austin pointed out (*see* 194) that in simple descriptive talk we are not bound by these priorities and that many simple descriptions operate with different priorities. Sometimes we have in mind a certain description which we are looking for an individual to fit, sometimes we have in view an individual and wonder to what class or kind it belongs: that is, how it should be described. From this an uncontroversial but trite principle of truth follows.

The Most General Criterion for a Successful Description

A statement can be said to be true, that is, to be a proper statement for the *conveying* of information, when the sentence used to make the statement is constructed in such a way and of such elements, as are conventionally used to locate and individuate something where the subject of the statement actually is, and to ascribe to it a property which it actually has. I have not said, be it specially noted, how it is possible for us to know that this criterion has been met on any particular occasion. I shall tackle this problem in the later part of this chapter. I shall argue:

1. That each of the other major traditional criteria that have been offered, that is, correspondence, coherence and pragmatic success, derive what plausibility they have from the fact that statements picked out in accordance with them also meet the above criterion.

2. That most common uses of 'true', as for example those elucidated by P. F. Strawson (*see* 193), can be explained in terms of the most general

criterion. I am unable to see that Tarski's theory of truth has anything to do with the epistemology of science. It seems to be a theory appropriate to the philosophy of mathematics.

Theories of Truth

The above account must not be taken for a definition of the 'sense', or 'a sense' of 'true'. In general, traditional theories of truth have set out criteria for judging statements on the basis of which we may say that a statement is true: that is, a successful statement. Confusing criteria with meaning would lead to an impasse, created by the fact that examples can be given of judgements of the success of a statement which are based on what seem to be quite different sorts of consideration from those of the general criterion. For instance, there are cases where what matters is whether a statement corresponds to the facts, but there are others where our assessment depends upon whether the statement under review coheres with other statements. It seems that we use different criteria on different occasions. In some contexts one seems appropriate, in other contexts another. If all the criteria are supposed to elucidate meanings of 'true' then 'true' must be a hopelessly ambiguous word, and it is not. In fact, all that such examples and counter-examples show is that in different circumstances different criteria are used to judge of the success of a statement, or perhaps it is better to say, of a piece of statement-making.

We do not usually, if ever, say or write anything like 'The cat is on the mat is true' or ' "The cat is on the mat" is true', forms of expression at one time frequently canvassed by logicians as typical examples of the use of 'true'. Nor do we often say 'It is true that the cat is on the mat'. Rather we respond to a statement by another statement using 'true' or 'untrue', 'lie', 'wrong', and very occasionally 'false', and referring to the original statement. You may say 'The cat is on the mat', and it is I who may say, I think that you are right, 'Yes, that's true'. A statement using the word 'true' typically is used

(i) to refer to another statement.
(ii) to pass judgement on it.

This passing of judgement on a statement may be according to a wide range of criteria, but these criteria must spring in the last analysis from the general set given above, for they represent the minimum conditions for the complex symbol, the declarative subject-predicate utterance or inscription, to convey a statement correctly, to convey information about the state of the world at some place.

We have already seen how we use the phrase 'That's true' in much the same way as we use 'That's correct' and other phrases, to register our agreement with and hence our confirmation of the statement referred to by the demonstrative. Confirming a statement involves both agreeing with the statement and lending our authority to it. 'That's true', like 'I know' has this special power of passing on authority. Its use suggests, at the very least, that the user also has reason to believe the statement (a hint at the coherence theory of truth), and at the most that he, personally, has verified it (a hint at the complementary, correspondence theory of truth). If I make a statement and you confirm it, I can round on you if it later turns out to be false. It is in this manner of use that 'That's true' has been classified as a performatory utterance, since in uttering it I confirm your statement, in contrast to say, merely repeating it. The statements of scientists should measure up to the highest standards to which statements can attain, and it is we who set those standards. To scientists we have set the standard of truth, that their words should meet such criteria as will justify us in accepting their authority as to the information conveyed in their statements. How do we know that the standards have been met? How is it possible for them to be met at all? To these questions we shall now turn.

Truth and Evidence

To tell whether what someone says is true, one attends to what they are talking about and examines it. In many cases it is quite easy to tell whether what has been said is true or false, right or wrong, accurate or inaccurate, exaggerated or moderate. In those cases where the subject of discussion can be examined without any special equipment or study, there is no question of finding evidence for the statement made about it. If one can *see that* the broom is in the corner then it would be odd to say that the broom's being in the corner is evidence for the truth of the statement that the broom is in the corner. The state of the face of an instrument should not be treated as evidence for the statement that the instrument has such and such a reading, since in observing the face of the instrument an instructed scientist can see that the instrument is registering 15 milliamps. The notion of evidence is appropriate to those cases where the subject of discussion is different from the things and processes which can be examined. Then when an investigator sets about assessing the statements made about the unexamined subject it is proper to speak of what is observed as evidence, because the alleged fact that is being checked is not what is observed, but is an inference from what is observed to be the case. Evidence must have 'logical distance' from the fact to which it is relevant.

How can we know that some group of words is a report of the way things are? The answer to this problem is the key to the whole group of

G 2

questions about truth. I shall try to provide a sketch for an answer along the lines developed by a number of writers recently, by developing the metaphor that learning to make reports, to use sentences to convey information, is learning to 'read' the world, in much the same sense as one learns to read words, signs and symbols. We have already had intimations of this idea earlier in the chapter when I argued that words cannot be given meaning in an empirical context without, at the same time, the development of a categorization of the world. This was the basis of the declaration that simple ostensive definition was a myth. There is a familiar distinction between the objects used to express a proposition, the external bearers of thought, one of which is the sentence, and the proposition so expressed. I shall assume this distinction in what follows, and attempt neither to examine its subtle variations, nor to justify it by reference to the metaphysics of language. When we grasp the idea of truth, what we have grasped, I believe, is the idea of arrangements and qualities of things expressing propositions to one who can 'read' them, that is, to one who has learned to observe; that is, to one who can do something other than just see. A statement is true when it is made by means of a sentence which can be understood to express the same proposition as the arrangement of the things, or state of the things to which attention has been directed, can be understood as showing to a sufficiently experienced observer. The function of the demonstrative conventions is to ensure that our attention is drawn to the right place at the appropriate time, that is to that place in which we are supposed to be able to *see that* such and such is the case, that is to observe the state of affairs. Some examples can illustrate the point.

Imagine a situation in which one wants to communicate with a friend without a third person present being aware of it. One might nod one's head, almost imperceptibly, towards the relevant state of affairs. The friend is brought to see that so and so is the case by having his attention drawn to the relevant state of affairs. If information can be 'conveyed' this way, why do we have language? The effort to communicate in this way *could* be just as successful when one draws attention to some arrangement of things as when it is a conventional sign or a written notice at which one looks. Why have descriptive conventions for carrying out our communicative intentions? Could we not simply point silently to this or that state of affairs in which our 'audience' would see that this or that was the case? *Propositions can be conveyed like that*, as the example shows. We have a language simply because adherence to its conventions puts categorial closure on our mutual understanding. There is less room for misreading a sentence than there is for misreading a state of affairs. A state of affairs is of much greater complexity as an object, than is a sentence, when recognized as such. The whole idea of truth arises from the fact that we can come to see that some thing is the case, and can express the same proposition verbally. There is no necessity to try to compare words as objects with things, as objects. The relevant relations between items of each class are very different, but there is an identity between what the words express, and what we observe to be the case in the world.

Telling the time is an excellent example of how things arranged in a certain way and words concatenated in a conventional pattern can express the same proposition. With both hands on twelve, the face of a clock expresses the proposition that it is twelve o'clock, which can also be expressed in words as 'It is twelve o'clock'. I speak truly when my words express the same proposition as the face of the clock, falsely when they express a different proposition. Now consider the sun in the sky. Its position can have significance, when we have learned about the sun's apparent movement through the sky. We see that the sun is at its zenith, and seeing that we not only see the sun at its zenith, but *that* it is at its zenith, that is, we observe the sun in a certain position. We come to see the sun being where it is as showing the same proposition as the clock face shows, and as is expressed by the sentence.

The idea of an arrangement of things being understood, as well as merely being seen or felt, has also come up in the discussion of models. It seemed that for certain objects to function as models they had to be understood as such, they had to be 'read', so that certain features could be seen as having significance and others ignored, though, in a sense, both relevant and irrelevant features would be seen.

The *old* controversy as to the nature of truth melts away if one adopts the view sketched above. However, new problems come into focus, particularly the problem of how understanding is possible at all, a problem which I do not intend to discuss. The traditional controversies lose their force since we can say that there is a correspondence of a sort, since sentences and states of affairs are matched through their both being capable of serving as the occasions for identical acts of understanding, that is, as conveying the same fact. Unless the things were of the right kind and in the correct arrangement for it to be possible to see by examining them that such and such is the case, the verbal expression of that proposition is false. There is coherence of a sort, too. If the statement made by an offending sentence is false, the proposition expressed by the use of that sentence is logically incompatible with the fact that can be seen in the things examined. The power of certain elements of language and of certain gestures and the like to draw people's attention to things and the states of affairs obtaining amongst them is what binds language to the world, since it is thus that our attention is drawn to those states of the world which we are required to observe, that is, to understand.

We have already studied these powers in considering demonstrative conventions in Chapter 3.

Perceiving and Observing

The points I have been making about the way things can be seen as expressing propositions can be brought out in another way. The peculiarity of perceptual verbs in either taking a direct object, or a 'that' clause as object, has often been pointed out. We can both see and see that . . ., notice and notice that . . ., observe and observe that . . ., even perceive and perceive that. . . . The direct objects of perceptual verbs are what we see, things, states, processes, and so on, that is what we perceive visually, while the corresponding 'that' clauses express what we come to know, or think we come to know by seeing, noticing, observing and perceiving. This is the contrast between non-epistemic seeing and epistemic seeing that is the basis of Dretske's argument (see 99). I shall argue below that in many cases there is no necessary connection between seeing something, and seeing that something is the case; that is, between sensing and observing, between visually perceiving and coming to know by looking, but I shall follow R. A. Putnam in arguing that in some cases there are strong connections between perceiving and observing, but they are not to be construed as the relations between evidence and conclusions based upon evidence. Seeing and observing are both perceptual skills, the model for the latter being reading.

This peculiarity is not confined to verbs associated with visual perception. I can feel the material and feel that you have used silk for the facings, but we are more inclined to say that while I can smell garlic I know by the smell that you have put garlic in the goulash, and that you have made a culinary error and the like. Whether we pass by a simple grammatical transformation from the verb of visual perception to the corresponding expression of the knowledge acquired by looking, or whether the verb of perception is explicitly epistemic, we can distinguish what we perceived from what we learnt in perceiving, that is, what we observed.*

I want to argue that this is exactly the same distinction as that between seeing the headline 'Chief Minister Resigns', and seeing that the Chief Minister has resigned by reading the headline. One might be inclined to say that reading is a very special kind of observing, one specific way of exercising the general perceptive-epistemic skill. But of course in the perceptual case what we learn is what we perceive, that is,

---

* The most complete account of the differences among, and relations between, non-epistemic perceiving (seeing, etc.) and epistemic perceiving (seeing that, etc.) is to be found in F. I. Dretske (see 99).

the content of the proposition learnt is the fact observed, and the fact observed is the relevant state of affairs as understood by an observer. Only in this way is seeing that the wall is fallen by looking out of the window at the ruins, different from seeing that the dollar has fallen by looking at the headline 'Dollar Eases Against Pound'. There is truth when the content of what is observed is identical with what is seen.

In a remarkably economical and perspicuous article, R. A. Putnam (*see* 198) develops the distinction I have been using between sensing and observing, in an examination of Hanson's curious but interesting remarks (*see* 186) about two astronomers of differing theoretical persuasions watching the sunrise. Both watch the same happening, but Tycho sees the sun rise up, according to Hanson, while Kepler sees the rim of the earth drop away. Mrs. Putnam rightly objects to this way of speaking, which seems to suggest either that the pair each saw different things, or that two senses of 'see' are at play in the anecdote. What is true is that both astronomers see the sun rise, that is, their visual experiences are the same, but each reads a different proposition off the world. While Tycho sees that the sun has moved up, Kepler sees that the earth's rim has dropped away, revealing the sun. It is what they make of their experience that reveals the differences in their astronomical point of view.

How is it that two men can see the same happening, but each observe something different, each read a different proposition off the world? This is a general problem in the philosophy of science, since it is not enough for a scientist just to see, he must observe, that is, he must have something to communicate when he tells what he sees. Mrs. Putnam puts the matter in this way: the upshot of looking for something is seeing or noticing it, when we have been successful in finding it. Thus, we must be ready with criteria of identification prior to our seeing, noticing or observing. In this case, 'I see a robin' uttered triumphantly, does entail 'I see that there is a robin in the garden'. It is also true that what we look for is partly determined by what we already know or believe, partly by our awareness of our own perceptual skills. This kind of seeing and noticing is observing, and hence what we observe to be the case, what we see to be true, is connected with our knowledge and beliefs, and fits within a propositional system. Whenever we are observing, we are doing that kind of perceiving which I have described metaphorically as 'reading the world', and the upshot of it is items of knowledge, in short truths.

On the other hand, seeing what we are not particularly looking for is an achievement without a prior task, therefore I can quite properly be said to see things, and their arrangements and so on, without seeing that anything is the case, that is, without observing or noticing anything. Of course, should I recognize something which I was not particularly looking for, I must have possessed the power to recognize it, and this might be represented as my having, somewhere in my conceptual equipment, the criteria for identifying it. So, if I notice something unexpectedly, I am nonetheless observing it, and I can see that it is a Great Crested Grebe, or the track of a positively charged particle. I can come to see that it is this or that for the first time, even after seeing the phenomenon many times, and not grasping what it is, that is, not observing that it is a positron. One might say that this way of looking at the relation between experience and knowledge, is in contrast to the idea that the sensory experience forms the basis for an inference to the knowledge, as if the senses provided 'raw data' which led to knowledge. But there is a gap between things and facts which no ingenuity can bridge. But looked at in the way I am suggesting, and following Hanson as reformed by Mrs. Putnam and the general line of Dretske's argument, we can see that there is no unbridgeable ontological gulf between things and knowledge of things, since it is one of our perceptual skills to be able to see what is the case.

Description as Classification

Broadly speaking then there are two kinds of items of knowledge of the world which can be acquired by human beings. One can find out what kinds of things there are, and one can find out how these things behave. Knowledge of how things behave is organized into the structure of explanatory theories, issuing in an understanding of natural mechanisms and their capacities for reaction, and their power to affect other things. This kind of knowledge we have discussed at length. So far I have taken for granted the capacity correctly to ascribe things to kinds, and so to describe, since to say what a thing is, is to say to what kind or kinds it belongs. Scientific description goes on to assign entities to taxa, and to develop systems of taxonomy, in which the taxa are themselves organized. The philosophy of taxonomy is as riddled with positivist myths as any part of the philosophy of science.

The central and most pervasive positivist myth is that classification is ultimately arbitrary, and that when classification is hardened into a taxonomy, it depends not upon the nature of things and substances, but upon the means and purposes of the classifier. I shall call this doctrine *the Myth of the Infinite Arbitrariness of Classification*. Opposed to this myth is the position of the realist, that things and substances differ by their internal constitutions and that these are real differences, discrete or continuous real differences, upon which the differences of manifested characteristics of things depend. So for the realist classification by manifested characteristics of things is only justified in that the manifested differences are a mark or sign of real differences in constitution. Once again the distinction of nominal essence, the set of manifested qualities whose appearance justifies the use of a common noun or adjective of the thing or substance, from real essence, the constitutional set-up responsible for the manifested qualities, must be invoked.

The origin of the myth of the infinite arbitrariness of classification can be understood with the help of a distinction I shall make between classes and kinds or taxa. Any aggregate of entities having at least one characteristic in common can constitute a class. A taxon or kind is a group of things having the same constitution; but not necessarily manifesting this common constitution in the possession of at least one characteristic in common, though usually each taxon has a nominal essence. Sometimes we speak of the distinguishing characteristic itself as the kind or sort. Not all classes are taxa, though taxa will usually be manifested as classes. The class, *red entities* is not a taxon, because the manifestation of redness in things may be brought about by a great many different constitutional set-ups of things. For instance one thing may be red because it reflects only red light, and another thing may be red because it transmits only red light, or yet another thing may be red, because, though it reflects light of all colours it is only seen in red light. So it cannot be inferred from an entity's manifested redness that it is of any particular constitution. *Red entities* is a class but not a taxon, because manifested redness is not a mark or sign of a common constitution of the things falling under the class.

It is clear why we should, in practice, make this distinction. While assignment to a taxon may be by reference to some one or a few manifested characteristics, constituting the nominal essence, the taxon, being a group of like constituted entities, has other capacities. Our belief in the continuing co-manifestation of qualities, and stability of behaviour traits, must be connected, as we discovered in earlier chapters, with knowledge of the causal mechanisms present. But assignment to a class gives no guarantee that any other characteristic than those making up the nominal essence will be manifested in all the members. The notion of a taxon is more powerful than the notion of a class. Science uses taxa, and does not use classes. Only if it used classes would the myth of the ultimate arbitrariness of all classifications be acceptable.

The classes to which a thing belongs are as multiple as its latent and manifested qualities. Indeed if there are k such qualities then there are $k!$ classes to which the thing might properly be said to belong, and its assignment to any one is connected only with manifestation of the quality involved. No reason can be offered for choosing one, or any, of these classes as being the natural kind to which the entity in question belongs. If we accept, at the same time, the myth of the Independence of Predicates, so that any of the $k!$ possible nominal essenses are admissible in science, we can only conclude that there are no natural kinds, since the entity in question will have an equal title to belong to each of the $k!$ classes. But if we suppose that the total set of manifested and latent characteristics of a thing is not arbitrary, not what it is *per accidens*, but is connected (as effect with cause) with the actual inner constitution of a body, with its real essence, we move directly to a rational theory of natural kinds. The different real essences are the basis of the different and discrete natural kinds, and are, of course, in their presence and power the reason why things manifest the actual set of determinate qualities that they do. The electron theory of atomic structure makes the chemical elements natural kinds.

The characteristics which constitute the empirical criteria of classification into taxa must be such as to be indicative of the generic constitutional differences in the things or substances classified. By and large, two main kinds of such criteria can be found. The constitution of a thing or substance may be assumed to be the same as another entity which is similarly descended or engendered. Such I shall call a 'generative taxonomy'. Proof or descent from a common ancestor constitutes a taxonomic criterion, only because it must be supposed to be indicative of similarity of constitution. But this generative taxonomy depends upon a prior classification of kinds among the ancestral entities which cannot itself be generative. It depends, too, on some such contingent principle as that like begets like, or the principle of the fixity of species, which we know is certainly not obeyed exactly in nature. The most fundamental criterion must then be qualitative difference. But since every perceivable individual has some qualitative difference from any other, the notion of a nominal essence must be resorted to again. A characteristic or set of characteristics which is the *differentium* of a species can be the basis of a differential taxonomy only on the assumption of the principle by which likeness of manifested qualities is related to likeness of inner constitution, in which qualities are seen as the manifestations of the causal powers of the essential constitution of a thing, and quality-ascriptions really ascriptions of powers.

It is this very principle, and the fact that a taxonomy is a classification into natural kinds, which allows the nominal essence to be used as the empirical criterion in a world of individuals, each of which can, by some qualitative difference, be differentiated from all others. The variation that actually occurs in determinate manifested qualities can be accommodated without disrupting the system, provided the variations can be put down to extraneous conditions causing the common constitution to be manifested in slightly different ways in different individuals. In this way the different allotropic forms of elements can be accommodated in taxonomy of chemistry. Diamond, black carbon and graphite manifest different characteristics, but they are all carbon because they are alike in the electronic structure of their atoms, that is they have identical constitutions, and belong to the same natural kind. Only on the realist view of the nature of science can this resolution make sense, because only the realist allows such hypothetical structures as the electronic constitution of atoms a place in the universe as real as, or indeed more real than the manifested qualities of the things classified.

The notion of the gene-pool is a nicely sophisticated device to allow a taxonomy of species which permits variation in the manifested qualities without upsetting the taxonomy. If the individuals in the species are endowed with somato-generative mechanisms which are drawn as selections from a group of components larger than the totality of components required for the individual somato-generative mechanisms, then the nominal essence of the species will be a disjunction of conjunctions. Suppose that the individuals of the population of some species are generated from 2-component somato-generative mechanisms, drawn from a pool of 3 components, a, b and c. Then the somato-generative mechanisms of the population will be either ab, or bc, or ca. Now if an individual is generated by somato-generative mechanism ab it will manifest characters N, M and P, if bc then N, M and Q, if ca then N, P and Q. N will become an indicative sign of the species, but the nominal essence of the species will be $(N \& M \& P) \vee (N \& M \& Q) \vee (N \& P \& Q)$. Any entity which satisfies this disjunction, that is, shows a combination of manifested qualities which is any one of the combinations that are disjoined, will be a member of the species. That is it will be a member of a taxon, because it will be an entity which is generated by one of the somato-generative mechanisms ab, bc or ca. These are the somato-generative mechanisms that are possible if the gene-pool is the set of components $\{a, b, c\}$. The concept of the species as the manifestation of a gene-pool only makes sense under the assumption that the species is a taxon and not just a class. Indeed, on this view, a species may not be a class, since it is logically possible that the species is unified by being related to a gene-pool, no single determinate character is manifested in all the members.

Another taxonomically significant kind of variation in manifested characters is also logically possible on this view of the nature of the taxon. Two individuals may be genetically alike, that is have somato-generative mechanisms of sufficient similarity to put them in the same taxon, but each develops in different circumstances and so shows different manifest characters. The resolution of the problem of the apparent evolution of a new species of primula in damp conditions, as merely a variety, can be legitimately achieved only by the supposition that the differentiation of manifested characters is due to the difference of circumstances in which the manifested characters are evoked, and not due to any difference in the somato-generative mechanisms, to which the phrase 'really the same species' must be taken to allude. The technical concept of 'phenocopy' is used for those cases where the somato-generative mechanisms are different, but the circumstances of development produce two phenotypically similar individuals.

Taxonomy is difficult in practice because of the long and complex process by which the somato-generative mechanisms, whose different kinds constitute the taxa, produce the individuals whose manifest characters can be observed, and according to which they must be classified in practice. Any individual can belong to infinitely many classes, but to only one taxon. Classification by manifest characters, if that is supposed to be the end of the matter, will not yield distinctions of natural kinds, only arbitrarily reclassifiable classes. The science of taxonomy only makes sense if the classification of individuals by their manifest characters is under the control of the theory that there are real differences in the internal constitution of things and substances, from which their manifested characters and their differences spring.

SUMMARY AND BIBLIOGRAPHY

Attempts to stabilize a vocabulary of descriptive terms, independent of theory and of wider linguistic ambience hinge upon the theory of ostensive definition. This theory has been destructively criticized in

173. L. WITTGENSTEIN, *Philosophical Investigations*, Blackwell, Oxford, 1953.
174. F. WAISMANN, *The Principles of Linguistic Philosophy*, Macmillan, London, 1964, Ch. V.

The theory that an adequate descriptive vocabulary for science can refer only to sensations was advocated by

175. E. MACH, *The Analysis of Sensations*, new edition, Dover, New York, 1959, and has been extensively criticized, cf.

176. A. C. LLOYD, 'Empiricism, Sense Date and Scientific Language', *Mind*, **59**, 57–70
177. P. ALEXANDER, *Sensationalism and Scientific Explanation*, London, 1963.

The operationist (-alist) method of creating a theory-independent vocabulary was first advocated by

178. B. C. BRODIE, 'A Calculus of Chemical Operations', *Phil. Trans.*, **166**, 781–859, **167**, 35–116.

A very similar view has been proposed by

179. P. W. BRIDGMAN, *The Logic of Modern Physics*, Macmillan, New York, 1954.
180. P. W. BRIDGMAN, 'The Nature of Some of our Physical Concepts', *BJPS*, **1**, 257–72, **2**, 25–44, 142–60.

For the main points of criticism of this theory see

181. A. C. BENJAMIN, *Operationism*, Thomas Springfield, 1955, particularly Ch. V.
182. MARY HESSE, 'Operational Definition and Analysis in Physical Theories', *BJPS*, **2**, 281–94.

For discussion of the more general question of whether theory and observation reports are independent in meaning see, for instance, the following:

183. MARY HESSE, 'Theories, Dictionaries and Observation', *BJPS*, **9**, 12–28.
184. P. ALEXANDER, 'Theory-Construction and Theory-Testing', *BJPS*, **9**, 29–38.
185. R. F. J. WITHERS, 'Epistemology and Scientific Strategy', *BJPS*, **10**, 89–102.
186. N. R. HANSON, *Patterns of Discovery*, Cambridge University Press, 1958, Chs. I and II.
187. P. ACHINSTEIN, 'Theoretical Terms and Partial Interpretation', *BJPS*, **14**, 89–105.
188. M. SPECTOR, 'Theory and Observation', *BJPS*, **17**, 1–20, 89–104.

Recently the discussion has taken a new turn as questions of the possibility of conceptual change, and change in meaning, have been raised. Some of the main papers on this topic are

189. P. K. FEYERABEND, *J. Phil.*, **62**, 266–74, sums up the early stages of this discussion.
190. D. SHAPERE, *Mind and Cosmos*, ed., R. Colodney, Pittsburgh University Press, 1966.
191. J. LEPLIN, 'Meaning Variance and the Comparability of Theories', *BJPS*, **20**, 69–80.

My own theory involves the treatment of quality descriptive statements as power attributions, in which the words for sensible qualities are used only metaphorically. Something like this theory appears in

192. W. JOSKE, *Material Objects*, Macmillan, London, 1967, Chs. 4 and 5.

It is, of course, close to Locke's theory for secondary qualities, cf.

J. LOCKE (*see* 11), Bk. II, Chs. 8, 24.

The question of truth was tackled with a brief survey of recent ideas, leading up to the conception of 'facts' as 'readings of the world'. An excellent collection of the most important recent papers on truth is

193. G. PITCHER, *Truth*, Prentice-Hall, New Jersey, 1964, particularly pp. 18–31.

For my purpose, see particularly

194. J. L. AUSTIN, *Philosophical Papers*, Clarendon Press, Oxford, 1961, Chs. 4 and 5.

Also see

195. R. M. MARTIN, 'Facts: What they are and What they are not', *APQ*, **4**, 269–80.
196. I. SCHEFFLER, *The Anatomy of Inquiry*, Alfred A. Knopf, New York, 1963, p. 57 ff.

The 'reading the world' view depends for its cogency, in part upon the propriety of speaking of propositional meaning for things and material structures. A number of papers have appeared discussing this question, among which are:

197. C. A. BAYLIS, 'Facts, Propositions, Exemplification and Truth', *Mind*, **57**, 459–479.
198. R. A. PUTNAM, 'Seeing and Observing', *Mind*, **78**, 493–500.
199. V. C. ALDRICH, 'Pictorial Meaning, Picture-thinking, and Wittgenstein's Theory of Aspects', *Mind*, **67**, 70–9.
200. S. K. LANGER, *Philosophy in a New Key*, London, 1951; Harvard University Press, Cambridge, Mass., 1963, Chs. III, IV and X, which has been discussed by

201. P. WELSH, 'Discursive and Presentational Symbols', *Mind*, **64**, 181–99.
202. N. FLEMING, 'Recognizing and Seeing As', *Phil. Rev.*, **66**, 161–79.

The *philosophical* problems of the real essences of species, as genotypes, have not been greatly discussed, cf.

203. D. J. HALL, 'The Effect· of Essentialism on Taxonomy', *BJPS*, **15**, 314–26, **16**, 1–18.

8. Principles

The Argument

WHAT IS A PRINCIPLE?

1. (a) acceptance of principles determines
   (i) the way phenomena are identified: taxonomic principles;
   (ii) the way phenomena are explained: causal principles;

   (b) principles have no special logical form, but rather are *treated as* secure from falsification, in the short run;

   (c) often our awareness of principles is posterior to our having capabilities of identification and explanation for a certain field of phenomena;

   (d) confusion is created by taxonomic principles appearing in the guise of causal principles.

2. 'Necessity' as a metaphorical 'attribute' expressing attitude. Taxonomic and causal necessity: both are connected with causal mechanisms, genotypes and real essences.

3. Taxonomic principles are concerned with real essences. Taxonomic criteria are concerned with nominal essences.

4. Principles of individuation are not to be confused with criteria of identification of individuals, since, within a kind, all individuals have identical nominal and real essence, though they may be identifiable idiosyncratically.

II

AN EXAMINATION OF POSSIBLE PRINCIPLES FOR PSYCHO-PHYSIOLOGY TO ILLUSTRATE THE DISTINCTIONS IN I

1. In ordinary language there is already a taxonomy of psychic states, events, conditions, etc. Physiology also already possesses its own taxonomy, of objects, relations, states and events.

2. The reduction of psycho-taxonomy to physiotaxonomy could be
   (a) direct, which will be shown to be impossible, *or*
   (b) indirect, via outer features of the body, which depends for its plausibility on the discredited verification principle.

3. Physio-taxonomy will be said to classify 'brain-states', and psycho-taxonomy 'mind-states'; 'states' to be liberally conceived in both cases.
   Note. The *practice* of psycho-taxonomy is indisputable: its principles are obscure.

4. Numerical differentiation of mind-states is by
   (*a*) reference to person;
   (*b*) reference to object.
   Neither is reducible to the means of numerical differentiation (individuation) of accompanying brain-states.

5. Examination of 'The brain is the organ of thinking':
   (*a*) originally an empirical hypothesis;
   (*b*) can be treated as the grounds for a *taxonomic* principle (not a causal principle);
   (*c*) Reasons for such treatment:
       (i) *Locke*: no causal mechanism can be conceived;
       (ii) Physiologists' taxonomy of cells, etc., is irrelevant to the classification of mind-states, *a priori*;
       (iii) The fact that the brain's rôle had to be discovered shows that the taxonomies of brain-states and mind-states were originally absolutely independent;
   (*d*) Since mind-state identifications are in fact used to supplement the physio-taxonomy, it is natural to try to combine the taxonomies under some naturally necessary principle, falsifiable only in the long run, so that the physio-taxonomy is dependent upon psycho-taxonomy.

6. Sleep, unconsciousness and death preclude general statements (P_1 and P_2) correlating individual mind and brain-states from consideration as possible principles.

7. Examination of
   (*a*) 'For every kind of brain-state there is a kind of mind-state' (P_3);
   (*b*) 'For every kind of mind-state there is a kind of brain-state' (P_4).

   Case 1. Since there are kinds of brain-states during the existence of which no mind-state of any kind occurs, P_3 is false (and hence contingent).

   Case 2. *Either*
   (*a*) whenever mind-state-kind *f* occurs, brain-state-kind *g* occurs, *or*
   (*b*) on some occasions brain-state-kind *h* occurs with *f*.
       (i) if (*a*) happens, then P_4 can be treated as necessary;
       (ii) If (*b*) happens, P_4 can still be treated as necessary by reforming the physio-taxonomy so that *g or h* is a new taxon. Reasons will be advanced for (ii).

   Note. To treat P_4 as necessary is not to try to make it analytic or tautologous. It is to give it a certain status in our system.

8. It is shown that choice of Case 2, (ii) makes it impossible to discover 'free' mind-states, i.e. it is *not* a support for dualism.

9. Justification for the introduction of disjunctive taxa:
   (*a*) Note that in the long run the necessity to form ever-lengthening disjunctive taxa would count against the principle;

(*b*) The components in a short disjunction may be all consequences of a common hidden factor *X*.

10. Objections to making brain-state-kinds taxonomically ultimate.
  (*a*) (i) Many mind-state disjunctions are absurd as taxa;
       (ii) Examination of the intelligibility of the notion of hypothetical mind-state-kind
            (*a*) e.g. Freudian innovations not hidden causal factors but novel classifications introduced as metaphors;
            (*b*) Difficulty of conceiving of the *discovery* of a new mind-state-kind.
  Only intelligible possibilities seem to be *either*
  unacknowledged class of reasons, *e.g.* a common motive, but no empirical distinction between providing a reason, and discovering a reason, *or*
  previously unknown powers, which is absurd in disjunctive case.
  Note. The cases examined are clearly handled better with hypothetical brain-state-kinds.
  (*b*) The postulation of some mind-state-kind to go with the brain-state-kind or kinds during unconsciousness is objectionable.

11. (*a*) (i) Distinction between privileged *access* to mind-states, which is false; and privileged *authority* as to mind-states which is true;
       (ii) The vocabulary for describing mind-states is public in use, i.e. no privileged access;
      (iii) No privileged authority or access for brain-states.
  (*b*) (i), (ii) and (iii) of (*a*) do not entail a rejection of mind-state reports for a possible science, only that the reporter (i.e. the subject of any experiments) is himself a member of the scientific team.

12. Metaphysical implications of making P_4 a principle are briefly explored through consideration of two models:
  (*a*) 'aspects' model;
  (*b*) 'sentences and statements' model;
  (*c*) the modified identity thesis is consistent with P_4 as a principle.

13. Physiology and psychology are not, on this view, steps in a hierarchy, but parallel. P_4 is consistent with psychology as a human powers and human nature investigation.
  Human Powers retain the priority of mind-state-kinds, but Human Nature investigations can aim at brain-state-kind reduction.

III

1. Other examples of taxonomic principles masquerading as causal laws;
  (*a*) force and mass-acceleration;
  (*b*) Optical activity and molecular structure.

2. In (*b*) a common cause of optical activity does exist, no common cause in (*a*).

IV

RESOLUTION OF RESIDUAL PROBLEMS ABOUT REALIST THEORY
OF CAUSALITY

1. Explanation of 'fundamentally different kind' distinction.

2. The problem of trans-categorial causality.

3. The modified identity thesis resolution for mind-states and brain-states is not available for, e.g. fields and things.

4. Resolution is by the conceptual device of recategorization, that is, e.g. things are declared to be fields of a certain kind.
   (a) Recategorization is not available in mind-state/brain-state case.
   (b) The misleading propaganda force of expressing recategorization existentially is examined.
   (c) But recategorizations do have different existential *consequences*, under the final constraint that spatial and temporal changes, considered in themselves, are not causally efficacious.

A principle is a general statement, adherence to which determines the way we view the phenomena we study. Just how this determination takes place will emerge in this chapter. Furthermore, a principle is adhered to: we speak of adopting principles. A principle is a statement whose falsity we are not likely lightly to admit. In this it differs from other statements. I hope to show, however, that this difference is neither ultimate nor intrinsic. The difference is rather in the attitude which we adopt towards a statement of principle. A statement becomes a principle not because of some special structural feature or because of any special kind of meaning, but because it comes to play a certain rôle in our thinking. We may not consciously decide to pick some statement for the rôle of a principle, but we may come to see that our attitudes to certain statements are such that they must be functioning as principles: that is, functioning as determiners of our way of thinking.

Two groups of principles seem to be needed for a science. There are those in which are found the bases for the criteria according to which phenomena are to be individuated, classified and described as being of this or that kind, in short, the taxonomic principles of the science. Taxonomic principles must be carefully distinguished from those statements which prescribe the cause-effect structure for the field of phenomena under study, and which determine in advance the general style of the pictures of the hidden mechanisms which we imagine to be responsible for the patterns of phenomena in the field of study. These are concerned with the theoretical concepts of the science and their origin and critique has been discussed in earlier chapters. One of the points with which I shall be concerned in this chapter is the detection of pseudo-causal principles, which, though having the air of causal principles, actually form part of the principles of the taxonomy of a science.

We shall find that taxonomic principles and classifying criteria worked out in accordance with them, are distinguished from other general statements by an unwillingness on the part of taxonomists to abandon them in the face of some apparent counter-evidence. As in earlier discussions of causality and of laws, we shall mark this feature of their use by using the word 'necessary' of them. As a preliminary move, we distinguish 'taxonomic necessity' as a metaphorical 'attribute' of taxonomic principles and classifying criteria that marks them off as being immune from sudden falsification, remembering that it is we who refuse to discard them.

How does 'if it is a lamprey then it *must* have a notochord' – an example of taxonomic necessity – compare with 'if a gas is compressed then its pressure must increase'? The necessity of the latter stems from the way our idea of the essential nature of the gas shows how the mechanism of pressure production is such that only one outcome can follow from a decrease in volume, that is a decrease in the room for molecules to move. Taxonomic necessity rests upon exactly the same structure. A raven must be black because the essential nature of ravens is such that there is no other colour possible for a thing which has the other raven characteristics, i.e. is a manifestation of the raven genotype. A lamprey has an essential nature such that it must have a notochord. This is not only because, as a matter of fact, the predicates 'being a lamprey' and 'having a notochord' are truly predicated at the same time of the same thing. Such an observed clustering of attributes would lead only to a conception of the nominal essence of lampreys. It is because we think that the real essence or essential nature of lampreys, i.e. the genotypes of the species are such that in normal circumstances they lead to the development of that structure. The genotype is involved both in the production of the set of attributes under 'lamprey', and in the generation of the notochord. The essential nature of a plant or animal is its genotype: the set of genes from which its observable characters flow. Taxonomic necessity derives from the mechanism of character production which ensures the regular co-presence of those characters of the adult organisms which form both their nominal essences, the set of characters we use to identify them, and the other properties which they usually have.

Taxonomic principles must be very carefully distinguished from classifying criteria. The former have to do with the *way* we are going to classify, that is, in accordance with what *kind* of criteria. Classifying criteria, such as those discussed above are concerned with what particular classifications we are going to make. I would not include among taxonomic principles the criterion of interspecific sterility which allows taxonomists to differentiate biological species by the absence of interbreeding, but rather the principle which sees 'species' as related to a unique genotype, from which idea the criterion of interspecific sterility follows. In short, taxonomic principles are concerned with the distinctions of the real essence of entities, i.e. with the principles which lie behind the distinctions we make of natural kinds, while criteria of classification are concerned with the nominal essences which things must have, given their natural kinds. In practice this makes the criteria of classification derivative from the phenotypical distinctions between things.

An example of those principles which masquerade as causal laws while actually serving as the basis of a classification of phenomena *a priori* would be Newton's First Law, which seems to state that changes in velocity or rest are to be attributed to the action of one or more forces. But on closer examination of the science of mechanics it turns out that this 'Law' is not a descriptive generalization nor a causal law in the sense of Chapter 4. The way in which we know that a force is acting, in a great many important cases, celestial orbits for instance, is by observing that the state of rest or motion in a straight line of a body has changed, and not by observing an agent exerting a force. Newton's Law then, one may say colloquially, allows us to say that this phenomenon *should be treated as* one in which a force acts, i.e. classified among dynamic phenomena. Newton's First Law then can be seen as a taxonomic principle, laying down a further set of classifying criteria for the class of force-caused motions.

Not only are the principles of classification into kinds, and the actual criteria, relatively *a priori*, but so are the principles of individuation for the entities with which a science deals, since they will depend upon the principles of the taxonomy of that science through its criteria of classification. To identify an individual one has to distinguish it from all other individuals both of its own kind, and of other kinds. Within a kind, distinctions are idiosyncratic. For instance, within the kind 'man' or the kind 'planet', the distinctions which allow the application of proper names are individual peculiarities. With classes of non-idiosyncratic individuals like electrons or the cells of some organ, there is not even the possibility of singling out an individual because individuals cannot be reidentified and hence cannot be given names. Furthermore, even for groups of individuals which do differ sufficiently amongst themselves to be nameable, since there is nothing intrinsic to the man which determines what name he gets there can be no science of men as the bearers of names, and no scientific taxonomy of men as having this or that name.

Science is concerned with essences, not accidents. The taxonomic principles express the way in which it has come to be decided (and here we adopt the metaphor of 'decision') that that distinction will be made for that particular science. These are 'decisions', not discoveries. They call for reasons in their justification, not for evidence. The sciences can produce no theories to account for numerical identity and difference, their interest can lie only in qualititative identity and difference. Relatively *a priori* taxonomic principles are required by every science because there always are differences between individuals of a certain level of complexity, and it is a theoretical decision as to whether any particular differences are to be treated as flowing from differences in essential nature, and so forming taxonomic differentia. From the standpoint of taxonomy every individual within a taxon is an atom, because atoms have to be exactly alike. We know that in many taxa individual differences can be discerned and the individuals named, but because these differences cannot be recognized within the classificatory system of the science, they cannot be given a scientific explanation. It is true that we have to 'decide' where difference in quality shall cease to count as qualitative difference, but once this has been 'decided', the further differentiation of individuals ceases to be scientific because which qualitatively identical individual is which, is a matter of designation by convention, and not a matter of discovery by some scientific investigation.

To illustrate these and other points in detail I want to discuss the possibility of setting up principles for a taxonomy with the putative science of psycho-physiology as an example. This science, could it get going, would take both psychic and physiological phenomena as its province. If there were relations between phenomena of these two broad kinds they would be studied in such a science. Currently, two main views of the relation (and so of the nature of psycho-physiology) persist. One is that psychic phenomena are really physiological phenomena, and the other that physiological phenomena cause psychic phenomena. Both of these views are mistaken, I believe, and I shall be offering a third view, not to my knowledge offered before, though it has strong connections with one strand in contemporary thinking about these problems.

I am going to assume for the moment that there is a tolerably adequate taxonomy in use by all of us, psychologists or not, for identifying and classifying the events, states and conditions of our mental life, including a wide variety of kinds of entities: for example, deliberate actions and contemplative thoughts. We know that there is an adequate taxonomy already in existence in the sciences of bio-chemistry, physiology and anatomy. I want to explore the relations between these taxonomies, partly with the possibility in mind of developing a psycho-physiological taxonomy ,and partly with the more general question of the possibility of any sort of science of psycho-physiology in mind. A science of psycho-physiology would deal with the events and states and activities commonly lumped together as the mental life, and with changes and enduring states of the body as dis-covered and classified by physiologists and anatomists. One way of putting the question as to the possibility of psycho-physiology would be to ask whether a monistic taxonomy were possible, that is whether for example both mental and physiological events, states and so on could be classified, employing only the taxonomic principles of physiology. If this could be shown to be possible then we might be inclined to say that for a scientific study of human beings physiology was enough. Were this to be admitted on all hands we would come to learn to classify our feelings as adrenalin-based fears versus triptamine-based fears, our thoughts as prefrontal thoughts versus medullary thoughts, our emotions as spinal versus visceral emotions, and so on. To some extent we do this already, witness the growing use of the word 'visceral' in the vocabulary of emotion. But I shall prove, not only that psycho-physiology could not have begun in this style, but that there are power-ful reasons for thinking that it could not continue in this style either. This discussion can be connected up with the traditional problem of the mind and the body. I hope to show as a corollary to my main line of argument that much of the difficulty in the traditional way of con-sidering these matters arises from confusing two different kinds of scientific principles, the taxonomic and the causal. But this will emerge later.

Some monistic systems have been proposed as answers to the question of whether a unified psycho-physiology is possible. For instance Carnap argued long ago that a physicalist reduction of psychological sentences is possible.* His argument will serve as a warning, since it is one of the classical slides by which evidence for a statement is taken to be identical with the content of the statement. This argument is a particular case of the notorious verification principle. Carnap's theory depends upon the idea that Rudolph's protocol sentences cannot be used to report upon Otto's state of mind, because all that can be stated in protocol sentences are reports on Otto's behaviour, expressions of face and so on. Then, using the verification principle, we conclude that the actual con-tent of Rudolph's protocol sentences, which seemed to be about Otto's

* R. Carnap, 'Psychology in Physical Language', Ch. 8 in *Logical Positivism*, edited A. J. Ayer, Free Press, Glencoe, Ill., 1960.

state of mind, is the facts of Otto's behaviour, facial expressions and so on. So the only public, intersubjective protocol language, that is 'vocabulary specially suited to reporting facts', for a science of psychophysiology would actually contain nothing but reports on the physical state of corporeal organisms. This would also have to be true, on this theory, of those statements by which one reported one's own state of mind. Apart from many other objections, it is obvious now that this theory confuses behavioural evidence *for* psychological statements with mind-state content *of* psychological statements, and so refuses to allow a man's own statements about his state of mind to be taken seriously. 'I'm tired' tells how I feel, and it is identical in meaning with 'He's tired' said by someone else of me. Any theory which denies these two facts must be wrong.

Brains and nervous systems are among the objects studied by physiologists and anatomists. They have a system of classificatory concepts which allow a wide variety of entities to be differentiated, particularly cells, and organs: that is cell structures with an identifiable function. A wide variety of chemical and physical states of cells and cell structures are also able to be differentiated. Finally, the physiologists' and anatomists' taxonomy provides criteria for distinguishing very many kinds of events, states and processes, for instance chemical changes in synapses, progressive changes in the electro-chemical state of nerve cells, and so on. I shall use the generic term 'brain-state' to refer to any state, momentary or enduring, of any part or the whole of the brain and central nervous system of a man or an animal, which is individuated and identified (that is classified) by means of the criteria derived from the taxonomic scheme of anatomy and physiology.

The taxonomy of thoughts and feelings currently used by quite ordinary people is understood widely enough to make novels and biographies intelligible. Narratives which include references to the thoughts and feelings of characters, real or imaginary, can be intelligible to someone other than their authors. We can all understand the difference, for example, between remorse and guilt, and see what an author or autobiographer is driving at when he classifies a feeling under one or other of these heads. This system of classification is wide enough to allow probation officers' and psychiatrists' reports to be intelligible to magistrates and others. It has been affected here and there by psychological investigations and theories. It is beyond dispute that thoughts and feelings are qualitatively identified with a public vocabulary, and that under these classifications what people think and feel can be discussed without utter confusion. Furthermore, not only are we capable of identifying thoughts and feelings, but individual thoughts and feelings can successfully be localized through their being the thoughts and feelings of individual people, and their temporal characteristics, like moment of conception for thoughts, or endurance for feelings, fairly readily determined. This is not to say that the laying down of the principles under which determinations of the numerical identity of thoughts and feelings are made, is not without considerable difficulties. However, despite the obscurity of the principles, the practice is easily instanced, e.g. 'I suddenly realized he was trying to borrow money'; 'I felt dizzy most of the morning'; 'It was only some hours after the murder that he began to feel that combination of guilt and fear that stayed with him throughout the chase that followed', and so on. For those mental predicates which ascribe powers to people, capacities, liabilities and capabilities, a quite different account of identification is required. We shall return to that problem later in the general discussion of powers in Chapter 10.

Any state which can be identified and classified by means of the thinker's taxonomy I shall call a 'mind-state'. I do not intend that this phrase should in any way suggest that this taxonomy should be taken to be identifying 'states of mind', that is qualitative differentiations of mind-stuff. One upshot of my argument will be yet another demonstration, on novel lines, that the conception of a mental substance is absurd.

If we follow Hampshire,* Strawson (see 95) and others at this point we would be inclined to offer two main dimensions upon which numerical differentiation of mind-states can be made. For a great many mental attributes and occurrences, the person whose attributes they are and to whom the occurrences occur is the source of the numerical individuation. The arguments that have been put forward to support the idea that feelings, for example, must be individuated as this or that person's feelings, seem to me inescapable. The other dimension of individuation is through intensionality, the thought is of this or that tree, the plan to bring about this change. Thus a particular thought of a tree is individuated by reference both to whose thought it is, and which tree is being thought of. But for the purposes of my argument I am under no necessity to produce anything more than enough of a sketch of a theory of individuation to show that the principles of individuation of thoughts and feelings are not, as a matter of fact, the principles of individuation of their identifiable physiological accompaniments, nor rest upon such principles.

What is it to discover that the brain is the organ of thinking? It is, in part, to discover that certain kinds of brain-state are regularly present in the organism when certain kinds of mind-states are reported, or are shown by performances which provide logically adequate criteria for

* S. Hampshire, *Thought and Action*, Chatto and Windus, London, 1959, Ch. 1.

their ascription, or are, in some rare cases, inferred from outward signs. This was once just an empirical discovery, but as I shall show, it can be made the grounds for a statement which can be treated as expressing a conceptual decision. But if we do that we must give reasons. We must be prepared to answer the question 'Why should we treat the discovery this way?' Locke gave one reason. He said in effect that it was just this correlation between body-states and simple ideas which had to be taken as fundamental in psycho-physiology. This was because *a fortiori*, no corpuscularian account of the mechanism by which brain-states produced mind-states (or ideas) could be given, since motion produced more motion, not ideas. So no causal account is to the point. Another obvious and inclining reason is that the discovery that the brain is the organ of thought *could not* have been made by anatomical and physiological investigations alone, however detailed, prolonged or sophisticated. And this is for the simple but profoundly significant reason that the physiologists' taxonomy does not provide any basis for identifying kinds of thoughts, feelings or intentional behaviour, let alone a basis for individuating particular examples of these classes of phenomena, on the basis of which an empirical correlation might be observed. For all that anatomy and physiology can tell us, it is logically possible that any, some, all or no organ, identified anatomico-physiologically, could have been the organ of thought.

I draw attention to the historical fact that the discovery that the brain and central nervous system is the centre of thought and feeling was made long after a considerable amount of anatomical and rather less physiological information was available about them. I remind the reader that Aristotle seems to have thought the brain was an organ for dissipating excessive heat, and that for long the heart was held to be the organ of emotion. It follows that it is *not at all obvious* that the brain is the organ of thought. So far as I know the hypothesis was first enunciated by Cabanis circa 1780, though somewhat similar ideas had been held by Galen. The discovery was made by finding that particular kinds of damage to the brain were associated with particular disturbances of mental functioning. In order for this discovery to be possible it is logically necessary that both the brain damage and the disturbance of mental function, including intentional behaviour, must be independently identifiable. This independence derives at this stage from the independence of the criteria of identification, which in turn derives from the fact that the taxonomic principles are independent in each realm.

Once one has decided to consider the hypothesis that the brain is the organ of thought seriously, it is both easy, and I believe fruitful in a very special way, to treat the hypothesis as a principle, and so to have something of the character of a necessary truth, but this necessity is natural necessity, that is it will remain possible that the progress of discovery might lead to a change of attitude to the hypothesis, under which it would become possible to falsify it, though the conditions under which this is possible are complicated and very special. I shall turn to them in a moment. Their special character, as we shall see, makes it very natural to treat the correlation of brain-state-kinds and mind-state-kinds as a taxonomic principle. The kernel of my subsequent argument is this: the selection from among the myriad possibilities left open by the anatomico-physiological way of identifying elements in the body as to which kinds of brain-states are to be associated with kinds of mind-states, involves a taxonomy in which the criteria for identifying mind-state kinds are already intimately involved. The taxonomy of mind-states, I argue, provides a crucial part of the taxonomy of brain-states. This is because any investigator of how brain and mind-state kinds are correlated is under the inescapable necessity of taking and accepting reports from his subjects of study as to their mind-states on the occasions that he determines their brain-states. The same point could be argued for the correlation between intentional actions and brain-states. These facts, I shall argue, can best be accommodated by treating the discovery that certain instances of mind-state kinds are correlated with instances of brain-state kinds in a regular way, not just as an empirical discovery but as the occasion for the invention of a new science, psycho-physiology. This is created by announcing that the newly discovered correlation is to be treated as grounds for a taxonomic principle which employs mind-state kinds in differentiating brain-state kinds, thus creating criteria for the identification of a specific group of kinds of entities. I want to explore some of the consequences of making this move, a move which I believe has in fact been made, in neurophysiology.

The first, and most obvious consequence, about which I shall say no more here, is that this move leaves open the possibility of psychology as an independent science, but makes physiology in some of its areas of study, dependent upon psychology. This is perhaps the most dramatic, but for my purposes of illustrating the logic of principles, the least interesting consequence.

Since the mental and physiological life of individual human beings run, at least for around seventy years, concurrently, there are always some brain-states occurring when any mind state occurs. But since, as a matter of fact, the mental life of people is discontinuous and their physiological life continuous, there is not always some mind-state when any brain-state occurs. So we already know from these elementary considerations alone that the proposition

'For every individual brain-state there is a mind-state' (P1) is contingent and false.

We also know from the same considerations that the proposition

'For every individual mind-state there is a brain-state' (P2) is contingent and true.

These propositions are of very little interest since they say nothing about kinds. The latter, P2, would be true were there never any repetition of a like contemporaneous pair of brain-state and mind-state.

Consider now another pair of propositions, this time a pair which state a correlation between kinds of states.

'For every kind of brain-state there is a kind of mind-state' (P3)
'For every kind of mind-state there is a kind of brain-state' (P4)

Case 1. Suppose the physiologists' taxonomy is used for differentiating kinds of brain-states. P3 could be tested something like this: N's brain is put into, or observed to be in a certain state, a physiologically identified state of kind a, and then N is asked to report his mind-state. For example an electrode is brought into contact with N's brain just under the temple and he is asked to report on what he hears or sees, as in the experiments of Penfield.*

Either N can report his mind-state, or he cannot because he is dead or unconscious. It is therefore contingent that N makes any report at all. Let us suppose that if he does report he says that his mind-state is of b-kind. For example he reports hearing Little Bo Peep played on the recorder. Clearly P3 is contingent. N might have experienced nothing at all, or nothing correlated with that particular kind of interference with his brain.

Case 2. Suppose now that the thinkers' taxonomy is used for identifying kinds of mind-states. Proposition P4, that for every kind of mind-state there is a kind of brain-state, could be tested something like this: N is asked to report his mind-state and he reports it to be of kind f. A physiologist using his taxonomy, reports that N's brain is in a state of kind g. To test the hypothesis, based on this result, that f-kind mind states are associated with g-kind brain-states, we wait until N again reports, using the thinkers' taxonomy, that he is in a mind-state of kind f. There are now two possibilities: either the physiologist, using his taxonomy, reports N's brain to be in a state of kind g, or reports it to be in a state of kind h. It is the second disjunct that is of interest. If that is what occurs there is a choice.

(i) We can abandon the principle that for every kind of mind-state

* S. Mullan and W. Penfield, 'Illusions of Comparative Interpretation and Emotion', *Arch Neurol Psychiat.*, **81**, 296 ff.

H

there is a kind of brain-state, and with it the possibility of psycho-physiology.

(ii) We can form a new *brain* taxonomy in which the disjunction g v h defines a kind of brain-state. The adoption of this alternative gives us a taxonomy for identifying parts and states of the brain and nervous system which selects the parts and states that it does because kinds of mind-states have become the ultimate taxonomic kinds. g v h is a kind in the new taxonomy only because both g and h are associated on different occasions with f.

Since, on alternative (i) the appearance of a new correlate for a mind-state kind is taken to falsify the hypothesis that the correlation of kinds is general, to consider the hypothesis under that alternative is to treat it as contingent, even if the falsifying instance, h, has not yet turned up. Alternative (ii) is then the only alternative which would allow us to adhere to the interpretation of P4 as a taxonomic principle, but adherence to this alternative requires that we be prepared to form disjunctive neuro-physiological taxa. We shall discuss the significance of such taxa later. I will also show that for a number of compelling reasons we are driven to adopt the second alternative. I want to ask now what the effect would be of adopting P4, the proposition that for every kind of mind-state there is a kind of brain-state, as a taxonomic principle.

The first point to notice is that the kind of necessity that accrues to P4 by our coming to treat it as a principle does not give any ground for classifying the principle as an analytic statement, whether that means either that the statement is a substitution instance of a tautology, or that its truth follows from the rules governing the use of the expressions that appear in it. But the converse principle that for every kind of brain-state there is a mind-state remains contingent, and is still false, whatever attitude is adopted to P4. To treat the statement that for every kind of mind-state there is a kind of brain-state as a principle makes it immune from falsification. It can then be used in support of the definition of a kind of brain-state not previously recognized by the physiologists' taxonomy. Thus, the adoption of the apodeictic attitude to a proposition can be an independent, creative act. Since P4 is certainly not analytic, and since there is no suggestion that the essential nature of mind-states is such that brain-states must 'accompany' them, i.e. exist when they do, it might seem odd to treat P4 as naturally necessary. But by calling a proposition naturally 'necessary' I mean that it should be treated as a proposition immune from falsification for the time being. I shall try to show that choosing to treat this proposition this way has peculiar advantages, in that it binds two sets of taxonomic principles together.

One of the important results which follows from our taking up the apodeictic attitude to P4 is that if we do so, *it becomes impossible to discover free mind-states*. If mind-state kinds are chosen as the ultimate sources of identifying criteria, for both thoughts and brain-states, so forming a unified psycho-physiological taxonomy, then mind-state kinds not associated with brain-state kinds cannot be discovered. But could, perhaps, individual cases of mind-states be discovered which were not associated with individual brain-states? No, because there is always some brain-state for each mind-state. The proof of this proposition hinges on arguments such as those of Strawson and Hampshire, to the effect that anything which can be identified as having any mind-state must be a kind of thing; arguments I do not intend to repeat here. We cannot discover mind-state kinds which are not associated with a brain-state kind, because under the alternative (ii) above, which makes the statement of the correlation of kinds into a principle, should a particular brain-state turn up which is different in kind from inductive expectation, then whatever kind it is, it simply joins the string of disjunctions which define in physiological terms a brain-state kind, but whose falling together into that disjunction depends upon their having been from time to time associated with instances of a particular mind-state kind.

The next step in our argument must be to look into the reasons which we might have for adopting this alternative, under which counter-instances to the general correlation are simply not admitted to have negative force. At one time this alternative was widely canvassed as a support for the possibility of dualism. It was argued that if there are instances of different brain-state kinds associated at different times with instances of the same mind-state kind, then brain-states and mind-states must be attributes of different substances. I have just shown that far from being a support for dualism, making mind-state kinds taxonomically ultimate makes it impossible to discover independent mind-state kinds: that is mind-state kinds which do not have a brain-state kind associated with them, provided that the formation of disjunctive taxa can be justified. In so far as there are independent arguments against dualism (cf. Ryle, (*see* 251) Strawson, Hampshire *et al.*) a theory whose adoption would make it logically impossible to make one of the kinds of discoveries that might give support to dualism, is a good theory.

We must now turn to the justification of the introduction of disjunctive taxa. In setting up the mind-state kind based taxonomy by disjoining apparent counter-instances to the originally associated brain-state kinds, there is of course the assumption that the disjunction will not only be finite but relatively short. We would surely be forced to abandon a mind-state kind based taxonomy if *every* time an instance of f, some well-differentiated mind-state kind instance occurred, a brain-state of different kind occurred. If we attempted to preserve the principle by disjunction in the face of indefinitely long disjunctive taxa of brain-state kinds we would be perhaps attempting something logically possible, but of no utility. For instance on the occurrence of a member of one of the disjunctive sets of kinds which form the new brain taxonomy there would be next to zero probability that f was the mind-state kind, since each time f has so far been reported an instance of a different kind of brain-state has been observed. It would be less unwise, but still unsound, to make any predictions from the occurrence of a mind-state of kind f to the likelihood of a brain-state occurring of the same kind as any of the kinds in the disjunction so far.

But what about even the short disjunction? Is any disjunctive taxon any good? Is not this theory based upon an implausible idea? Can I really be said to be *defining* a kind of brain-state which appears as now this and now that? Only, I think, if it is plausible to introduce a hypothetical brain-state kind, instances of which are responsible under differing conditions, for the manifest brain-states, of kinds $g\ v\ h\ v\ .\ .\ .$, which are observed to be the correlates of f, the mind-state kind. I think that this move is not only plausible but precisely the move by which decisive scientific advances take place. It is the same move by which a disjunction of symptoms can be definitive of the syndrome of a disease when we believe, or have reason to believe, or know that the same micro-organism is the cause of each alternative manifested set of symptoms. The value of this suggestion would not depend upon whether an instance of the postulated hypothetical brain-state kind could be identified here and now, independently of some member of the disjunctive taxon. It might be that all that could be identified was some difference in *other* brain-states, constituting a difference in the environment between that in which g occurs and that in which h occurs, a difference which explains why the hidden factor x is sometimes the cause of g-kind phenomena, sometimes of h-kind phenomena, while leaving the question of the identification of x quite open. Even in the absence of any empirical differences of this sort it is still open to an investigator to advocate a physiological theory which postulates a hypothetical common cause of the members of the disjunction.

A slightly different example of the combination of the logical priority of mind-state-kind taxonomy and the consequential postulation of unknown physiological factors is Schachter's discovery that the emotions experienced by subjects given a dose of adrenalin depend upon the situations they believe themselves to be in, and can differ as widely as fear and joy.* Physiologists have responded to this discovery by postulating

* S. Schachter and J. E. Singer, 'Cognitive, Social and Physiological determinants of emotional state', *Psychol. Rev.*, **69**, 379–99.

a further physiological difference, so far undiscovered, which differentiates the joy-adrenalin state from the fear-adrenalin state. In this
way mind-state-kind taxonomy is ultimate.

We are now in a position to see why a choice of brain-state kinds as
taxonomically ultimate for a psycho-physiology is the less rewarding
move. To make brain-state kinds taxonomically ultimate would involve
the possibility of our having to form a mind-state taxonomy by disjoining mind-state kinds of the thinkers' taxonomy when counter-
instances have to be accommodated. It would be incumbent upon an
investigator who finds that whenever a certain sort of electrical dis-
charge occurs in the brain of his subject, he sometimes reports himself
happy, sometimes sad. Does it make the slightest sense to say that this
can be explained by the hypothesis of a hitherto undiscovered mind-
state-kind, instances of which are occuring each time he reports himself
happy or sad. The rationale which made such a move respectable in
the brain-state taxonomy is not so clearly available in this case. It is
not at all clear that hypothetical mind-states of a hitherto unknown
kind is even an intelligible conception. Freud perhaps offers us the
nearest thing to this when he forms disjunctive taxa of differently identifiable phenomena through such concepts as 'infant sexuality'. But
Freud offers an extensive justification for the move from within the
mind-state kind system of thought. The case we are considering is very
different, since it involves the postulation of the new kind of mind-state
on the sole grounds of there being two or more different kinds of mind-
state present when a given brain-state kind is present.

Why does the idea of a hypothetical mind-state kind seem dubious?
The postulation of hypothetical entities is an essential feature of the
dynamics of most sciences. To examine this question we have to ask
first, whether the hypothetical mind-state kind is offered as a fiction,
perhaps as a metaphor which illuminates without explaining the facts.
It is in this sense that Freud's ideas seem to be most profitably taken.
The idea of infant sexuality is not so much an explanation of a wide
range of behaviour, but a device by which a wide range of apparently
disparate phenomena are drawn together, and in terms of which they
are seen to have an essential similarity. Such a move is like 'See it this
way . . .' rather than 'This is why it happens . . .'. In terms of the
problem of this chapter such a use of hypothetical entities simply leads
to more or less picturesque affirmations of the disjunction of taxa, and
is not parallel to the postulation of a hypothetical brain-state kind in
explaining a disjunctive taxon in physiology. On the other hand if the
hypothetical mind-state kind *is* offered as part of a possible causal
mechanism it must be possible to envisage the situations in which it
would be correct to say that an instance of that mind-state kind had

been discovered. What would it be like to discover an instance of a new mind-state kind? It might be the case that at least some people have had this mind-state, or been in this mind-state, but have never identified it. Then the introduction of the term for the new mind-state could be justified as a useful extension to the vocabulary. But what if a new kind of feeling, a new kind of emotion, a new kind of reason, a new kind of action, a new kind of power or capacity, i.e. something which had not previously been experienced by anybody were proposed? Some proposals can be ruled out *a priori*. To say we had had but not experienced certain feelings and emotions will not do, since unfelt feelings, unexperienced emotions, and, we can add, unperformed acts are all clearly self-contradictory notions, so hypothetical feelings, emotions and acts do not seem to be available to provide the background identity to create disjunctive taxa of mind-state kinds. Only two obvious candidates remain, unacknowledged reasons and unknown powers.

Let me remind you of the situation with which we are dealing. A physiologist is supposed to have discovered that when a subject reports, or is seen to be in, or experiencing mind-state kind a, *and* when he is in mind-state kind b, his brain-state is of kind k. If brain-state kinds are supposed to be taxonomically ultimate, then a new taxon for the classification of mind-states must be created, namely $a \: v \: b$. Following the model of the physical sciences this peculiar situation would have to be resolved by the theory that both states of kind a and states of kind b are products, effects or results of the hitherto undiscovered state of kind c. Thus, k marks the operation of a hitherto undiscovered mental power whose exercise produces a state of kind a sometimes, and a state of kind b at other times. Or k marks the presence of a hitherto unacknowledged motive, or unformulated reason, which lies behind each of a and b when they severally occur. We now have to consider what would recommend such moves in preference to those which make the mind-state kinds taxonomically ultimate and which, as we have seen, force the physiologist to postulate hitherto unknown physiological phenomena.

Of the remaining alternatives, it is again clear that the idea of 'exercising a hitherto unknown mental power' has certain difficulties, though science-fiction writers have made the concept intelligible. Powers are defined partly in terms of their exercise, so the power to produce either a or b is not an unknown power. Compare this with the genuine novelty of suddenly discovering that one can do something one did not know one could. But in this case it is not an unknown power that is discovered, but that one personally has a known power. We are being asked in the supposition I am considering to suppose that I can know what I can do, namely a and b, and not know that I have the power to do them, which is absurd.

Only the hypothetical common motive remains. Apart from the fact that this restricts the possible mind-state kinds where disjunctive taxa might be formed to those in the genesis of which reasons and motives might figure, we run into an empirically unresolvable problem when we try to theorize in this way. The physiologist has found that when the subject does, or experiences, *a* and when he does, or experiences, *b* a brain-state of kind *k* occurs. Of course the physiological chain of causation from *k* to each of the different movements that distinguish *a* from *b* will be different. Our hypothesis is that the subject has thought of *a* and *b* as distinct kinds of acts or experiences, and hence as being performed or experienced for distinct reasons or motives. The psycho-physiologist now wishes to form a disjunctive taxon *a v b* which defines a kind of act or experience. Asked to justify this he offers the theory that both *a* and *b* stem from the same motive, are done for the same reason. The empirical discovery of this common motive or reason is then pursued in the only way possible, the subject is asked to consider whether on reflection he might have done *a* and *b* for the same reason.* But notice: what empirical distinction is there between the subject *providing* a reason which is common to both, and *discovering* that he had the same reason for both acts. It might be very difficult to find one. We can now bring this long and rather elaborate argument to an end, since the most favourable situation in which a move parallel to the move which creates disjunctive physiological taxa might be expected to be helpful is such that every alleged discovery could be reinterpreted as a justification after the fact. Of course this does not amount to a conclusive argument, but here we are concerned with inclining reasons.

It is worth noticing that the example we have just considered offers rather better scope for the hypothesis of a hitherto unknown brain-state kind than it does for the mind-state kind hypothesis. It seems clear to me that, in practice, in such a situation it is precisely a brain-state hypothesis which would be used to resolve the situation, thus retaining the taxonomic primacy of mind-states. Surely we would be inclined to say that *k* and *x* occur with *a*, and *k* and *y* go with *b*, and leave later generations of physiologists to identify brain-states of kinds *x* and *y*.

Since bodily life is continuous and mental life discontinuous, there are physiologically identifiable brain-states independent of any mind-state. Making brain-states taxonomically ultimate would require the postulation for the intervening periods of unconsciousness e.g. dreamless sleep, either of unconscious mind-states of kinds already acknowledged, or of a special *ad hoc* class of mind-states, the unconscious ones. Under the first alternative it would be impossible to verify which kind

---

* The only case I know, is L. Festinger's concept of 'cognitive dissonance', which fails for just the reason given.

of mind-state one was having, when unconscious. Since, by hypothesis, neither a contemporaneous nor a retrospective report can be given by the subject, this supposition does not differ empirically from the supposition that he is in no kind of mind-state. This case is actually more complicated than I have represented it, since only if the brain-states occurring during unconsciousness are of a kind with those regularly found with certain kinds of mind-states during consciousness, can the hypothesis of unconscious mind-states of already recognized kinds be formulated. And then it is hard to see how unconsciousness and consciousness are distinguished physiologically. On the other hand if the brain-states occuring during unconsciousness are characteristically different from those which are known to occur during consciousness we have little to go on, except perhaps some analogies in postulating the existence of any kind of mind-state. That is to say, in response to the logical possibility that a variety of brain-state kinds occur during unconsciousness, which prevent the simple introduction of unconsciousness itself as a mind-state kind, one would be forced to postulate a variety of unconscious mind-state kinds, and this makes neither more nor less empirical sense than to say that only one kind of mind-state or no kind is occuring. The second alternative is the idea of unconsciousness as a mind-state kind. Here it seems improper to speak of a mind-state at all. To do so would seem to involve the slide from the absence of a state of kind A, to the presence of a state of *non-A*. But either *non-A* is a disjunction of all possible states of the universe except A, which is pointless as a taxon, or *non-A* is just nothing at all.

Thus the alternative of making brain-state kinds taxonomically ultimate seems to me to be subject to very serious objections.

But not everything involved in these arguments has yet been brought to light. I have been assuming in my argument some of the important epistemological points made in recent philosophy of mind.

1. It is not true that only A can know what A's thoughts, feelings, intentions and motives are. Usually what A is thinking, feeling and intending can be known to other people quite well. But whenever a dispute arises, in most cases, A is himself the final authority. This is quite different from the idea that only A knows what he is thinking and feeling. It seems to me again that a conclusive argument has been offered in contemporary philosophy of mind in favour of the idea that the things that A does, including especially what he says, are not properly treated as *evidence* from which other people infer his mental states, but as *showing* his mental states, and as *examples* of his capacities. If we hear someone complaining this is not evidence for an inductive ascription of discontent to him, it is the way his discontent is shown, it is an example of discontent. This idea has been modified and adapted

very successfully in recent work in the philosophy of mind. But we must beware of moving too far in the opposite direction from the ancient errors. It does not follow from the dethronement of 'privileged access' that a man loses his authority as to his motives, feelings, thoughts and aspirations. That a man is, on most occasions and in most cases, the best authority on such matters is a central part of my argument. It is one of the reasons why the taxonomy of mind-state kinds is essentially involved in determining what are the relevant kinds of physiological states and conditions when we are seeking the physiological basis of human thought and action. It is also worth noticing that the exceptions to the Principle of First Person Authority are rather special. It has to be A's wife or his psychiatrist who can properly claim sometimes to know A's mind-states better than A. The significance of A's words and deeds is fully evident only within the context of his life. This is an important point for philosophical psychology but need not detain us here.

2. Usually what anyone says in report of his mind-states, for instance what emotional state he is in, what aspirations he has, what he is feeling and where he feels it, is perfectly intelligible to other people. The vocabulary in which we report our states of mind is a public vocabulary.

3. Anybody, suitably trained and equipped, including A, can find out A's brain-states within the limitations of the current state of science. This is the Principle of All-Person Authority as to Brain-states.

Since we like the facts of a science to be ascertainable by anyone with equal right, for a variety of familiar reasons, there naturally exists a strong temptation to try to found a science of psycho-physiology on the Principle of All-Person Authority as to Brain-states. But to carry this through it would be necessary to make brain-state kinds taxonomically ultimate, so that behaviour and feeling, speech and thought would all have to be classified by reference to the brain-states accompanying instances, with all the attendant difficulties with which we have already dealt. The final step in my argument is to show that the necessity we are under to take mind-state kinds as taxonomically ultimate, that is that we are obliged to classify physiological states and conditions by reference to the acts, thoughts, feelings and so on that accompany them, does not oblige us to abandon a possible science of psycho-physiology. The fact that the vocabulary for mind-state description, is public and understood by most people ensures that A's remarks about, and reports of, his mind-states are remarks and reports in the public domain. The Principle of First-Person Authority is no impediment to including reports of the occurrence of kinds of mind-states in a scientific psycho-physiology. It is no more an impediment to a scientific psycho-physiology than is the fact that sometimes the only report as to the shape of the interior of a cave is brought out by a particularly enterprising

H 2

speiliologist just before it collapses, is an impediment to a scientific speiliology. *A*'s privileged authority can only be taken as an impediment if it is confused with the entirely unfounded supposition that *A* is privileged as to access to his mind-states, and as to understanding what he means by his reports. In doing experiments to ascertain the nervous seat of various mental powers one depends upon the fact that the mind-state kind taxonomy operates through a public language, and one literally *recruits the subject of study to the scientific team*, and his job is to report on *his* state of mind, which may involve the ascription to himself of any of the whole gamut of predicates appropriate to people, including what he is doing, feeling, thinking, hoping and so on. Reports about a man's mind-state are not unverifiable, though sometimes they are verifiable only by him. Since we can understand such reports there is no impediment to incorporating a man's mind-state reports into a scientific investigation, and thus no impediment to a scientific psycho-physiology. But it will be characterized by the fact that the mind-state kind taxonomy is determinative of what kinds of physiological states are to be identified.

The adoption of particular taxonomic principles in any part of science is not innocent of metaphysical implications. Choosing to classify the chemical elements according to their relative weights is already to shift from one metaphysical system to another. It is to think of chemistry as the science of materials maintaining their identity through qualitative change *à la* Lavoisier, and not to think of it as the science of qualities which pass from material to material *à la* Lamarck.* I must now sketch an answer to the question as to what is the most economical metaphysical picture under which the adoption of the principle that mind-state kinds are taxonomically ultimate, can be made intelligible. It seems that the theory of ontological dualism, that mind-states and brain-states are states of two independent substances, is not only not required to make the choice of mind-state kinds as ultimate for classification intelligible, but positively ruled out as a possible metaphysical picture, since strong arguments have been produced to show that making mind-state kinds taxonomically ultimate makes it impossible to identify, and hence to discover, free mind-states. If mind-states and brain-states are not qualifications of two substances, how shall we understand their differences in a way which preserves the intelligibility of the system of psycho-physiological taxonomy to which we seem to be driven? I explicate this by a pair of models, neither of which are wholly adequate to the relation between mind-states and brain-states, since that relation in which these states of an organism stand to each other is a unique phenomenon in the Universe.

* For a discussion of this issue, see C. C. Gillispie, *The Edge of Objectivity*, Oxford and Princeton, 1960, pp. 273–6.

Model I – 'Aspects': Mind-states and brain-states differ in a way similar to that in which different aspects of the same thing differ. One can see an adding machine as a system of gears or as a device for totting up sums. It is the same machine looked at in different ways, and under each aspect it makes sense as a machine independently of the other aspect. Furthermore our identification of certain actions of the gear trains of the machine as 'adding', imports into the mechanical taxa, criteria of classification that come from the taxonomy of numerical operations. I dare say a fairly complete analogue of the arguments of this chapter could be constructed which would show that the taxonomy of numerical operations was ultimate in the identification of the parts and operations of the machine, if it were being considered as an adding machine.

However, attractive and fairly thorough-going as this model is, it will not quite do, since the two system whose interlocking criteria of classification we have been studying, differ not as two external aspects of the machine, but as internal from external. An adding machine's addings are not thus classified because that is how the machine classifies them. To try to catch this internal/external distinction we can supplement the 'aspects' model with another.

Model II – 'Sentences and statements': Mind-states and brain-states differ as a sentence heard by someone who understands a language, and a sentence heard by someone who does not. There is just the one utterance, and he who understands it, the insider, does not usually attend to its phonemic structure, while he who does not, the outsider, can attend only to its phonemic structure. Furthermore the classification of phonemic structures as linguistic elements depends upon the insider's identification of this or that sentence as having this or that meaning: that is playing this or that rôle in his life. This model has a further virtue. The insider, can, if he will, attend both to the meaning and to the phonemic structure, just as the subject of a physiologico-psychological study can attend both to the feeling of anxiety he is experiencing and to the concentration of adrenalin and other substances in his blood. In the case of feelings the subject directs his attention to the very same bodily spot, in many instances.

Each of these models seems to express the idea that our awareness of mind-states and of brain-states differ as two different ways of knowing about the same thing. There is a dualism, but it is epistemic. Each of these ways of knowing involves a taxonomy, and the criteria of classification in one of them dominates the other. But the identification of a mind-state and of a brain-state, and of a brain-state as identified through a mind-state, are each and every one of them predications of states to more or less specific places. The point of reference, the subject, to which any of these predicates are applied, is no wider than a human being, though it may be narrower: that is, as for example, in the case of feelings and specific sensations it may to be to a point or circumscribed region, on or within the person. This connects on to a strand in contemporary philosophy of mind. This view is not without its difficulties since awareness of mind-states and awareness of brain-states are both mind-states.

The view that the sense of mind-state predicates and brain-state predicates are different but their point of application is the same, seems pretty convincing for such psychic phenomena as feelings. If the stab of fear is a mind-state, and the physiological goings on in the nervous system are, in my extended sense of the term, brain-states, then the very fact that the feeling is felt where the nervous system is excited can lead us into thinking that there is an identity of substance. Even when one begins to look at this in a somewhat more sophisticated way, and to be concerned about the rôle the brain itself plays in the feeling of feelings, this is a case where the sensory and the physiological are confined within the one envelope, the spatial confines of the person himself. But what of perception? I do the seeing of the yellow daisy, and whatever occurs, occurs in and to me. Yet while the physiology of perception deals with states and occurrences in me, the yellow is in the daisy. The point of application of the predicates '. . . is yellow' and '. . . sees a yellow daisy' are different. I do not know of any adequate view on this matter, though there are metaphors enough in the literature. Perhaps this difficulty cannot be resolved in a way that would lead to the formulation of a cast-iron argument for epistemic dualism and ontological monism.

If the arguments in this chapter have been correct, then there are consequences for our view of the relation between the sciences of physiology and psychology. It is often said that the sciences form a hierarchy with respect to explanation, so that the causes of the phenomena recognized and studied in one science are to be found among the phenomena recognized and studied in some other. Indeed in the earlier parts of this book I have laid considerable emphasis on the importance of recognizing the stratification of scientific knowledge. We recognize that the explanation of chemical phenomena is to be sought in the physics of atoms and their components. We recognize that the explanation of biological phenomena is to be sought in the chemistry and physics of the components of organisms. We might be inclined to think that the explanation of psychic phenomena, which are the subject matter of psychology (and that, of course, includes all that class of behaviour too in which a man may be said to be doing anything), must be looked for in a science which stands to it as does physics to chemistry, or biochemistry to biology. But this analogy will not do. The identification and classification of the entities with which physics deals is entirely independent of the taxonomy of chemistry. There could be a physics of electrons, protons and neutrons, had these entities never come to form the structures which are the atomic components of the chemical elements. There could have been the chemistry of carbon had there never been organisms. But there could not have been the study of the physiology of perception were there no perceiving. Psychological individuating criteria play an indispensable rôle in individuating physiological states and processes. Inside a man's head are billions of cells, and trillions of chemical and physical phemonena are occurring. What a biochemist, a physiologist or an anatomist pick out as entities could bear only accidental relations to psychic states if they used their given taxonomies. Only subjects, and conscious subjects at that, can report on their mind-states, and as we have seen, the identification of brain-state kinds proceeds through the identification of mind-state kinds.

The tenor of these considerations leads one to say that the relation between physiology and psychology is not of explaining science to explained science. Rather than steps in a hierarchy they stand side by side. They are related taxonomically, not causally. When we say such things as 'Mind-states are correlated with brain-states' we must understand this, so I have argued, not as a sketch of a causal relation, but as a sketch of a principle of taxonomy. The correlation is not the kind of correlation that exists between smoking and lung cancer, since these are separately identifiable phenomena and the question of their connection is the problem of finding the causal mechanism. In the case I am discussing we cannot identify the physiological states involved in human behaviour independently. There is no causal problem to solve. Only if the relation between psychology and physiology is mistakenly assimilated to relations of the smoking/cancer type is there a causal mind/body problem. If we see their relation as a principle of taxomony the mind/body problem takes on a different aspect.

The detection of pseudo-causal laws is of very great importance in deciding which existential questions to pursue in any branch of natural science. If Newton's Second Law is treated as a pseudo-causal law then it points to two aspects of mass acceleration, not to two contingently related and existentially distinct phenomena. As a pseudo-causal law it cannot serve as the basis of an inference to the hypothesis of the existence of forces, as independent and distinct entities in the universe. So the attempt to identify forces independently of mass-accelerations would be pointless. While a law is treated as a genuine causal law the existential questions continue to be pursuable. It continues to be sensible to try to think of other manifestations of force, for instance, than mass-acceleration, or as in the previous example it continues to be sensible to look for kinds of mind-states not regularly correlated with kinds of brain-states, and of kinds of brain-states not regularly correlated with kinds of mind-states. In short the question of whether some term, in a two term relation, refers to an entity whose existence might be demonstrated independently of the existence of the entity referred to by the other, is, in part, an aspect of the question as to whether a certain law-like statement is a contingent general statement of a causal relation, or actually to be treated as a necessary statement about the different ways the same entity manifests itself in experience.

This needs careful elaboration. It is a matter of fact that an entity manifests itself in two or more observationally distinct ways. Take for example the structure of a fairly elaborate molecule. The structure may manifest itself in the fact that certain derivatives can be obtained from it and not others. It may also manifest itself in optical activity. But the fact that a solution of tartaric acid is dextro-rotatory *and* forms such and such is not a candidate for a causal relation. This case is perfectly clear and no one would be tempted to gloss the fact otherwise than as follows 'Because the molecule has such and such a structure, S, and S is responsible for dextro-rotation, and S is responsible for derivative a being formed, rather than compound b, then, if it is dextro-rotatory, it *must* form compound a and not compound b!' From whence the 'must'? It derives from the necessity, at least in classical chemistry before the idea of resonant structures was introduced, that any given substance can have only one structure, because to have that structure is part of what counts as being that substance. The other part of what counts as being some definite substance is, in classical chemistry, having been prepared in a certain manner, from certain other materials. So while it is the substance it is, it must have the structure it has, and since it is the structure which is being manifested in the phenomena, the phenomena are necessarily related. Thus dextro-rotation is an effect of the structure appearing under some circumstances, and that such and such a derivative is formed is an effect of the structure appearing under other circumstances. If, however, one treats the optical activity and chemical powers as independent of the structure, and thus the effects rather of the circumstances that the structure, as one might be tempted to, following a Humean style of thought. Of course, there is no necessity whatever that optical activity and the obtaining of such and such derivatives go together. The same explosion manifests itself in a visible cloud of dust and debris, and in a loud bang, but, in this case, it is perfectly clear that the explosion is neither a cloud of dust, nor a loud noise. So cases where there is a single cause and two different effects, three independent phenomena, have to be carefully distinguished from the case in point, where the effects are themselves the manifestations of the entity in question, be it thing, event or quality.

Only if one is prepared to *take* the two effect phenomena, each as manifestations of the 'third thing', that is, as causally related to the third thing, can one take the two effects as themselves not causally related. And no empirical evidence of a conclusive sort could be produced to decide which way to take them. As I tried to show in the mind-body example, something like weight of advantage or inclining reasons can be brought, but one can never do better than that, and for a very simple reason. Any case where we wish to take the correlated phenomena as aspects of one original phenomenon, could, for all we can ever tell observationally, be a case like the optical activity and chemical derivatives, that is a case where there is a cause existing independently of these effects and which only produces the effects in the situation of the appropriate conditions and circumstances obtaining. Can we decide *a priori* which cases of an apparent correlation to treat as two or more aspects of one phenomenon? No, but in deciding one way or another we express our conceptual decisions. It is in this and other ways that we express our beliefs about what kinds of entities there are. And in another way this shows too that necessity is not a property of propositions which they might have in independence of who particularly entertains them. To say that a proposition in science admits of no true alternative is to express one's attitude to that proposition, not to describe the proposition itself. Those propositions which do not have the attribute of being analytic in structure, and which yet demand such an attitude on the part of one who entertains them for the reasons given in this chapter, may still come to be treated in the same way. Their inviolability then serves as the mark and expression of some conceptual 'decision'.

In earlier chapters it became clear that the full development of the realist position, demanding, it seemed, at least the consideration of existential hypotheses for all kinds of hypothetical referents of theoretical terms, created difficulties of a fundamental kind in the universal deployment of the realist notion of cause. It is now time to resolve these difficulties, and, in showing them to be merely apparent, to advance our understanding of hypothetical entities to its completion. The problem for the realist view of causality comes out of the need that sometimes arises for the introduction of hypothetical entities which are of a fundamentally different kind from the phenomena in the explanation of which they are intended to figure. By a 'fundamentally different kind' I mean that two entities A and B are of a fundamentally different kind, if, for every property characteristic of kind A, kind B has a property which excludes the possession by the entity of a property of kind A. For instance, if kind B is characterized by being of a continuous substance, kind A is atomic; if kind B is coloured, kind A is of such a kind that colour predicates do not apply to it, that it is, in the strictest sense, colourless; if kind B has finite density, then kind A has either no density at all or infinite density, and so on.

The history of science provides two very beautiful examples of this kind of difference. Cartesian mind and Cartesian matter differ in the way described as fundamentally different in kind. The Faraday-Maxwell field, and charged bodies and magnets also differ in this way. The test for two entities being of a fundamentally different kind is whether they can be said to be in the same place at the same time. Then either there are not really two different entities there, but the same entity differently described, or the two entities are of a fundamentally different kind. In so far as minds are located, they are where a person's body, including his brain is, and fields are certainly in the same place as the particles they are supposed to act upon.

The realist view of causation advances the view that concomitance is not enough to satisfy the requirements that a relation be causal. There must also be the hypothesis, or best of all, the knowledge, of the existence of a mechanism capable of producing phenomena of the types of the effect, on being stimulated by the cause. Now suppose the state of the continuous, insubstantial field is offered as the cause of the acceleration of the particulate, massy, substantial corpuscle. Then of what nature is the causal mechanism? If it is material, of what nature is the mechanism by which the insubstantial field acts upon it? If it is field-like, of what nature is the mechanism by which the field-like mechanism produces the effect in a substantial corpuscle? To answer these questions either with a field-like account, or a material-thing account, is clearly no help whatever. On the other hand one might try to resolve the difficulty by saying that at this point a new kind of causal relation has appeared, reducible to simple concomitance, without generative or productive connection, which is effectively to abandon the realist theory of causation. Indeed it concedes victory, in the final analysis, to the regularity theorists.

A resolution of the mind-body problem has already been achieved, for scientific purposes, by the insight elaborated in the first part of this chapter. That is, while not denying the genuineness of the epistemological distinction between the ways 'mind' and 'matter' manifested themselves, a way was found of denying any ontological distinction between them. Minds and bodies can be in the same place at the same time because a pair of one of each, constitutes, not two things, but only one. The problem of the relation between fields and things cannot be handled so simply.

What, though, if the claim that fields exist is treated not as the claim the fields exist *as well as things*, but that fields exist *instead of* things? Any attempt at such a move in the mind-body case leads either to the absurdities of materialism or of idealism. To work out the consequences of this suggestion in physics a new notion is needed, that of *recategorization*. It happens from time to time that the whole world comes to be seen in a new light. In a way everything remains the same, but in a way everything is different. The experience of the convert, when suddenly everything comes to be thought of as the acts of the hand of God, or of the child, on first learning that adults can lie, provide us with models of recategorization. Everything which existed, that is which could be referred to and which was potentially demonstrable, remains. Instead of being incorporations of forms in prime substance, the very same things can come to be thought of, and even to be seen, as clusters of corpuscles. To decide whether the forms exist, or whether corpuscles do, is only partly an empirical question. It is only partly an empirical question because, if something is shown to be a cluster of corpuscles, a believer in the theory of substantial forms has but to say that here is prime substance incorporating corpuscularian forms. If you had asked an Aristotelian and a Corpuscularian what there was in the world, each would begin their answer in the same way: 'Men, animals, minerals, earths, plants, planets, . . .' and the Aristotelian would add 'and Forms and Prime Matter'; and the Corpuscularian 'and Corpuscules'. But the addenda are different sorts of claim. Carnap has noticed this difference and distinguished them as internal and external existence claims. They are not settled by demonstration, as for instance is the question of the existence of a certain metal or of a certain tribe or even of a certain person. To assent to the addendum is to assent to the claim that corpuscles exist instead of forms in matter, and that everything in the early part of the list is in fact corpuscularian. One might be tempted to say that the claims and counter-claims of Aristotelians, Corpuscularians, and adherents of the field, should not be expressed as existential claims at all. One might want to say that the existential form is mere propaganda, inviting the assumption that demonstrative proof for, say, the existence of corpuscles, might one day be forthcoming. But it is clear that whatever can be brought into a demonstrative relation with a man, either in the senses or in the extended senses, can be either informed matter, or a cluster of corpuscles, or a singularity in the field, according to taste. So the existential form of the claim conceals rather a claim that whatever exists should be redescribed, in the terms proposed in the addendum. In short it is not, it seems, an existential claim at all. Candidly put, the Aristotelian world view would seem to sum up as 'Men, planets, metals and so on exist, and *they* are all manifestations of different forms in prime matter.' This re-wording of the addendum does make it into a general descriptive statement, and not an existential claim at all. But this tempting simplification will not really do.

There is an existential difference as well as a descriptive difference between the three positions. It becomes particularly pronounced in the case of the field. On the realist view, the universe cannot contain both fields and corpuscles. No way can exist by which they could interact. So either the field is a metaphorical conception, heuristically valuable in understanding the distinction of the possibilities of action at a distance from the corpuscularian agents, in which case fields are no more candidates for existence than are say the elastic 'skins' of drops. Alternatively corpuscles could become nothing but specially distinguished volumes of the field. The full development of this last idea will be the subject of the last chapter of this book. But if the corpuscularian view is adopted there is nothing whatever between the corpuscles. Descarte's vortices are surely the *reductio ad absurdum* of the idea that there can be a corpuscularian plenum. The field view demands that we say that the field exists where there are no corpuscles, namely between the corpuscles. And there the claim is not reducible to a claim to merely redescribe whatever is there, because by hypothesis, nothing is there. Clearly the move from corpuscles acting at a distance, to continuous fields, filling all space, involves both recategorization of the corpuscles and an additional claim to existence.

But what of the suggestion that in saying that the field exists between corpuscles *space* is being redescribed, recategorized? This will not do. Space has no causal powers, and, in the corpuscularian world view, all causal powers are located in the corpuscles. But in the field view, the mediate action at a distance of the powers of the corpuscles are turned into the immediate action of the field. So the field must have causal powers at every point at which it can act, that is everywhere. So the field cannot be recategorized space. It must fill space.

In general terms, where does the existential difference lie? It seems to be in the causal mechanisms. He who denies the existence of substantial forms is not denying that there are red things and that they differ in colour from yellow things. He is denying that the difference is to be explained by saying that the former is an actualization of the form of redness and the latter an actualization of the form of yellowness. There is, he might claim, a difference in structure of the surface clusters of corpuscles between the kinds of things. Looked at in this way the difference between the protagonists may come to look very like empirical differences. How then is it that we can also regard these differences as ones of a free choice of concepts? It lies in the truth-conditions for the choice. This will be explored in detail in the last chapter, where the way the decision between these options must be exercised will be expounded.

SUMMARY AND BIBLIOGRAPHY

Here we return to the question of general statements immune from falsification. The view advocated distinguishes between statements which are treated as immune (dubbed 'necessary') because of their form, and those treated as necessary to provide a stable conceptual structure. For necessity as a distinct concept from a certain logical form see

204. D. W. Hamlyn, 'On Necessary Truth', *Mind*, **70**, 514–25.
    H. Putnam, 'The Analytic and the Synthetic' (*see* 44), pp. 358–97.

For attributions of necessity as expressions of attitude see

205. R. Crawshay-Williams, *The Methods and Criteria of Reasoning*, Routledge and Kegan Paul, London, 1957.
206. H. G. Alexander, 'Necessary Truth', *Mind*, **66**, 507–21.

These points are illustrated by a discussion of principles for psycho-physiology. A similar view as to the nature of these principles, though not fully worked out has appeared in

207. J. J. Fodor, *Psychological Explanation*, Random House, New York, 1968.

For various versions of the 'identity-thesis', briefly referred to in the text as a suitable metaphysical position, given the theory about the principles, see

208. J. J. C. Smart, 'Sensations and Brain Processes', *Phil. Rev.*, **68**, 141–56.
209. J. J. C. Smart, *Philosophy and Scientific Realism*, Routledge and Kegan Paul, London, 1963, pp.92–105.
210. D. M. Armstrong, *A Materialist Theory of the Mind*, Routledge and Kegan Paul, London, 1968.
211. J. Teichmann, 'The Contingent Identity of Minds and Brains', *Mind*, **76**, 404–415.

For more general surveys of possible models see

212. T. Nagel, 'Physicalism', *Phil. Rev.*, **74**, 339–56.
    H. Feigl, 'The "Mental" and the "Physical" ' (117), pp. 370–497.
See pp. 484–97 for an extensive bibliography.

The identification of pseudo-causal laws has been discussed by

213. G. Buchdahl, 'Science and Logic', *BJPS*, **2**, 217–35.

What I have called 'recategorization' is discussed by

214. F. Sommers, 'Types and Ontology', *Phil. Rev.*, **79**, 327–63.
215. K. Baier, *Proc. Am. Soc.*, **61**, 19–40.
216. L. Sklar, 'Types of Inter-Theoretical Reduction', *BJPS*, **18**, 109–24.

9. Principles of Indifference

The Argument

Our general theory is that explaining is in terms of enduring mechanisms. The existence and behaviour of ultimate mechanisms cannot, therefore, be explained.

What are the principles of permanence?

I

SPACE

Change of location in space is not causally efficacious. This implies the principle of inertia.

1. *Proof of the Principle of Location Indifference*

Preliminary examination of the concepts of space.

(a) *Spaces*

    1. Enclosed volume;
    2. Place or occupiable volume;
    3. $3-D$ continuum of places;
    4. Idealized system of relations of $3-D$ continuum;
    5. Abstracts system of relations with a certain formal character.

(b) *Boundaries*

Physical space = that volume which can be occupied (physically possible locations);

Space of physics = that volume which could be occupied (logically possible locations).

(c) *Minimum Formal Conditions for Spatial Relations*

    $1-D$ space as illustration: conditions on objects a, b, c, d, e.

      1. $aSb = \quad a \neq b$ & $(MEc)(a \neq c, b \neq c, cBa, b)$
      2. $cBa, b = -aBb, c$ & $-bBa, c$
      3. $-cBa, b = \quad aBb, c$ v bBa, c
      4. $cBa, b = \quad (MEd)(MEe)(dBa, c$ & $eBc, b)$

(MEx) means 'There might be an x'.

This system of concepts is predynamic so resistance to incursion is not included. 'Shoving-in' will therefore be ambiguous between Latin and Teutonic queue jumping.

2. *Translation as Space-defining*

Using time and two objects there is a spatial interval between them if a third object takes time to change from contact with one to contact with another. This is post-dynamical because contact is a dynamic notion, otherwise the test body could push the limit body along indefinitely.

(a) Add metric to II by

    (i) comparison of times (clock);
    (ii) dynamic assumption of same cause of motion;

(b) Add metric to I by
    (i) adding dynamic concept to distinguish gap-filling from displacement;
    (ii) idea of counting standard units filling gaps.

3. *Uniqueness of Desiderata*

(A) The visible spectrum of colour though linear, is not continuous; not even of the order of the rationals; 'shoving-in . . .' axiom ensures for space an order of infinity of at least the rationals, though kinematics requires space of the order of the real numbers;

(B) Translation generation of space has no colour analogue because colours, though uniquely related to other colours, do not involve the notion of formerly this or that colour, so no analogue of place.

Space is the totality of places as defined by I, 1, (c).

Change in spatial relations is not causal, because it does not depend upon particular things, only things in general. So there can be no qualitative distinction of nexuses of spatial relations. Change in spatial relations cannot be causal.

4. *Cusa's Principle*

Change in location has, of itself, no physical effects.

Note. Ratios of distances do not seem to be arbitrary. At least changes in ratios require a dynamic account.

5. *Principle of Isotropy*

Change in orientation is not, by itself, causally efficacious. (If changes occur during reorientations, they are put down to the effects of bodies or fields.)

II
TIME

1. *Principle of Epoch Indifference*

'There are no physical consequences of existing at one time rather than another.' Change in time is explained by processes and causes other than the effluxion of time.

2. *Events or Durations*

Both required for intelligible system for our world. Heraclitean world must be modified to give some endurance to individuals. Parmenidean world must be modified to give some change to individuals.

3. *Converse Principle*

(a) Enduring is in no need of explanation.

(b) No problem of induction if change sequences are referred to permanent or semi-permanent mechanisms.

Note. Continuous causality required for practical maintenance of things because of existence of Manichean forces. This *follows* from the principles.

III
PARITY

Left-handedness and right-handedness have no physical consequences. Non-conservation of parity seems to be experimental refutation of a principle.

2. The principle seems to be connected *a priori* with non-causality of spatial relations. Alternative interpretations of second cause are possible.

IV
NUMERICAL INDIFFERENCE

Raw number is not a cause.

1. *Derivation of Principle from Range Theory of Probability*
$$\text{prob } \alpha/\beta = \frac{\text{no. of subclasses for event of kind } \beta}{\text{no. of } \alpha \text{ subclasses.}}$$
Statistical evidence for probability statements would make numerically different sized classes, different ultimate subclasses.

The idea that an n-membered class is a different natural kind from an $n+1$ membered class is absurd.

2. *Possible Resolution*
Laws of nature operate upon initial conditions. Raw number is a feature of the initial conditions.

Change forces itself upon our attention. We are inclined to overlook the quiet persistence in existence of much that does not change. In the sciences, assumptions of principles of indifference, of unchange, play an important, and insufficiently noticed part. These are not the conservation principles which import persistence into change, but the principles of non-change: of endurance. Philosophers of science have commonly overlooked these principles in their efforts to unearth the principles involved in our understanding of change. And yet, if my general account of laws, theories and of causation is correct, the understanding of the principles of change actually depends upon the persistence of the structures and mechanisms of nature within which and by which change is brought about. What principles do the changeless aspects of existence obey? There are, I believe, two classes of such principles. There are the principles of spatial and of temporal indifference, including the principles of inertia and induction. The latter is merely the principle that lapse of time does not, of itself, constitute a cause. There is the principle of numerical indifference, a principle concerned with the raw number of things of each kind that there are.

(A) Spatial and Temporal Indifference

In the classical world view it is a principle that there are no physical consequences of being in one place rather than another. Often, by moving something, we change it. Does not this fact refute the principle? Indeed not, because when this seems to happen it is the relocation of the thing with respect to other specific things and their influences, not the spatial relocation, which is held to be the cause of change. In the absence of all but one body, rest and motion of that body would be indistinguishable. As a test body, is further and further removed from all other bodies, its being in motion will be more and more difficult to distinguish from a state of rest: that is, its state will be unchanging. Motion is defined only within a system of other bodies, while change in motion is now naturally attributed to the influence of some, or all, of those other bodies. One wonders, indeed, what is the logical status of the principle of location indifference. Are there any conceivable circumstances in which it could be true that mere spatial translation was responsible for a physical change? To answer the question, it is necessary first to give a general account of those relations between things and materials which we distinguish as spatial relations, by trying to identify the minimal conditions which a set of relations between things must satisfy to be spatial relations.

It is possible to identify at least five distinguishable notions in the spatial conceptual system, roughly comprehended in various contexts under the general term 'space'. (1) There is 'enclosed volume or displacement', (2) there is 'place', (3) there is 'three-dimensional continuum of the material world'. The theory of this is geometry interpreted as a branch of physics. (4) There is space in the geometer's sense, 'a system of relations among bodies, obeying certain very general conditions'. The theory of this is geometry in Euclid's sense. Finally, (5) there is space as the system of relations obeying certain formal conditions, between a set of abstract entities, having no conceptual connection with things or places. The theory of these relations and entities is geometry as pure mathematics: geometry as conceived by Hilbert.* Euclid's Geometry might be looked upon either as a physics book or as a mathematical treatise. Under each view different aspects come to the fore.

I suppose the most central concept and the one with which to begin a discussion of this fivefold bouquet of notions, is that of space as a system of relations of a certain sort among material bodies.

I shall now try to formulate the characteristics that a set of relations between bodies have to exhibit to be distinguished as spatial relations. There are, it seems to me, two quite distinct ways in which this might be done. One method would be to use the notion of coexistence and to exploit the relation of betweenness. The other method would be to use the notion of translation. We are aiming at conditions which do not presuppose any particular dynamics.

The conditions for a set of objects to constitute a pre-dynamical linear

* D. Hilbert, *The Foundations of Geometry*, trans. E. J. Townsend, Open Court, Chicago, 1902.

spatial sequence, according to the first of these methods, can be set out as follows:

> Let *a*, *b* and *c* be the names of reidentifiable objects.
> Let *S* be a relation to be defined – the spatial relation.
> Let *B* be the relation 'between'.
> Let (*MEx*) be the modal quantifier 'there might be an *x*'.

The sense of *cBa*, *b* can be elaborated with some axioms which it must obey, though these do not exclude other interpretations.

> 1. $aSb = a \neq b$ & (MEc) $(a \neq c, b \neq c, cBa, b)$
> 2. $cBa, b = -aBb, c$ & $-bBa, c$
> 3. $-cBa, b = aBb, c$ v bBa, c

The continuous nature usually attributed to space cannot be represented in any refined way in this system. The best that can be done is the 'shoving in' principle, derivable from 1 above.

> 4. $cBa, b = (MEd)$ (MEe) $(dBa, c$ & $eBc, b)$

This principle is acceptable in a predynamic account of spatial relations since *a*, *b*, *c*, etc. are not 'fixed' with respect to any other object, nor yet with respect to any abstract frame abstracted from all actual objects. With our superior wisdom we can see that there are circumstances, where, because of the size of *a*, *b*, *c*, *d* and *e*, for *c*, *d* and *e* to shove in would involve the displacement of *a* and *b*. But at this stage the concepts of size and displacement have not been introduced. The concept of size involves a metric, even if only a comparative measure, which does not yet exist, and the concept of displacement involves the selection of one set of objects as the spatial referents. Actually, this step usually involves the introduction of some imaginary objects to carry the permanent basis of reference, such as the geographer's equator. 'Shoving in' is conceived here as a very primitive notion. Its use does not allow us to distinguish between a third object pushing in between two objects by displacing the others forward and backwards, like the method of entry used by an Italian to get into a bus queue, and an extraneous object slipping into a gap, queue-jumping in the German style. Thus, the concept of shoving in does not yet permit the distinction between occupied and unoccupied place, though it depends upon the distinction between occupied and unoccupied *space*. To achieve the former distinction some dynamic concept, properly a part of physics, like resistance to motion would be required, though this might be concealed by expressing the difference between two kinds of shoving in in terms of the different times taken to move into an occupied and an unoccupied place.

To put this another way: the modal quantifier used above, 'there

might be an x such that . . .', is used here in the sense of logical possibility, not physical possibility. It should be noticed, however, that,
though the quantifier is modal, it carries the sense in the definition of
coexistence, that *if* there is anything, it coexists with anything else said
to exist. So the set of relations defined by axioms (1) to (3) defines space
without time. But it is another aspect of the absence of a temporal concept that Italianate queue-jumping cannot be distinguished from
Teutonic. If all translations were instantaneous, we should need no
dynamic concepts, since the concept of more or less quick motion, and
hence of the differential causes of more or less quick motion, would not
be required. We shall see, too, that the introduction of a metric is also
impossible in total independence from some physical, that is, dynamic
concept.

Is there a second method of arriving at the principles governing the
minimal spatial concepts? I depend upon the notion of a translation. I
shall try to see how far the spatial concepts can be developed using the
notion of possible degrees of freedom of translation of just one object.
But even in its most primitive form the method of translation involves the
addition of some dynamic concept. One object alone, such as the whole
universe, cannot be said to have moved or be moving in any direction.
For a solitary object to be said to have moved, it must, for instance, be
known to have been acted upon by another object or by a force. By
hypothesis there is no other object. But how can the case of a force acting impotently on an immoveable object be distinguished from the case
of the one object in the world moving off under its influence? Surely we
cannot do without some other objects, *not under the influence of the force*, to
provide a frame of reference. This point has been made by Reichenbach
in terms of the redundancy of universal forces.* And these other objects
must be related by at least those relations which define a linear spatial
sequence. However, by the introduction of the concept of temporal interval, and by supposing at least three objects to exist, a conception of
linear space can be generated, without using the principles of the first
method above. A linear space can be generated by requiring that the
following conditions be met:

There is a spatial interval between the bodies a and b, if and only if
a third body, c, in moving from contact with a to contact with b, takes
some time. Later we shall see that simple contact is not an entirely
satisfactory notion, but for the purposes of this argument the unsophisticated notion will do. There is no need of a dynamic concept to
help distinguish the case of c moving from a to b from a and b moving
relative to c, because in either case, for the change of contact to take
time, there must be a spatial interval. This method of generating space

* H. Reichenbach, **6**, Ch. I, 225.

is *not*, however, predynamic. The notion of contact between bodies is properly a dynamic concept, related to the concepts of impenetrability and inertia. The notion of a translation taking some time does involve some minimal dynamic notion too, otherwise the third body could just push the limiting body along for an indefinite time or reach it instantaneously. There is not a second, predynamic method.

To generate a metric two further concepts must be added. The comparison of times will enable an investigator to say that c has passed from a to b in longer, shorter or equal time to that in which it has passed from d to e. But to be significant for a spatial metric, c must in each case come under the same cause of motion. So there is need for a dynamic concept as well, and for either a method or a convention by which it will be determined that it is the same force acting in each case, and that inertial mass of the test bodies is conserved. Only then do the times of translation become the measures of distance, and true in relativistic or absolute space and time.

A metric can be grafted on to the simple linear definition of space by adding a number of concepts. A dynamic concept is required to ensure that the metrical body enters only gaps. We can then define two bodies a and b as being the same distance apart, if identical test bodies $cl \ldots cn$ can, without encountering resistance, be placed between a and b, but to insert n + 1 such bodies between a and b would meet with resistance. By reapplying a similar procedure to the c's, we can both ensure that all c's are the same size, and that our estimates of distance can be made progressively as accurate as we please by the progressive reduction in the size of the c's. By a further extension of these procedures we can justify the technique of using a single test body, c, inserted n times between a and b, under the physical condition that c satisfies some convention and some laws as to remaining the same size, having the same inertial mass, and so on.

The final step in this section will be to ask whether the conditions set out lay down unique desiderata with which only spaces agree. Colours in the spectrum seem obvious candidates for an alternative interpretation of the terms a, b, etc. in the principles of the first, non-dynamic, method. *A* and *b* are different spectrum colours if a colour can be found between them, and this seems to meet the conditions of the first principle. Similarly, the conditions of the second and third principles can be met by a concept of 'betweenness' for hues. But with the essential 'shoving in' condition a difficulty appears. Human colour discrimination is not infinite so that a situation can be reached in which, when a hue is alleged to be inserted between two discriminable hues, the inserted hue will not be discriminable from either of its neighbours. There are similar difficulties about the discrimination of exact place in visual space too, but space can be divided in the imagination in a way in which the spectrum of colours cannot. In other words, to demand that the 'shoving in' condition is obeyed ensures that the space so generated is a continuum of a sort, but only of the lowest order of infinity. Such a space would not contain sufficient points to serve as the basis for classical kinematics. The reason is this. A set of points formed by division of intervals leads only to a set which has the order of infinity of the rational numbers, that is, of the integers and rational fractions. But it is easy to show that, in, say, the interval 0 to 2, there are points, say, that marked off by the diagonal of a square of side 1, which cannot be represented by any rational fraction. Kinematics requires a continuum of the order of the real numbers, and our desiderata will not yield such a one. Nevertheless, it will be a space, and the 'shoving in' condition will ensure that is of the order of infinity of the rational numbers.

The translation generation of space might also seem to find some sort of an analogue with colours. Colours could change. But here we do encounter a philosophical problem. It concerns criteria of identity. In the generation of spatial concepts the objects were individually discriminable, and so the notion of 'different object in the same place' had application. By being in the same place as another object, that is in the place another object had formerly occupied, that is, by coming to bear the same spatial relations to another object, an object does not become the original place holder. Thus, places are distinguished from objects occupying them. However, hues and their relationship to the other colours in the spectrum cannot be similarly distinguished. To be such and such a shade of yellow is, *inter alia*, to be in such and such relations with other colours. An analogue with space could only be set up if it could be said of some yellow that it was different from another in that it was formerly red, while the other, seemingly identical in hue, was formerly blue. But there would have to be no visible trace, in the yellowness of each, of their origin. Such a distinction is empty. The democracy of colours contrasts with the aristocracy of objects in space, whose current place and whose existence as separate objects with distinct ancestries are logically distinct.

Finally, let us compare 'having a certain colour' with 'being in the same place'. In the previous arguments we have been testing the idea that colours might satisfy formally similar conditions to those whose satisfaction by objects constitutes space. Our arguments showed that the relations between hues were not formally isomorphic with the relations between objects. But this still leaves the possibility that some other attribute of objects might show a formal similarity to place. The argument against this supposition is very simple. It is characteristic of place that two objects cannot occupy the same place at the same time.

But clearly, potentially infinitely many objects could have the same colour at the same time. In this way all genuine attributes are distinguished from place. We should, though, be obliged to say that there was a space when the shover-in was not a *material* object and occupied the same place as some other object, just because it would vacuously satisfy the betweenness requirements. The place of an object is the set of relations it bears to other objects, which are spatial relations. Since these relations can be satisfied by indefinitely many distinct objects we come to treat the place as having an existence independent of the objects that from time to time occupy it. It is an easy step then to the error that there is a place independent of any object at all, that is, to the idea of absolute place. The only objects from which place is truly independent are the particular objects occupying it. The objects which are the other terms of the betweenness relations cannot be entirely dispensed with, though, since any of these occupies a place contingently, that it is this or that particular object which is the term of the spatial relation, is neither here nor there. The space of physics now becomes the totality of places whether occupied or not. As Gassendi once put it, 'Space is not just the places where objects are, but also the places where they might be and are not'.* Space is not only the totality of places where things can be, but also the totality of places where things could be, were the laws of nature different from what they are. The consideration of the spatial relations independently of the particular bodies of the material world gives yet another sense of 'space', that of which geometries of the Euclid kind are the theories. Finally, the consideration of these relations merely in their formal character gives the mathematical sense of 'space', that of geometries of the abstract kind.

So the method of generating spaces does seem to ensure that only spaces are so defined. The five spatial concepts are now readily distinguishable. If the parts of an actual object are themselves objects satisfying either set of conditions, then the object has a volume, a spatial extension. If they do not, it has not. One of the principal requirements for a kind of object to be a material thing, that no two objects of that kind be in the same place at the same time, is an additional requirement on an object having spatial extension. This is another way of making the point that the 'shoving in' condition is pre-dynamic and so more general than any system of physics. The problem which prompted this investigation into the conditions under which a relation is properly to be called spatial, was the question of the status of the principle that mere spatial relocation could not lead to consequential physical change. Is this a matter of contingent fact to which exceptions might be discovered, or

* P. Gassendi, *Syntagma*, Lyon, 1658, Vol. I, Part II, Section I, Book 2, '*De loco et tempore*'.

is it a consequence of the way spatial relations among things are marked off from all other relations? It should be clear that it is possible to lay down conditions for a relation which intuition admits to be spatial (what else would count as a proof?), which do not involve any dynamic concepts. So change in these relations alone is not a cause and need not be an effect.

A place in space can be identified by reference either to the spatio-temporal framework, which depends upon the existence of material things in general, or if independent of the spatio-temporal framework, by reference to one or more qualitatively identifiable and numerically individuatable particular things. A place in space cannot be independently identified, if by this is meant a location which is identifiable in independence both of material things in general, and of any particular material thing. Now space, that well-defined system of relations, though it is dependent on material things in general, being a certain subset of all possible relations between things, is not dependent on any *particular* material thing. Hence the location of some special spatial point, point of space, such as an origin for three-dimensional Cartesian axes, is quite arbitrary. The system of relations that is space, is constituted by just those relations which are not dynamic, and are not associated with causes. So no one nexus of this set of relations can be qualitatively distinguished from any other. Hence, any material thing can constitute the basis of the choice of origin for a spatial system. This, to be historically fair, should be called Cusa's Principle.* Nicolaus Cusa argued that there could be no centre to the universe, since from whatever point one chose to view the Universe there would be things above, below and around one, so that there would be no empirical difference between standing on the Earth and looking up at the Zenith, and standing on the Zenith and looking 'up' at the Earth. Any point is capable of serving as the centre of the Universe, that is, as the origin of the spatial system.

The Principle of Location Indifference does have corollaries of considerable importance for the philosophy of physics. The principle that it is impossible to locate oneself independently of things follows directly from it. If there were physical consequences of mere spatial relocation, that is, if there were mutual variation between some observable and measurable parameter and an object's spatial co-ordinates, then merely by noting the value of the parameter, one's position could be determined, and by maintaining that parameter constant one's position could be maintained, in absolute space! Temperature and expansion are often mutual covariants so expansion can be used as a measure of

* Nicolaus Cusa, *Of Learned Ignorance* (translated by G. Heron), Routledge and Kegan Paul, London, 1954, Bk. II, Ch. XI.

temperature. As we have seen there *can* be no parameter which varies merely with the values of the spatial co-ordinates, because their values are arbitrarily dependent on a choice of metric and, it follows from Cusa's Principle, can be increased or decreased merely by the arbitrary will of the co-ordinator.

However, ratios of distances are independent of units and metric. Since there is no co-ordinate parameter that varies with the place of a body, independently of what other particular bodies there are about, simple location can only be with respect to the spatial, noncausal relations with other bodies. But, perhaps ratios of distances are exempt from this, e.g. Kepler's Third Law. Two questions can be distinguished here. Is there an absolute spatial measure? Is there an absolute spatial location? To the latter, our answer must be no, there could be no absolute spatial location, given the distinction of the space relations above. But to the former the answer is, surely, yes, there is. It is the space ratio. The simplest case is with three bodies A, B and C in a linear space. The value of that ratio is a constant, however the chosen metric of that space may be changed, though the values assigned to measures of the intervals AB and BC will not be constant. Indeed, the character of a space is defined, in part, by the exact form which the invariant takes, for instance, in special relativity *the* invariant is not a simple space ratio. It would follow from the general tenor of the argument, that changes in, say, the simple ratio of intervals in a linear space ought to be accompanied by a covariant physical change. In short, though to accomplish changes in intervals no dynamic interference is required, changes in the ratios of intervals would, and indeed on this line of argument, must require dynamic interference. Changes in the value of the ratios of spatial intervals in simple linear space calls for a causal explanation, and changes in the appropriate parameter in other more complex spaces similarly must be explained.

Mere position then has of itself no correlated property or disposition of bodies. This has not been a feature of all scientific systems, and for instance in Aristotelian physics mere position was correlated with a particular degree of power to move towards the centre of the universe, with downward appetition.

To Cusa's Principle of Location Indifference another must be added, the Principle of Isotropy, that there are no physical consequences of mere orientation. Direction does not by itself, and without reference to other properties of bodies than their spatial position, figure in the antecedent of a causal law. Merely to turn from Jerusalem to Mecca has, of itself, no physical consequences, though the spiritual consequences may be profound. Of course, there are physical changes consequent upon such a deviation, but influences such as forces and field are brought in to account for them. It is not merely that a magnetic compass is in a different orientation with respect to the Earth's axis that causes it to deviate from its north-seeking orientation, but distortions in the magnetic field are postulated to explain this. This principle is perhaps of even deeper significance than the Principle of Location Indifference. Cusa, as it is no doubt clear from the sketch of his argument for the relatively of spatial origins above, depended upon it.

In actual science, the Principle of Isotropy seems to function straightforwardly as a methodological principle. By laying it down that mere orientation is not a cause it puts an onus upon a practising scientist to invent or detect an influence which operates causally and differentially with orientation. In fact, it imposes the need to invent spatially oriented influences. It is not, though, a constitutive principle, because it does not lay down in advance just what kind of influences these must be. For this reason I prefer to use the word 'influence' rather than the word 'force' in these and similar contexts. 'Force', narrowly understood, is a strictly mechanical concept, and more widely understood is still, to this day, heavily infected with anthropomorphic notions of power and effort exerted. Its use inclines one to a push-pull view of the interaction of matter, a shover and hauler point of view, and this, as will emerge in the last chapter, prejudices metaphysics in the direction of a kind of influence which is not typical amongst bodies.

Why should the Principle of Isotropy be accepted? Arguments in favour of the Principle of Isotropy turn on the same point as do arguments for Cusa's Principle, namely that space is just that set of relations among things which have no dynamic character, and changes in which have no effects. Our conceptual system is so structured that a change in relative position is never, of itself, responsible for any physical change. Should a physical change occur on a change of position, that is, for the Principle of Isotropy, a change of orientation, a non-uniform force or influence is invoked to explain the change. Since space has no dynamic character, a change of orientation of a body, with respect to space, that is, with respect to the system of relations considered in isolation from any *particular* material things, could have no physical effects, since all physical effects are caused by influences from particular material things. So orientation with respect to space is as arbitrary as is position with respect to space. But, considered with respect to particular material things, change in position, and change in orientation, will surely be accompanied by effects. The purity of space is preserved by the introduction of influences, forces and fields, in association with the particular things with respect to which the physical changes can now be explained.

I turn now to consider somewhat similar temporal principles. What I shall call 'The Principle of Epoch Indifference' could be formulated as

'There are no physical consequences of existing at one time rather than another'. Upon this principle ultimately hangs most of the rationale of the account of scientific thinking I have been giving in this book. In one way, it is the deepest scientific principle, that the mere effluxion of time brings about no changes, or, as Kant put it, there is no differentiation of empty time.*

There are two important time concepts with which our temporal experience is comprehended. They are 'event' and 'duration' .In the early part of this book I was concerned to show how a certain account of the logic of theories, of the nature of laws and of their relation to the evidence which counted for or against them, was connected with a certain metaphysical predilection for events. I was concerned to show, too, how epistemological phenomenalism, a theory closely connected to the logical theory, led to the belief that events are what are the real and only proper subject matter of science. Thus, the world is held to be ultimately analyzable as a Heraclitean flux, objects being nothing but bundles of changes, and durations nothing but sequences of events. The abandonment of this epistemology, and the logic that goes with it, allows one to abandon the metaphysics too, and to elevate the concept of duration to an equal status with change. The Principles of Indifference for Time are really none other than the laws of endurance, just as the Principles of Causality are the laws of change.

What are events and durations? Here it will be well to pause to consider the nature of time. The lapse of time is characteristically known to us in two ways: by our perception of duration, and by our power to be aware that the present state of the world, and of ourselves, is different from what it previously was, that is, by our awareness of change. We can be, and indeed often are, aware of duration when there is no change, but not of change when there is no duration. It is a common mistake to suppose that memory is involved in the awareness of change. All we need to know, to know that a change has occurred, is *that* the present state of the world is different, not in what way it is different. We no more need to compare our present experience with our memory of the past, to be aware that change has occurred, than we do to compare our colour perception with remembered samples to know what colour we are seeing. It is perhaps this mistaken view about the recognition of change that has led philosophers in recent years to overlook the vitally important temporal concept of duration – temporal persistence without change. It might, perhaps, have been unthinkingly assumed, that if the past and present state of the world are the same, then their comparison will yield no awareness of difference. Consequently, it might be supposed that they would not be temporally

* I. Kant, (*see* 241), Transcendental Dialectic, Bk. II, Ch. II, Section 2, A427.

distinct and, indeed, that a world in which no change occurred would not properly be said to persist in time. It might be argued that, in fact, duration can only be known by reference to successive change, so that the paradoxical conclusion could be reached that unless something changed, nothing else could endure. But this, like so many similar arguments, is based upon a confusion between the fact of duration and the conditions under which duration can become known, and in particular with the measure of duration. Of course, one cannot compare the duration of things without successive change to act as measure. But it does not follow from this that duration is successive change.

On the other hand, it used to be thought that change was less fundamental than duration, that while duration was possible without change, change was not possible without duration. It was argued, for instance, by Kant,* that to know that a change had occurred, and not merely that one thing had been substituted for another, an enduring entity had to be known which remained the same throughout successive change. But this seems to me clearly fallacious. I am quite unable to see why one sort of change should not be the background for the awareness of another sort. There does not have to be persistence in the same state to enable a change in some other state to be noticed. All that is required is that the reference state shall change in some standard manner.

Parallel to the epistemological argument, and its modern analogue, the belief in the necessity for a constant subject of predication, there is a metaphysical argument. Again it hinges on the distinction between the *substitution* of something different for a given thing, and the *change* of a thing from one state into another. If there is to be change, rather than continuous substitution, then enduring entities unchanging in certain characteristics must be the bearers of those that change. But one might well ask, why should the conceptual system of change and enduring entities be preferable to a conceptual system in which the world is treated as a succession of fleeting individuals, each unchanging but being generated and passing out of existence in a flash? This is not the Heraclitean world. In that world there are individuals which endure but they are all undergoing modifications at all times in all respects. The Heraclitean world derives its individuals from the application of practical rules as to the limits of change within which individuality is preserved. Such a rule might forbid reference to the ashes in a funerary urn by the use of the name of the former person, of whose cremation they are the product. This, and similar rules, would be applied under the general rubric that any individual would have to be spatio-temporally continuous to count as an individual. But in the conceptual system envisaged above, individuals are not modified, new individuals replace

* I. Kant, *op. cit.*, Transcendental Analytic, Bk. II, Ch. II, 3A; A180–5.

I

the old. Such individuals need not necessarily be of instantial duration only. The system might even be sufficiently liberal to allow the same individual to enter into different relations with other individuals, provided there were no changes in its constitution consequent on the change of relations, that is provided the relations which changed were purely spatial and temporal. So far as we known there are no individuals of which we can say for certain that they are quite without internal changes in the world we are trying to understand, so the substitution-of-individuals scheme would seem to be less preferable than our own, in which both fleeting individuals (events) and enduring individuals* (things) figure fundamentally.

At this point, I come to the most fundamental and the most powerful of methodological principles. It is this.

Enduring is in no need of explanation.

We are not required to explain the fact that something remains the same; only if there is a change is an explanation called for. Philosophers have understood that changes require causal explanations, but misled by the event metaphysics, they have misunderstood enduring. They have treated it, tacitly or explicitly, as a set of like or identical events, whose likeness or identity require explanation.

Philosophy of science is prone to pointless scholasticisms. The present morass of deductivism is the successor to the even more scholastic morass of inductivism, and its great problem. The 'Problem of Induction' arises because it is supposed that we need an explanation of evidence for, or guarantee of sameness, of unchange. The problem is generated by asking 'What guarantees have we that the regularities and existences we have so far discovered will persist in the future?' It is then suggested that anything might succeed any previous state, and in the metaphysics of events (each independent of the previous one) this would indeed follow. But sequences of events are the products of generative mechanisms, regularities of appearances are the consequences of the endurance of internal structures. Persistence, endurance and unchange require no explanation and particular cases of them have no causes. Consequently, no evidence for belief in persistence, in endurance, is required; only identification of that which is enduring. Because enduring has no causes there can be no call for evidence to support belief in the continuance of endurance, other than that something has endured, and that is enough. Only if things decay, if processes change their direction, if regularities go awry can we ask for a cause, and so only for the likelihood of future change can there be evidence – unless a body is acted upon it endures in its state. There is no spontaneous

* Notice the conceptual connection between existing as a thing, and enduring.

generation or decay. That is the basis of our conceptual system. There is no problem of induction since no evidence is required to support the belief that regularities persist. There are no physical causes for the maintenance in existence of a physically isolated structure, and there are no physical causes for the continuation of any processes which depend upon and issue from a continuously existing structure.

But there are processes which seem to require continuous causality to persist in the same manner and there are plenty of structures which seem to require maintenance to prevent them collapsing. Does the repairing of a house, or the flight of an aeroplane, violate the principle of indifference? Indeed not. The fact that the existence, unchanged, of a structure is the result of an equilibrium of forces shows, not that the structure had a spontaneous tendency to decay, but that a destructive influence was operating upon it. The most friable sandstone will not need maintenance where wind and weather cannot reach it. Of course, one maintains a permanent structure by putting on cement as fast as it flakes off, but the putting on of cement is not an action taken against the spontaneous decay of a house, but against the forces which tend to destroy it. Thrust is required to maintain an aircraft in flight only because it is subject to drag. There is no paradox in the fact that in the practical application of our conceptual system, things that endure are usually maintained. Our understanding of the necessity of maintenance derives precisely from our understanding that decay is not spontaneous, but is the effect of the action of causes upon the thing to be maintained. In the absence of such causal influences the structure would remain unchanged.

'Change and decay in all around I see.' Is there nothing which is not subject to Manichean influence? Are there in the world no structures which endure because no causes operate upon them?

There are. The Principle of Endurance governs the structures, entities and processes which we imagine for the purposes of explanation. The changing is explained in terms of the unchanging. A sequence of changes is given scientific treatment when it is shown to be explicable as the effects of the interactions of changing structures of unchanging things. Whatever is conserved in such explanations is not conserved by a balance of influences. It is conserved without causes. In the older physics it was not possible to ask for the cause of the conservation of mass. It was a subsidiary principle of endurance that mass could neither be created nor destroyed, so consequently its conservation was not a phenomenon which had a cause. The essence of modern genetics is the idea that the multiplicity of manifested characteristics of the individuals of a species are the product of the combination and recombination of fixed hereditary units, the genes, which do not change in the process of reproduction, but are redistributed in successive generations. It turns out that what is actually conserved in the generation of plants and animals is patterns, i.e. structures among individuals, not individuals. Chemical change is explained by unchanging atoms, atomic change by unchanging fundamental particles.

It is now possible to see how the principle of the non-causality of time, and the acausality of endurance are connected. To endure is to be in a mode of existence for which the only change is in the temporal parameter. Change in the temporal parameter is itself marked, but not constituted by, change in the manifested qualities of other things. If there is a change only in the temporal parameter of a thing, then other things must be changing and not the thing which endures. Since, as will emerge in the next chapter, everything must be regarded as influencing in some degree everything else, true unchanging endurance is more a theoretical ideal than a matter of fact. It constrains our conceptions.

Just as there was a principle of isotropy so there might seem to be a principle of temporal symmetry, and physicists have sometimes talked as if this were the case, as if direction in time was as arbitrary a determination as direction in space, as if then and now were as arbitrarily ordered as up and down. The fact is, however, that for a great many laws of nature the substitution of $-t$ for t leaves the law unchanged. In practice the significance of this fact is confined to the unexciting conclusion that most processes are reversible.

(B) *Parity*

Recently, one of the most fundamental principles of indifference has moved from the status of a principle to that of an empirical hypothesis, widely held to be false. The principle asserts that for all screws and coils and like things, and cyclic or rotatory phenomena, in general phenomena having 'sense', there are no physical consequences of being right- of left-handed. In short it was once taken as a principle that the laws of nature are the same for right- as for left-handed screws, for entities with a right- or a left-handed spin. And indeed it was widely held that as a matter of principle it could be supposed that a universe in which the sense of all things and processes was the reverse of our universe, would be indistinguishable from it. Sense, left- or right-handedness, it was held, could not by itself act as a cause. Which spin a thing has leads to no differentiating physical consequences, by which left- or right-handed senses could be differentiated. For instance, there is no predominant colour shown by screws having a left-handed sense. This seems a very obvious and indeed simple principle of indifference in keeping with the Principles of Location Indifference and of Isotropy. A corresponding principle would lay down that the cause of a thing being right- or left-handed must be looked for in the sense of its generating mechanism, i.e. only from sense can sense arise. The sense of a screw is inherited from some sense-feature of the initial conditions of its production, so, for instance, the colour of the causal mechansim cannot be responsible for the sense of its products.

It is worth noticing that both the principle of location indifference, Cusa's Principle, and the Principle of Isotropy have been held to be false empirical hypotheses. In Aristotelian physics it was supposed that changing a body's place changed its appetition for the centre of the universe. Indeed, the whole theory of a natural place runs counter to the Principle of Spatial Indifference. In the non-spherical universes of some early cosmologies, orientation was not without its physical consequences, but in our conceptual scheme one can only describe oneself as finding one's location with respect to other bodies; not by means of internal changes brought about by mere changes in location in space. One can only orient oneself with respect to other bodies and changes they bring about. One cannot orient oneself by means of the observation of internal changes caused by mere change in orientation in space. These are not matters of fact but depend, as I have shown, on how we demarcate that set of relations among bodies we call 'spatial'.

It was thought that the screw-sense of a single entity could not be determined in isolation from other bodies because it was thought that there were no physical consequences of rotation or organization in a different sense, in the absence of other bodies to influence and to be influenced by. In short, the sense of a screw, like its location and orientation could only be determined with respect to other bodies: in particular with respect to their sense. Lately there have appeared experiments which do seem to show, at first glance at least, that there are physical consequences of differences in sense which are not further instances of sense. The atomic physicists' property of 'spin' is differentiated by left- and right-handedness. An experiment should have shown, had left- and right-handedness been without physically discernible effects, that if the spins were all the same, other physical non-screw-sense phenomena would be randomly distributed. But an experiment was designed in which Cobalt 60 atoms were lined up for spin, and their products, beta-particles and neutrinos should have shown no preferred direction. But they did show directional differentiation.

The interpretation of this experiment was suggested by the hypothesis that had led to its being performed, namely the differential emission of beta-particles and neutrinos was held to be a physical consequence of the particular screw sense imposed on the Cobalt 60. Of course, no such interpretation is mandatory. That it seems to be so derives, I think, from a confusion between the Principles of Screw Indifference, namely that differential screw-sense has no differentiable physical effects, and the empirical hypothesis that right- and left-handedness of screws do not derive from differentiable physical causes. The latter hypothesis, it seems, should be regarded as falsified by this experiment, and so should others like it. As an alternative interpretation for the experiment, intuition would suggest a hidden cause, having both a sense- and non-sense component, the former component producing the differentiation of screw-sense, which, according to the principle of indifference, has no further consequences, and the latter the differentiable effects. By introducing hidden parameters we can preserve the conservation of parity. But clearly we are under no obligation to do so, since the former status of the Principle of Parity as a necessary truth, reflects the attitude of scientists, and not any analytic, logical structure in the proposition itself.

Numerical Indifference

The principle states that the number of entities of a kind is not subject to the necessity of law. This principle emerges in a number of contexts. It seems that the fact that there are nine planets is an entirely accidental matter, but the fact that their orbits are elliptical is not. At least this is the way we are inclined to think about the matter. The principle can also be made to emerge, in what is for philosophers, perhaps a more interesting way than the mere fact that scientists tacitly assume it. It can be connected with the range theory of probability. The range theory avoids certain inductive problems by shifting the grounds for probability statements from the ratio of numbers of individuals to the ratio of the number of subclasses into which an individual can properly be said to fit, to all possible subclasses of individuals of that kind. Constraints on the number of possible ultimate subclasses in the universe are derived from the known laws of nature. The class of bodies which are both material and *repel* each other according to an inverse square law, is not a possible ultimate subclass. The problem situation arises in this way: suppose one finds that a certain individual in a certain state F, is classifiable into three out of seven ultimate subclasses, each of which contains a G component, the seven ultimate subclasses being appropriate to that kind of individual. The probability that another F individual will be in a state G is then $3/7$, according to the range theory. But suppose only one other individual having the characteristic G falls into one of the F subclasses, while millions that are G-ish fall into one of the four subclasses that are not-F-ish. Surely the preponderance of the number of individuals will affect our estimate of probability? To suppose that it does not so affect the estimate seems to me to presuppose that the number of individuals in a class is not subject to causal law, for if they were subject to causal law then a five-membered class of Fs would be a different class from a six-membered class of Fs, in the appropriate sense, that is would constitute a different natural kind, which is absurd. For the purposes of the application of probability theory the answer has already been given in Chapter 8, where it was seen that a probability estimate depends upon two quite different kinds of evidence: the nature of the individuals and their natural kinds, and statistics. For the range theory to *include* statistical evidence in the considerations determining what ultimate subclasses are to be considered, would be to be involved in the absurdity of treating the number of members in a class as a property of the members. Is it then merely the danger of the fallacy of division that prevents us treating number as a characteristic, subject to causal law? The other alternative is to suppose that all ultimate subclasses, admitted by our current knowledge, have the same number of members, which is just plain false.

Galileo put number among the characteristics inseparable from the notion of matter. Atomism requires that the number of ultimate individuals remains the same, since if atoms are truly indivisible they cannot be destroyed by the ordinary physical processes of dissolution and decay. The vanishing of an atom would be as much a miracle as its creation, in the view of the Corpuscularian philosophy. This is the short way with the principle. If there is no question of a change in a characteristic then there can be no question of the characteristic being subject to causal law. Is number then something which does not change, and is it therefore subject to a principle of indifference? There are two cases.

If anything is constant in number it would be the Parmenidean One, for example a concourse of unchanging atoms. Whatever is not constant in number must consist of individuals of other types. But, perhaps even the numbers of Aristotelian individuals, which are subject to generation and corruption, are subject to causal law? One suggestion of how Aristotelian individuals might still obey a principle of numerical indifference, I owe to Wilfrid Sellars. It is that number may be transitive through the initial conditions, that is it never appears in laws, but is transferred through the particulars which express the initial and final states of systems, in short it is a feature of the boundary conditions of systems. Consider the application of a law of binary fission. The law is not a conjunction of assertions like 'Four individuals must become eight' and 'Eight individuals must become sixteen'. It applies to any number of individuals, and its application allows one to pass from any numerical initial condition to the final state which a system, say of amoebae, will be in, which is characterized by that initial number of individuals. Thus the cause of their being sixteen, is not that there were eight, but that each, of however many individuals, has divided.

But this does seem to need a measure of qualification. Demographers and ecologists are prepared to assign causes to increases and decreases in populations. If number was wholly immune from causality how could one seek to find means for reducing or improving numbers, for instance seek for means to control the world's population? Clearly, here, what is under causal law, is again not raw numbers of individuals, but differences in raw numbers. Further, in contexts where generation and dissolution of individuals is admitted (the pediatrics and geriatrics of individuals), in a certain sense to be investigated, we do seem to allow for a causal account of numbers. But in most contexts, to be an individual, subject to change, is to have some degree of permanence. It is, at least, to be enduring for a little time, and so to be an Aristotelian individual. And then the numbers of individuals are not subject to causal law, because they do not change. They are transmitted whole, through the series of initial and final conditions that constitute the world process.

There are two possible kinds of individuals not subject to change. There are genuine atoms, Parmenidean individuals, which persist through change but do not themselves change. Complexes of these can be formed and they are the usual individuals of our common experience, and these can change, that is be the subjects of incompatible predicates. Of course for any individual to be properly the subject of predication of two or more incompatible predicates, it must endure in time, and so have some degree of permanence. We have already noticed the existence of such Aristotelian individuals. But each of them had its moment of conception, and each of them will have its dissolution. So for the *change* in numbers of such individuals we may ask the cause, and seek the law of reproduction. The other kind of individuals not subject to change are events. Being ephemeral and themselves changes they are not subject to change.

The very same Parmenidean individuals to which I have been alluding in this section are none other than those which were identified above as the basis of explanation: the final and ultimate things. It is to a deeper account of them, that I turn in the rest of this book.

SUMMARY AND BIBLIOGRAPHY

It is argued that predynamic and non-causal concepts of space and time can be created, thus setting up certain principles of indifference, from which the necessity of an inertial concept can be derived. The very same principles are shown to lie behind the scientists' way out of the problem of induction.

For the relation between principles of indifference and explanation see

217. A. I. FINE, 'Explaining the Behaviour of Entities', *Phil. Rev.*, **75**, 496–509.

For a discussion of the viability of a set of spatial relations extracted from the actual relations between things, see

218. D. M. ARMSTRONG, 'Absolute and Relative Motion', *Mind*, **72**, 209–23.

For the ordinary notions of space as in Concepts 1 and 2 in the text

219. D. LODNEY, 'The Concept of Space', *Phil. Rev.*, **64**, 590–603.

The very general question of boundaries between spaces is discussed by

220. A. M. QUINTON, 'Spaces and Times', *Philosophy*,
221. M. HOLLIS, 'Times and Spaces', *Mind*, **76**, 524–36.

A set of general conditions for time, consonant with my predynamic account, but more fully worked out as a time system is

222. M. BUNGE, 'Physical Time: The Objective and Relational Theory', *Phil. of Science*, **35**, 355–88. This paper has a very valuable bibliography.

For the logic of the general sense of 'event' as used in this book, see

223. F. I. DRETSKE, 'Can Events Move?', *Mind*, **76**, 479–92.

and for the general conditions of order on temporal sequences

224. K. W. RANKIN, 'Order and Disorder in Time', *Mind*, **66**, 363–78.

I am not concerned, in this book, with general questions in the philosophy of space and time. Valuable recent books are

225. H. REICHENBACH, *The Philosophy of Space and Time*, Dover, New York, 1958.
226. A. GRÜNBAUM, *Philosophical Problems of Space and Time*, Alfred A. Knopf, New York, 1963.
227. G. J. WHITROW, *The Natural Philosophy of Time*, Nelson, London, 1961.
228. R. SWINBURNE, *Space and Time*, Macmillan, London, 1968, particularly Chs. 1, 2, 10, 12, 13 and 15.

For the derivation of the principle of inertia from principles of indifference

229. J. C.-MAXWELL, *Matter and Motion*, Longmans, London, 1925, p. 28.
230. G. J. WHITROW, 'On the Foundations of Dynamics', *BJPS*, **1**, 92–107.
231. G. W. SCOTT-BLAIR, 'Some Aspects of the Search for Invariants', *BJPS*, **1**, 230–44.

is a general discussion of the problem of invariants.

Limitations on the concept of the basic individuals under the demands of endurance and transience are set out in

232. V. C. CHAPPELL, 'Sameness and Change', *Phil. Rev.*, **69**, 351–62.

I 2

The philosophical discussion of parity can be found in

233. I. KANT, *Prolegomena*, translated by P. G. Lucas, Manchester University Press, 1953, §13. For discussion of Kant's views see

234. N. KEMP-SMITH, *A Commentary to Kant's Critique of Pure Reason*, Macmillan, London, 1923, 161–6.

235. H. WEYL, *Symmetry*, Princeton University Press, 1952, pp.3–38.

A particularly clear account of the physics involved in the non-conservation of parity is

236. P. MORRISON, 'The Overthrow of Parity', *Scientific American*, **196**, no. 4, 45–53.

The problem of raw numbers has not been much discussed. For the view that there are causes of the raw numbers of entities see

237. B. SPINOZA, *Ethics*, Part I, Prop VIII, Note II, Premise 4, and for the view that certain raw numbers are not accidental, but necessary *a priori* see

238. A. EDDINGTON, *New Pathways in Science*, Cambridge University Press, 1935, Ch. XI.

10. The Ultimate Structure of the World

When we look at a thing, we must examine its essence and treat its appearance merely as an usher at the threshold, and once we cross the threshold, we must grasp the essence of a thing; this is the only reliable and scientific method of analysis.
'*The Works of Chairman Mao*', I, 119

The Argument

I

REGRESSES OF EXPLANATION

1. *Access to the Ultimate is by Regression*

(*a*) Regress of explanatory mechanisms $\Big\langle$ microexplanation / macroexplanation

(*b*) Regress of causes $\Big\langle$ past / future

2. *Epistemic Antimony* against knowledge of ultimate ends and beginnings.

(*a*) If *e* is the first event, and *k*-kind events are those we could know; either *e* is *k*-kind, then regress of causes from the present state of the universe to *e*, leads beyond *e*, i.e. it is not knowable that *e* is first,

(*b*) Or *e* is non *k*-kind, then the regress of causes from now cannot lead to *e*, therefore *e* is not knowable.

3. *Regress of Microexplanations*

(*a*) *Problem.* How are parts related to wholes?
Comment (i) Attribution of perceived qualities to parts will not do, because of the secondary character of many if not all qualities.
Comment (ii) Primary quality theory depends upon forming a conception of ultimate entities as teleiomorphs of perceived things.
(*b*) *Proof of Unsatisfactoriness*
Lemma. Definition of power.
To say *A* has power to *B* is to say *A* is of such a nature that, in the appropriate circumstances, it will do *B*. Can be analysed as an unspecific attribution of a certain nature to *A*, and a specific attribution of a disposition *B*.

II

ANALYSIS OF NOTION OF 'POWER'

1. *Two Paradigms of Action:*

(*a*) Cartesian: the origins of action are wholly extrinsic to the actor;

(b) Helmontian: the origins of action are wholly intrinsic to the actor.
    (i) Neither paradigm is purely exemplified in nature, nor do the applications of these paradigms match the inorganic/organic distinction;
    (ii) The Cartesian paradigm is an intimate part of positivism.

2. (a) How powers are distinguished from dispositions;

  (b) Analysis of power-attribution: (antecedent is stimulus conditions and not enabling conditions)
    (i) specific dispositional conditional;
    (ii) unspecific nature categorical,
  (i) and (ii) are conjoined by the as yet unanalysed connective, 'in virtue of'.

  (c) The distinction between stimulus conditions and enabling conditions is explained, the latter leading to knowledge of natures, and change in natures;

  (d) Liabilities have the same analysis as powers, but differ from them in falling under the Cartesian rather than Helmontian paradigm, i.e. stimulus conditions are extrinsic.

3. Connection between powers, natural necessity and nomological character of law statements.

4. (a) The critique of powers-ascriptions as *ad hoc*, trades upon Humean concomittance;

  (b) A powers ascription leads to
    (i) a further explanation of dispositions, i.e. the study of overt reactions under changing conditions;
    (ii) an account of the essence or nature of the substances involved.

5. *Criteria for Ascription of Powers*
  (a) Instances of the disposition must be known (i.e. some *tests* favourable);

  (b) There must be a ground for belief in the rôle of the nature of the materials involved in the manifestation of the disposition. (b) is not usually sufficient, e.g. there is a need for clinical trials of drugs of 'correct' composition.
Rebuttal of power-ascriptions may involve either (a) or (b), i.e. apparently successful tests are compatible with the materials not having the right nature.

6. *Relations of Powers and Natures*
  I. (a) The permanent possession of a power follows from the temporal self-identity or endurance of a nature;
    (b) (i) Cases of the relation of power and nature being necessary, i.e. where loss of power implies change of nature;
      (ii) Capabilities are powers where the power/nature identity connection is not so strict.
    (c) Tendencies are powers which
      (i) can be exercised (reference to enduring nature);
      (ii) likely to be exercised (reference to enduring enabling conditions).

7. *Identity of Powers*
  In time.
  (a) *Constant Powers*
    (i) The same power can be exercised again and again;

(ii) Differences in manifestations are often attributed to circumstances and not to a change in the power.

(*b*) *Variable Powers*

(i) Some powers are said to decline or augment.

Query 1. Is the distinction between constant and variable powers based upon a distinction between constant and variable *natures*?

Answer. Yes. In particular ultimate powers are constant in virtue of enduring natures.

Query 2. If the dispositions are different and the nature is the same, is each a different exercise of the same power?

Answer. Usage does not give an entirely clear reply.

(*c*) Generally, identity of power depends upon endurance of nature, and this sanctions the ascription of powers when they are not being exercised. Sometimes the nature of the material, in virtue of which it has a power, can be ascertained without evoking the power.

(*d*) *In Different Circumstances*

These are cases where a power is manifested differently in different circumstances. Here identity of power must depend upon identity of nature.

Conclusion. In general, identity of powers implies endurance of nature even when the nature is unknown, cf. particularly ultimate entities.

8. *Rôle of the Concept*

(*a*) to close regresses of explanation temporarily, because it is proper to ascribe specific powers even when the nature, in virtue of which they exist, is unknown;

(*b*) to avoid the ever-mysterious attributions of qualities.

III

CRITIQUE OF THE THEORY OF CORPUSCLES AS TELEIOMORPHS
OF PERCEIVED THINGS

1. Action and change in motion.

2. Action by contact is impossible for ultimates if they are material.

(*a*) *Proof.*

(i) Action by contact for non-ultimates requires deformation;

(ii) Ultimates, if material, are not deformable;

(iii) Either action, by contact, among ultimates is discontinuous, or ultimates are deformable;

(iv) Either action by contact is impossible, or action is discontinuous.

(*b*) *Lemma.* To prove the Law of Continuity.

(i) Inductive proof is provided when there are favourable instances, and when all apparent exceptions are explicable in terms of the law;

(ii) If time is continuous, and infinite forces are impossible, then action is continuous;

(iii) To prove (ii).

(*c*) *Definition of Continuity.*

(i) Consider a point Q, if Q is in the upper segment it is the least upper bound of lower segment, or Q is in the lower segment and the greatest lower bound of the upper segment;

(ii) If discontinuous change of state occurs in a certain interval of time, then at some time Q, a test body, is both at S_1 and S_2, which is impossible, or at neither S_1 nor S_2, which is contrary to hypothesis.

(*d*) *Proof that Time is Continuous.*

Consider the motion of real particles in the following worlds:

    (i) Space discontinuous and time continuous: then either there are some times at which body does not exist, contrary to the hypothesis of its reality; or it is at either S_1 or S_2, indifferently at the same time, which is impossible, because simultaneous reference to some body at S_1 and S_2 would prove the existence of two bodies.

    (ii) Space continuous but time discontinuous: either the body is at some places at no time, contrary to the hypothesis of its existence, or it is at S_1 *and* S_2 at same time, which is enough for proof of the existence of two particles.

    (iii) Space discontinuous and time discontinuous: not so easily shown to be impossible. Consideration of the idea of a less than mathematically dense continuum suggests the impossibility of discontinuous time.

    (iv) The supposition that both space and time are continuous is the only unproblematic case, and if time is continuous, so must action be.

The Search for the Ultimate

Access to the ultimate entities and structures of the world can only be by that combination of reason and imagination that has been identified throughout this book as the main intellectual tool of creative science. The ultimate entities must lie 'behind' whatever can be observed or detected. I have emphasized that scientific reasoning, that is the organization of knowledge of nature by thinking, involves the joint use of sentences, under the control of deductive logic, and of models and pictures, under the control of the principles of analogy. Scientific thinking is done, in psychological terms, both by propositional thinking and by imagining things and structures, processes and unobserved events. The use of reason and imagination to arrive at knowledge of the ultimate structure of the world is possible only by the exploitation of regresses, sequences of stations of rational thought, that lead on into the final ultimates of reality. Such regresses must begin with observations of the world as we see it, hear it, touch it, in short from the world as manifested in perception.

Two great systems of regressive stations of thought seem to lead towards the ultimate. There are the regresses of explanatory mechanisms, and there are the regresses of causes into the past and of effects into the future. Two main kinds of regresses of explanatory mechanisms appear to be possible: there are those which arise from the use of the principle that the behaviour and properties of wholes are to be explained by the interacting behaviour and properties of their parts. Then there are those which arise from reference to the characteristics of wholes in explanation of the characteristics of their parts. Though the former kind of regress is a millionfold more common than the latter, regresses to wholes may be of equal significance, when considered as guides to the ultimate structures of the world.

To explain a phenomenon is to identify its antecedents and to identify or imagine the mechanism by which the antecedents produce or generate the phenomenon. Sometimes the mechanism can be observed, but usually it cannot, and in the first instance an iconic model of it must be made or imagined. In favourable circumstances the model may come to be treated as the hypothetical mechanism itself. Sometimes technical advances may permit an extension of the senses by which the true mechanism is revealed. The components and mode of operation of the causal mechanisms are, in general, different in kind from the phenomena that they explain, which are either things in various relations, or things undergoing qualitative change. So explanatory mechanisms become a new subject for scientific study and the explanation of their principles of operation calls for the hypothesis of further explanatory mechanisms, new model building and so on. Here a regress opens out, which is, for all we can know, endless. I shall call it the 'regress of microexplanation', but it can, and will be asked what for *us*, scientist and plain man alike, would bring an end to the regress? What, for us, would be the ultimate mechanisms? I propose to develop the answer to that question in this and the following chapter. The ultimate, in this case, will be the ultimate parts of things. Enduring things, which are the mechanisms of causal change, are structures of parts, and these parts are themselves structures of parts. But it should not be assumed lightly that the answer to the question 'What are the ultimate mechanisms?' is the same as the answer to the question, 'What are the ultimate constituent parts of things?'

Sometimes the way a thing behaves is to be explained, in part, by reference to the larger structures of which it is itself a part. All explanation involves reference to the behaviour of mechanisms under external stimuli of some sort. The first explanatory regress arises by concentrating on the explanation of the behaviour of mechanisms with reference to their internal structure and the behaviour of constituent parts. The second kind of explanatory regress arises by concentrating on the explanation of the behaviour of mechanisms with reference to the system in which they are embedded, and the influences which are brought to bear upon them and stimulate them. Perhaps what really differentiates biology from physics is that both the structure of an organ, and the influences upon it which derive from its place in a larger structure, are given equal prominence *from the start*. Not only is this true of biology; physical chemistry began by paying attention to the environment in which chemical reactions take place. Cosmology refers to the totality of which the known astronomical systems are but a part. We have some reason for thinking that inertia is an endowment from the system to the parts of the system, not an intrinsic quality of the parts. So another regress opens: what can we infer about the progressively more embracing wholes of which the things we can study form parts? The method of investigation is the same. A model stands in for the unknown hypothetical whole. Like other models this can pass through the stages of being treated as a hypothetical mechanism, to perhaps being held to be reality itself. This is exactly the history of the helio-centric model of the universe. To explain the behaviour and character-istics of the parts by reference to the whole opens up a regress because, for all we can know, structure is embedded in structure, of a complexity and scale without end. I shall call this a 'regress of macroexplana-tion'.

Can it be said that there is an end to this kind of regress for *us*? I am quite unable to see how it can. In distinction from the regress of micro-explanation which aims at using fewer and simpler kinds of things, the regress of macroexplanation leads to entities of greater and greater complexity, and no way seems conceivable by which a principle of paucity could be applied to the macroconstituents of the universe. As I will show, in the regress of causes, the method of seeking a paucity of conditions for the existence of the universe is also closed to us. No model, I believe, could be recognized as successfully depicting the ultimate macrostructure of the universe.

In a world of enduring causal mechanisms another pair of regresses are possible. The world process consists of the mutual interaction of the mechanisms influencing each other in innumerable ways. Temporal concepts allows us to order the influences as causes and effects. Each cause is an influence exerted on some mechanisms from without, and so is itself produced by some other mechanism; that is, is itself an effect. The stimulus or stimuli which brought it into being are causes, and to come into existence in a world of enduring mechanisms must them-selves be effects. Effects become causes of further effects, and causes are the effects of antecedent causes. A regress of causes opens out towards the past and a regress of effects opens out towards the future. Can these regresses be followed into ultimate ends and beginnings?

Kant seems to have thought that equally good reasons could be given both for supposing that the world has a beginning in time, and that it has not such a beginning. I think that it can be shown that had the world had a beginning in time we could not know it, and similarly, were it to have an end, we could not anticipate it. To retrodict a first event, or to predict a last event, the present and recently past states of the universe are the only data. A description of these data must serve as a premise, and then by supposing *all mechanisms to retain the same causal powers*, the antecedent and subsequent states of the universe can be inferred by using the knowledge of causes that has been acquired by contemporary study of the world. Consider a possible first event. It can *either* be an event of the kind we know now, that is an event which is generated or produced by some known kind of mechanism, on the stimulation of a known kind of cause, *or* an event not of the kind we know now; that is, an event not supposed to be produced by any known mechanism. If it is an event of the kind we know now, then, if we can retrodict to it, we must suppose the natural mechanisms enduring thereto, because retrodiction supposes the actual course of events to be as they are inferred to be, by the retrodiction. Since all events of the kind we are familiar with at present are produced by the action of causes on generative mechanisms, we can infer a cause for the supposed first event, whether or not it is actually the first event. So if it were an event of a familiar kind, we could retrodict to it all right, but we could not know that it was the first event. On the other hand, if the first event was an event of a wholly unfamiliar kind, we could never retrodict to it from our knowledge of the present state of the universe, and of the behaviour of its natural mechanisms.

Therefore, whatever kind the first event was, we could never know it as the first event, since either we could retrodict to it but would have no means of telling that it was the first event, or we could not retrodict to it, so it would have no place of any kind in our knowledge.

What happens is the product of the responses of enduring mechanisms. Events do not produce other events directly, they stimulate fairly permanent structures of things to react in certain ways, consonant with their natures. A permanent atom absorbs energy and later emits it as radiation; the atom endures through the process. Many times throughout this book it has become plain that a prime condition for rational thought about nature is the introduction of notions of permanent enduring entities by creative acts of imagination which are the ultimate elements of everything relatively enduring. So if a rational system of thought must contain concepts of permanent entities, those entities must be supposed to exist prior to any possible particular event we could think of, and after any possible event we could envisage as the end of the world. So in no rational system of thought can knowledge of the end and the beginning of the world be contained. Endurance is prior to event temporality. The relatively permanent mechanisms of the world are not immortal but must endure through change. So we cannot think, rationally, of the beginning and the end of the world if these are supposed to be events in the natural order, discoverable by the application of scientific method. If we do think of the beginning and the end of the world, it is a thought or thoughts which do not belong in a rational natural enquiry, and is of another kind.

This argument is not entirely satisfactory as it stands, since there are two cases which seem to escape the dilemma. There is the case where all retrodictions from the current state of the universe converge upon a single state, similar to some situation we are currently aware of, in which there seems to be an uncaused event. For instance if all retrodictions converge on a state of the universe in which all the material is located in the one small volume, in a primaeval atom, so to speak, then the beginning of the disintegration of this atom might be held to be an event which escapes the force of the dilemma, on the grounds that it is an event of the same kind as the sudden, an apparently uncaused, disintegration of radioactive atoms. We are familiar with events of this kind, and we can devise no explanation for them. But our conceptual system is such that the mere lapse of time does not constitute a cause, so that our failure to find a cause for the disintegration of radioactive atoms need not be construed as the discovery that these are events which have no causes. Rather, we are constrained to say that their causes have not yet been found. All that the experiments show is that the cause must be looked for in internal changes in the state of material atoms, and not in external circumstances. This can be nicely accommodated in our conceptual scheme by ascribing to radioactive atoms the power or liability to disintegrate. This leaves open the question as to what changes have taken place in the natures of the atoms which are responsible for their disintegration.

The other case which may seem to escape the dilemma is the convergence of all predictions on a final state which is such that no further effects are possible. The redistribution of energy until a final uniform state is reached might be thought to be an example of such a convergence, once dramatically described as the 'Heat Death of the Universe'. Arguments of this sort are not in favour nowadays, partly because of the difficulty of transferring the concepts of thermodynamics, initially defined for finite, isolated systems, to whatever 'The Universe' might be; partly because of the uncertain state of the science of cosmology. Even in their heyday the arguments tended to show, not so much a final state from which further change was impossible, but only an assymptotic convergence upon that state, which would never actually be reached, that is the date of the Heat Death of the Universe could not be announced in advance.

In discussing the regress of explanation, the question of how the properties of parts are related to the properties of wholes was not raised. If the main source of knowledge of ultimate things is to be the regress of explanatory mechanisms this question must be faced. Otherwise how can it be discovered what are the characteristics of the final things, for human thought? Newton believed that the way to knowledge of the ultimate things lay through the principle that the primary properties of the parts are the same as the primary properties of the whole, and that the quantitative measure of such, in the wholes, is simply an additive function of the quantitative measures of these properties in each of the parts. This is a deceptively simple principle, and raises some difficult and complicated questions.

If we distinguish qualities, that is the characteristics objects are perceived to have, from properties, the characteristics they have intrinsically and essentially, then it is a feature of science to assume that properties and qualities may not be identical. It is not proper to speak of electrons as coloured or colourless, but some state of the electronic structure of atoms and molecules is responsible for a substance being coloured. The colour of a structured aggregate of electrons is not an additive function of the colours of the electrons. Nor is the warmth or coldness of an aggregate of molecules an additive function of the warmths and coldnesses of constituent molecules. Some function of the motion-states of molecules, considered individually, is responsible for a warmth state of the aggregate. This is the familiar origin of the notion of secondary qualities, which are perceived as determinates of a different determinable from that whose determinates are the states of matter responsible for those qualities being perceived.

It was widely believed in the seventeenth century that perceived things and ultimate things, while not sharing all determinables, certainly shared some. In the terms of Chapter 2, ultimate things were conceived as teleiomorphic models of perceived things, being both idealizations – for instance perfectly elastic and not subject to decay – and abstractions; for instance not being coloured but being in motion, having volume, mass and other characteristics that were supposed to be evident in 'grosser' pieces of matter, that is in the things that could be perceived. It is quite clear from what Galileo (17), Newton (18) and Boyle (15) have to say that, at least from one point of view, the ultimate entities are conceived as very small versions of perceived things. Galileo elevates this opinion into a principle by treating the attribution of primary qualities to matter as necessary, while for Boyle and Newton it is the most general scientific fact, and their methodology is particularly accommodated to it. Newton lays it down as a principle of method that those characteristics of perceived things, which are both universal and invariant in perceived things, are to be ascribed to the ultimate entities too. His famous list of such characteristics runs as follows: 'extension, hardness, impenetrability, mobility and inertia'.

In one aspect of his thought Locke follows Boyle and Newton in believing in the identity of some determinables of perceived things with some or all of the determinables of ultimate entities. He expresses this in his assertion that ideas of primary qualities (some of the sensible qualities of what is perceived) resemble the primary qualities of bodies (properties of real things, among them the ultimate entities). But this is certainly not Locke's final doctrine and he offers a further exegesis of qualities from which, in the course of this and the next chapter, a radically different view of the nature of the ultimate entities will be derived. He goes on to gloss the notion of a quality as 'nothing but a *power*' to affect other bodies and ourselves. The introduction of the notion of 'power' transforms the entire scene. If the intrinsic characteristics, or properties of bodies, are conceived of as powers, Newton's Principle collapses, since powers cannot be perceived, though it does not follow from this that they cannot be known. So, *a fortiori*, the powers of the ultimate entities cannot be additive components of the qualities of perceived things. To establish Locke's intuition properly I shall have to show that the qualities on Newton's list, are, as attributed to bodies, and so as attributed to ultimate entities, not qualities at all, but actually powers. Boscovich and Faraday will be the chief allies in this enterprise.

I now turn to an analysis and development of the notion of power.

Two Paradigms of Action

Since the concept of 'power' is going to play a central rôle in the theory of ultimates, its logic must be studied with care. To begin this process, I want to describe and contrast two main paradigms of action.

Paradigm 1. *Descartes Paradigm*. A stationary ball resting on a smooth table is struck by a moving ball. It begins to move and continues moving. The ball which struck the stationary ball continues on but moves less quickly. We are strongly tempted to think of this happening as a transaction in which the totality of the effect is due to the originally moving ball, and to suppose that the stationary ball contributes nothing to the final situation. The second ball which originally was stationary has no intrinsic power of motion, though it can move. Looking deeper and further into this paradigm, we can ask about the origin of motion of the first ball: that which was originally moving. Did it possess its motion intrinsically, did it move of itself? By reapplication of the paradigm it can be inferred that the first ball must have received an impetus to motion from something other than itself and that all its motion must have been acquired from sources external to itself. By continual reapplication of the paradigm, it can be concluded that there is no ball, not even a First or

Primal Ball which has an intrinsic power of motion. This paradigm, while exercising, I believe, a powerful influence over our thought, is not a straight metaphysician's crib from the science of mechanics. In that science, the final state of a system of bodies is a product of two factors, an extrinsic factor which is the various states of motion of the balls before they collide, and an intrinsic factor, which is the inertia or mass of the balls. Inertia is, as it were, a negative power, the power to resist indefinite increments of motion. But this shadowy negative sort of power has little metaphysical excitement about it, and for those attracted by it, the paradigm is still much as Descartes expressed it. The problem of the origin of motion can be solved either by conceiving of the hand of God setting every piece of matter in motion in the beginning and so arranging the laws of nature that the total quantity of that commodity is conserved; or by conceiving of the Universe as infinitely old, and so refusing the application of any questions of origin. The application of the Cartesian paradigm requires us to suppose that in all action the effect is completely produced by the impressed or stimulating cause.

The exigencies of fact require the modification of the paradigm in practice. The buffetings to which things (including persons and animals, and magnetic tapes) are subjected, changes them intrinsically, as well as stimulating them to respond. These changes in the natures and constitutions of things subsequently have an effect on the responses which the things can make to impressed stimuli. Continually bashing away at Hume's billiard ball will eventually knock bits off it, so that if it acquires, for example, the same increment of momentum every time something collides with it, it will move off more quickly each time, leaving a few chips of ivory behind.

Such modifications are in the direction, but not very far in the direction of another paradigm.

Paradigm 2. *Van Helmont's Paradigm.* On a fine sunny afternoon, of only moderate heat, and with no breeze to speak of, a man dozes in a deck chair in a garden. There are no flies, nor mosquitoes, nor wasps, nor shouts from neighbours' children. Suddenly the man jumps up, walks smartly to the shed, takes out the lawn mower and begins to mow the lawn. Nothing extrinsic to him had changed. The subsequent changes, for example the smelly racket of the motor mower, are entirely the products of the action of the man, the ultimate causes of which are to be found among states intrinsic to him.

But most cases of human action are not quite like that. This paradigm too gets modified. The man in the garden had the power to act and the capacity to be stimulated to act. He might have begun to mow because he was asked to do so. But a stick of dynamite or a loaded gun do not fit this paradigm, since their powers, though in some way conditional on the existence of certain intrinsic states, are exercised only in the appropriate circumstances, and under suitable stimuli. To say that nitroglycerine is a stable explosive is not to denigrate its power to blast but to intimate that it will manifest it only when appropriately stimulated. To say that ammonium tri-iodide is an unstable explosive is to intimate that whatever care you may take to isolate it from interference from without, it has been known to go off, *by itself*, as we say. We begin to assimilate it to the paradigm of the sleeping gardener.

If we consider the totality of our knowledge of things, materials, and persons one would be much inclined to say that neither paradigm of action is exemplified without qualification, but that a wide range of different proportions of responsibility is assigned on the one hand to the constitution or nature or powers of things, materials, and persons, and on the other to the stimuli to which they respond, i.e. to intrinsic states versus extrinsic circumstances.

On the official metaphysics of positivism, what could we possibly make of spontaneity, of the permanent possession of powers by things and substances, of a kind of causality where the cause and the conditions in which it acts are very much subordinate in their responsibility for whatever effects take place, to the intrinsic natures and constitutions of things?

Clearly, the Cartesian paradigm is enshrined in the metaphysics of positivism. If things are nothing but the collocations of their manifested qualities, whether this is given a phenomenalist ring or not, and if causality is but regularity of sequence of like pairs of events, and events are what happens *to* things, there can be no place for active powers.

There has seemed to be something fishy and soft, occult and mysterious about the second paradigm, and something tough, scientific and empirical about the first paradigm. But what are things, materials and persons really like? Are they like sitting ducks and stationary billiard balls, or are they like loaded guns and sticks of dynamite? Why has the second paradigm seemed fishy when it seems to be so natural and so clearly forced upon us by the way things are? Part of the answer lies in a mistaken epistemology which confines the data, and thus the content, of science to simple truths about sensory qualities manifesting themselves to an observer in particular circumstances, so that all that science can really be about is the obvious and the overt, and all laws but the statistics of the obvious and the overt. It is also partly due to a mistaken metaphysics, in which 'power' is seen as a concept surviving from magic, an occult quality appealing only to those of too tender a mind to face the stern truths of empiricism. I shall show in this chapter that not only are powers indispensable in the epistemology of science, but are the very heart and key to the best metaphysics for science. In so doing, I shall show that the concept of power is neither magical nor occult, but is as empirical a concept as we could well ask for, yet richer in capacity than those concepts it succeeds, such as qualities and dispositions.

Powers and Dispositions

In widening the sources of paradigms for action from the Cartesian patient to the spectrum between patient and agent and in treating this as deriving from the difference between intrinsic conditions and extrinsic circumstances, I have assumed an intuitive understanding of the concept of power. This must be made good by an analysis or explanation of what it is that is being said of a thing or material when a power or powers are ascribed to it. And here we groom an old and familiar concept for a technical rôle. What sort of description of the thing or of the material is now being given? It is clearly *not* a simple qualitative description of the thing or material. Simple, qualitative descriptions such as 'This one is red' or 'That stuff is cold' are intended to ascribe just that sensory quality that is then and there manifested by the thing or substance. But to say that that thing is brittle, that stuff is explosive, and the like, are not that sort of ascription.

The analysis of the ascriptions of such properties have been the occasion for the introduction of the idea of a disposition. Similar notions such as propensity, trend, tendency, and liability have been treated in an essentially similar way.

Under the influence of Ryle (244), and others, a particular way of handling all these ascriptions has become widespread. We are recommended to treat all of these ascriptions not as genuine categorical assertions of the presence of a quality, but to analyse them all as hypotheticals or conditional statements, or conjunctions of such statements. Any fine differences in import are supposed to be expressed by the use of the indicative and subjunctive moods. The meaning of 'It is brittle' is to be 'if maltreated, it will break', together with a trail of subjunctive conditionals of varying degrees of specificity trailing after it, whose rôle is to capture the force of ascriptions of brittleness to those things which did not, or have not yet broken. In a similar spirit 'It is poisonous' is to mean 'if taken, it will kill or make ill', and 'It can crush a car' is to mean 'if it presses a car, the car will be reduced to the size of a suitcase'. 'He is intelligent' is to mean 'if he is presented with a problem, he will quickly produce the solution.' Since the kind of problem is not specified, such ascriptions are analysed as generic rather than as specific hypotheticals: that is to say an unspecified range of specific dispositions are implied by the generic hypothetical.

The problem of what the assertion of a power or disposition to a thing means when it is not exercising it is not really solved for this analysis. To say that to assert that a particular piece of glass *is* brittle is to make a prediction about how it *would* behave if certain conditions were fulfilled is not enough, since it leaves the intractable problem of the truth-conditions of the subjunctive conditional conjunct in the analysis unresolved. At best one might agree that the ascription of a disposition may perhaps be appropriately treated this way, but this technique cannot work with the concept of a power or liability. Things and materials *have* powers even when they are not exercising them, and that is a current fact about them, a way in which they are currently differentiated from other things and materials which lack and now lack these powers. Indeed, *the reason why we believe that a certain disposition can be asserted truly, of a thing or material, is that we think or indeed know that it currently has such and such powers.* One of our reasons, and sometimes it is the only reason, for believing that if certain conditions are fulfilled then a material or an individual will behave in a certain way, is that the thing or material now has the power to behave in that way, should the conditions obtain. The difference between something which has the power to behave in a certain way and something which does not have that power is not a difference between what they *will do*, since it is contingently the case that their powers are, in fact, ever elicited, but it is a difference in what they themselves now are. It is a difference in intrinsic nature.

The Analysis of Power Ascriptions

The proper analysis of the ascription of a power to a thing or material (and with some qualifications, also to a person) is this:

X has the power to A = if X is subject to stimuli or conditions of an appropriate kind, then X will do A, *in virtue of its intrinsic nature*. The last clause is vital, and marks the difference between the ascription of powers and any other kind of description. In ascribing powers to people 'can' must be substituted for 'will' in the apodosis of the hypothetical component. Whether he will or no is up to him.

From this follow two vital points, and these really explain why we must have the concept of power for making sense of science.

(i) To ascribe a power to a thing or material is to say something specific about what it *will* or *can do*, but to say something unspecific about what it *is*. That is, to ascribe a power to a thing asserts only that it can do what it does in virtue of its nature, whatever that is. It leaves open the question of the exact specification of the nature or constitution in virtue of which it has the power. Perhaps that can be discovered by empirical investigation later.

(ii) But to ascribe a power is to say that what the thing or material does or can do is to be understood as brought about not just by the stimuli to which it may be subject or the conditions which it finds itself in, but in some measure by the nature or constitution that the thing or material has. That is why the ascription of a power is a schematic explanation.

It might be argued that the use of the structure 'If a is C then a will be D in virtue of being N', for expressing the logical form of a power's attribution to a thing or material, with its puzzling 'in virtue of' clause, is unnecessary, on the grounds that the maintenance of the nature of the thing or material could be treated as one of the conditions which had to obtain for it to behave in the ascribed way. There are, however, three important distinctions which the simple conditional form obscures.

The distinction between *extrinsic circumstances*, changes in which are not properly to be considered to be changes in the thing or material *itself*, and *intrinsic conditions* which are a proper part or feature of the thing itself. There are certain subtle considerations, particularly important in handling the conceptual structure of human powers, by which this distinction is differentiated from that between *internal* and *external* conditions of the manifestation of a power. Internal conditions lie within the spatial envelope of the thing, and external conditions outside it. It is not always true that the whole of the conditions pertinent to the nature of a thing or material are wholly inside its physical envelope. For example, the state of the magnetic field external to a thing may be an intrinsic condition considered with respect to its behaviour, since the nature of the body, in virtue of which it has certain dispositions, may derive in part from the state of the whole field within which it lies. An intrinsic condition is not necessarily an internal state and to assume that it is must be is a fallacy.

Finally, there is the distinction between *enabling conditions*, the satisfaction of which ensures that a thing or material is in a state of readiness, and *stimulus conditions* which bring about a response, on condition that a thing or material is in a state of readiness. The satisfaction of the intrinsic enabling conditions ensures that a thing or material is of the right nature for the exercise of a certain power.

Generally speaking, the intrinsic enabling conditions are satisfied by a thing or material being of a certain constitution, for example, having a certain chemical composition, or having a certain crystalline structure, or being in a certain state, say having a certain electric charge, and relative to the exercise of any powers the thing or material may have, the

possession of such constitutions or states is permanent, that is, they endure. It is partly for this reason that I am suggesting that they should
be picked out and assembled as the nature of the thing or material in
question. These distinctions are of rather general application. To use
the example cited by Ayers (243), the presence of an engine is an intrinsic enabling condition for a car to have the power to move, and
must be included in the specification of what it is to be a proper car.
The presence of the driver is an extrinsic enabling condition, while the
actions performed by the driver to set the car in motion are extrinsic
stimulus conditions. All of these conditions are internal to the car.

To ascribe a power to a thing is to locate the thing or material, for
the purposes of understanding its behaviour, in certain contexts of
action, towards the agent end of the spectrum of paradigms of action,
and to remove it from the patient end. At the patient end, things and
materials do not have powers, because at that end of the spectrum of
action, effects are brought about through the agency of extrinsic stimuli
or conditions.

'Power' is a notion particularly associated with agency, with the
initiation of trains of events, with activity, and though philosophers in
the past used to speak of passive powers, it seems better to use a quite
different term for such attributions as imply that the thing or material
falls under the Helmontian paradigm. I want to maintain the everyday
connection of the concept of power with that of activity, of initiation,
of cause and so on, though *not* anthropomorphically, but through the
distinctions between the degree to which intrinsic nature and extrinsic
circumstances enter into an explanation of the behaviour referred to in
the hypothetical component of a power ascription. I shall use the term
'liability' for a 'passive power'. The failure to keep powers and liabilities distinct was another failure of the disposition theory. Most things
and materials not only have powers, but also have liabilities.

The structure of the concept of liability has to be given in terms of
dispositions and natures, like the analysis above of the concept of power,
but now a thing's or material's liabilities are its dispositions, to suffer
change in virtue of its essential nature, that is the stimulus which produces the change is part of the extrinsic circumstances. When we say
of someone that he is liable to catch cold, what we mean is that in the
extrinsic circumstance of infection he has a disposition to develop a cold
in virtue of his essential nature. We are speaking in a similar way when
we talk of someone as having a delicate constitution. Many properties
of substances are strictly speaking liabilities and not powers, though
their analysans is formally alike, e.g. all those old favourites like 'solu-
bility', 'inflammability', 'brittleness', 'magnetic', etc. In the days when
the picture of nature as a crowd of passive sufferers of external and

imposed causality was in vogue, these were the only dispositional properties that ever got mentioned. But these only have a rôle where the world is also full of things and materials with powers. Nothing could be brittle in a world where nothing could smash, nothing could be inflammable in a world where nothing had the power to ignite, and there could be no solutes where there were no solvents.

Not only were the analyses of such terms as 'brittle' defective in failing to acknowledge the rôle of the essential nature of a material or thing, but they were grossly misleading in their direction of attention to the sufferings and passive pronenesses of things, leaving out of account the powers and agencies of materials and individuals, i.e. they led one to overlook the rôle of the intrinsic natures and over-emphasize the rôle of the external circumstances. I want to emphasize that the concepts of power and liability, as I have distinguished them, are the poles of a spectrum of concepts, distinguished by the relative degree to which we assign responsibility for particular behaviour between intrinsic conditions and external circumstances. I am not sure that any naturally occurring entity can fall under the pure concept of a patient, since it must be qualitatively identifiable to be known to exist, and hence must have at least the power to manifest those qualities which are required to identify it. To ascribe a power is to say that when any of a certain specific kind of phenomenon occurs the intrinsic nature of the things and materials involved are in more or less measure responsible for that phenomenon occurring.

It follows from this account that it is always incorrect to analyze an expression used to attribute a power or a liability wholly in terms of hypotheticals. That there have sometimes been superficially satisfactory analyses of power expressions in terms of hypotheticals is a consequence of the fact that in these analyses there has been a subjunctive component. The basis of our acceptance of 'would have' and 'could do' statements is just that which differentiates power ascriptions from quality ascriptions, namely that those things to which powers are ascribed are of such a nature or constitution that that is the way they *must* behave. Our theory of natural necessity works well in this context, since in assigning part responsibility to the nature of the thing or material we are offering a reason, independent of the actual facts, why those facts are as they are. Similarly, if law statements are offered in part analysis of statements ascribing powers, their nomological character derives from the tacit embedding of such statements in a view about the responsibility of the nature of the things or materials in producing the phenomena with which the law is concerned. Within this view one may see their behaviour as flowing from their natures or constitutions as consequences of what they *are*. So they must behave in the specified way, or not be the things that they are. And so necessarily while they are the things they are, they behave in those ways or have a tendency to behave in those ways or are disposed to behave in those ways mentioned in the hypothetical statement. Being of the right nature endows a thing or material with the power to manifest itself in certain ways or to behave in certain ways in the appropriate circumstances. In the extreme case of the pure agent, it is to do something or to be capable of doing something whatever the circumstances, i.e. to be the initiator of an act. A pure agent is that thing or material in which a causal chain terminates, be it the resolute gardener or a radium atom. I doubt if there are really any pure agents, either organic or inorganic.

At this point nothing has yet been said about the meaning of the expressions 'in virtue of having a certain nature', 'flowing from the nature or constitution of a thing or material' or 'being endowed with a power', or of 'the nature of a thing being responsible for certain phenomena occurring'.

A very important consequence indeed follows too from all this: a consequence of critical importance for the epistemology and logic of scientific discourse and knowledge. If the patterns of phenomena are explained by reference to the powers of the things and materials concerned, and sentences used to ascribe these powers are analyzed into a conditional clause expressing the hypothetical and a categorical clause, referring to the natures of things and materials, then because the latter clause indicates that there is a reason why the predicates characterizing a thing or material are found together, power talk always involves sets of non-independent predicates. So neither Hempel's nor Goodman's paradox can arise, if predicates stand for powers. This is a special case of the resolutions of these paradoxes established in Chapters 1 and 4.

The Epistemology of Powers

Before we can discuss the criteria for deciding upon the truth or falsity of ascriptions of powers we must face the objection that has been tacitly or explicitly accepted by most philosophers, at least since Hume. It is that a sound, empiricist, no nonsense philosophy—the kind of philosophy which science ought to have—must not employ such concepts as 'power', since these introduce bad and occult characteristics which science, with its emphasis on ascertained fact and observation has been struggling to banish from our view of the world. The alleged absurdity of offering a *virtus dormitiva* as an explanation of the fact that taking opium leads to sleep is so taken for granted as to allow that very example to be trotted out in elementary philosophy books as an instance of fatuous pseudo-explanation. It is then supposed that this is to be replaced by the allegedly scientific explanation that this man has been put to sleep by opium because all cases or most cases of opium taking have been followed by sleep, and the mind anticipates the senses and expects further cases of ingestion of opium to be followed by drowsiness. But with our analysis of power up our sleeves we can turn the tables here. In fact, the Humean emptiness of the general concomitance statement contrasts pretty unfavourably with the *promise* of the power statement, which says that the cause of the sleepiness is not to be put down just to the fact of the ingestion of opium but is to be looked for in the nature of opium itself. This is important because two different sorts of empirical investigation branch off from here. Following Hume's account of the affair the man who wanted to know more could only proceed by collecting more cases of that and similar phenomena until he had enough to convince himself of the lawfulness of the statement of concomitance. He who ascribes the effect to a power of opium does not collect further statistics but begins an exhaustive analysis of opium trying to find out what sort of stuff it is, what is its chemical constitution. Then he tries to follow it through the body after ingestion, and tries to ascertain how it acts upon the central nervous system, the higher centres, and so on. He undertakes quite a different sort of investigation. *He does what scientists actually do*. Powers are not occult qualities at all, since they are not qualities. To ascribe them is to offer schematic explanations, and in the growth of knowledge of the natures and constitutions of things the explanations are filled out.

Things and materials are in fact structures of other things and materials. So the analysis of common salt into a cubical lattice of sodium and chloride ions, of steel into iron, iron carbide and carbon provides us with some idea of the nature of these materials, in virtue of which salt is, for example, soluble. The nature of salt, in virtue of which it has its powers, is that it is a cubical lattice of those ions. In this account, such powers of the ions as their electric charge are used. These must again be subjected to a similar analysis. Since this analysis, in its turn, refers to the nature of sodium and chloride ions it opens an empirical question of a specific account of their natures. Any entities like protons and electrons which might figure in such an account must be assigned powers which have to be similarly analyzed, and so on. But things might have been continuous media distinguished at different places by different states of strain. What are field potentials but powers of the field? If there is no way by which the nature of the field in virtue of which such potentials exist can be specified then we have reached what, for the time being, must be an ultimate level of scientific analysis.

We are now in a position to discuss the criteria by which we judge an ascription of a certain power to a thing or material to be true or false.

To ascribe a power is to assert a specific behaviour hypothetical, together with an unspecific categorical, referring to the nature or constitution of the thing or material concerned. Behavioural hypotheticals are judged by tests and trials, constitutional categoricals by analyses and observations, for instance, under a microscope. To test the ascription of a power some test or trial must be undertaken, because the hypothetical is specific, but because the categorical is unspecific, we do not have to spell out the nature or constitution of the subject in detail, that is, one is not called upon to perform the analysis of the thing or material to be justified in ascribing a power. In the ascription of powers the categorical component is like a promissory note, we need only believe that it is not logically impossible for it to be cashed. In order for it to be proper to ascribe a power we must believe only that it is in principle possible to ascertain the nature of the subject. We are not committed to being able to tell it then and there. Analysis of the subject need not be carried out along with the test in order for us to be able to say that such and such a thing or material has such and such a power, though to ascribe a power is to locate the cause or part of the cause of what something does or can do in the nature of that thing itself.

So the verification of the ascription of a power is a two-stage process. In the first stage a test or trial shows that the specific hypothetical is true, if dynamite is detonated, there is an explosion. In the second stage the nature or constitution of the material is investigated and it is seen that it is in virtue of its chemical nature that the dynamite can explode. The power to explode is properly attributed to the dynamite after the trials have shown that it does go off, on condition that we believe that the explosion is a consequence of the nature of stuff, even though at that stage we may not know what its nature is, that is, that the intrinsic nature of dynamite is a largish component in the scientific account of the explosion. It was discovered by trial that fresh lime juice had the power to prevent scurvy, long before anything was known of the nature and manner of working of Vitamin C.

Similarly, if I want to rebut the ascription of a power to a thing or material, it may be that it does not even pass the first stage of the test, as when we suppose that some new synthetic may have the power to cure arthritis, but after a trial on some hopeful, but by now cynical hospital inmates we find that as has happened only too often in the past, it does not work. Sometimes, however, a trial of something may seem to give the required result and one might say that it has the power because the hypothetical component seems to be true. But, alas, on pursuing the second stage of the investigation, on looking into the nature of the stuff and its manner of working, it turns out that it could not have been responsible, it was not the right sort of stuff. For example, a placebo may pass a test designed for a drug, though the investigators know it is neutral, because it is not the kind of substance, is not of the chemical constitution, for the observed disposition to flow from its nature.

A final epistemological point: the muddle between dispositions and powers derives in part from the fact that statistics of behaviour do provide a *prima facie* reason for the ascription of a power, and we are properly entitled to ascribe a power on the empirical basis of that sort of fact alone, because we do not have to have *specific* knowledge of the nature or constitution of the stuff before so doing. But knowledge of the natures of things is usually not sufficient grounds for predicting how they will behave, we need empirical trials as well. It is just this point that lies behind the demands of governments that aeroplanes be tested for airworthiness, however studiously they have been designed. We shall see later that this itself follows from a quasi-logical feature of the metaphysics of the concept of power, namely the stratification it imposes on scientific knowledge.

Powers and Natures

Powers must be possessed relatively permanently because in attributing a power we are asserting that the behaviour is a consequence of the thing having such and such a nature, that is, we are asserting that this behaviour is not entirely explicable by the stimulus to which the thing or material has been subjected. If powers do depend in this way upon the natures of the things, then, since things must continue to possess their natures (to be themselves) whether they are being stimulated in this way or that, they must permanently have their powers. Thus the quiescent stick of dynamite can explode, the packet of arsenic can poison, the carbon dioxide can form hydroxyl ions in solution, the acid can turn the litmus paper red, the loaded gun can go off, the somnolent gardener can spring into action, and what we mean by these remarks is that they are of that nature. Dynamite is an explosive, arsenic is a poison, carbon dioxide is soluble, litmus paper is sensitive to acid, the gun is loaded, and the sleeper is a keen gardener. To say that dynamite is explosive, to go over to the adjectival form asserts, in one short sentence both what it *will* or *could* do and that in explanation of that the intrinsic natures of the things or materials concerned, are involved.

At this point it will be useful to contrast 'power' with two somewhat similar notions, that of 'capability' and that of 'tendency'. Each is differently related to the thing or material which possesses it. Powers have been closely related to natures, and a loss of powers or a decline of powers is evidence for a change in nature. Indeed, in some non-scientific contexts the connection between continuity of power and numerical identity is a necessary one. To strip a deputy of his powers is just what is required to make him cease to be a deputy. A hangman without the power of execution is a hangman no longer, he is but a former hangman. After a time calcium chloride loses its power to absorb water, and we are inclined to make this out as a change in nature going so far as to distinguish, for example, in the case of lime, between anhydrous calcium oxide and hydrated lime, treating them as different substances.

Capabilities do not seem to be related so tightly to natures or constitutions. So that loss of capability may be put down to the thing or material now being in pretty unfavourable circumstances for the exercise of them. In non-scientific contexts this is fairly clear, as for instance the use of 'incapable' for those cases where one's nautre has not changed but one is drunk or has a broken leg or some disability. In such a case an external circumstance interferes with the normal manifestation of one's nature as a judicious bystander or a ballroom dancer. These ordinary language considerations are useful as indicating that we might have here a family of concepts, ways in which the basic notion of power, as I have developed it, is qualified in various ways for different purposes. So far then we have loss of power associated with a change of nature and with the beginnings of a loss of identity. This will be studied in more detail below. 'Loss of capability' seems to be a concept whose ascription implies a certain permanence or constancy of nature, since the qualification of power in such cases is put down to interference with its normal manifestation through external circumstances.

'Capability' as a positive concept expresses something of this idea: a capability is a power which can be acquired (or lost) without there being a change in the fundamental nature of the thing or material in question. Such a concept, if we are to be able to bring it into use, requires a consequential qualification of the concept of 'nature' or 'constitution' of a thing or material. Since to acquire a capability is to make some change in one's nature we need the idea of a change of nature which preserves self-identity as compared with a change of nature which destroys self-identity. It is the whole burden of the supplemented dispositional analysis of power that natures and constitutions are neither just the totality of manifested qualities nor the totality of dispositions, which a thing exhibits in its actual behaviour. I shall return to this very important point in a later section.

Examination of the notion of tendency leads to another qualification of the notion of power. A tendency is a power in abeyance, a power in the course of coming to be exercised, a power about to be exercised or manifested. So, and here we have the strong part of the case against

Ryle and Hume, powers may be in opposition in such a way that though much is going on, nothing actually happens. The powers of greed are not abrogated in the case of Balaam's Ass, they are exercised, but still nothing happens, and the explanation of the fact that nothing happens is given just in terms of the exercise of power.

The Identity of Powers through Time

A power can be exercised over and over again or not exercised at all. So the possession of a power is different from the exercise of a power. Provided the constitution or nature in virtue of which the power is possessed remains unimpaired, then some divergence in outcome attributable to divergencies in the conditions under which the power is exercised is compatible with power identity through time. Alcohol retains the power to intoxicate despite the divergent effects it has upon different people, since these can plausibly be attributed to differences in the people, i.e. to circumstances extrinsic to the nature of alcohol. However, one of the species of power is capacity and we are familiar with the idea of 'impairment of capacities' and even of 'being incapacitated'. These are cases where we wish to say that a power has been lost or diminished through some temporary or permanent change in the *nature* of the individual involved. So the divergence of outcome from normal, or even failure to have the proper outcome at all cannot be unequivocally construed without reference to the identity or loss of identity of the material or individual involved.

Some powers may be diminished or augmented without losing identity. A drug may lose its effectiveness after a while, but this may happen slowly. We may still attribute to it the power to stop pain, but we may want to say that it does this less efficiently. The power of a car engine may decline. A man may slowly lose the power to move his limbs, his capacity to remember poetry, his capability as a teacher. Catalysts lose their power of catalysis. Powers can also retain their identity while they are augmented.

Are there powers which have to be possessed in full or not possessed at all? By this I mean 'Does our concept of power cover such cases?' Well, as we have already noticed, there are cases of legal or constitutional powers where a certain person either has the power to sentence, the power of attorney, etc., or he just hasn't. Then there is the case of the priest, who has the power to forgive sins. While he has it, this power can neither be diminished nor augmented, but is co-terminal with his being a priest. I do think that the way we use the concept of power covers both powers which can diminish or be augmented, and powers which are either had or not had in their entirety.

K

I shall distinguish these as 'variable powers' and 'constant powers'. As always with powers, we must now consider how they are related to the natures in virtue of which they are possessed. Is an invariable nature associated with a constant power? Might we conclude from a belief that the nature of the material has not changed that it has an undiminished power, say to turn litmus paper blue? And is the variability of a power ultimately dependent upon the variability of a nature, and what then is the criterion of identity under variability of that kind of nature? Does our distinction of two kinds of powers indicate a distinction between two kinds of natures, and thus ultimately two kinds of individuals? Are the bearers of constant powers Parmenidean individuals whose natures are unchanging under identity, and are the bearers of variable powers Aristotelian individuals whose identity is compatible with considerable changes in their nature? I am inclined to think that the answers to all these questions are affirmative.

Scientists make use of both kinds of individual, devising concepts for ultimate individuals that are, as far as possible, Parmenidean. But through being forced to acknowledge the increasing fragility of the structure of things and substances from tough atoms, through fairly robust molecules to fragile crystals and delicate organisms, there is a tendency to devise concepts for more complex individuals that are Aristotelian. So we should expect ultimate individuals to have their full powers while they retain their identities, that is, while they exist at all. These powers could neither be diminished nor augmented without a change in the nature of the individuals who have them. This is reflected, for instance, in the use of positive and negative charge, as a classificatory concept for distinguishing *kinds* of ultimate individuals. While the electron is playing the rôle of an ultimate entity, it is that individual which has a constant elementary negative charge, i.e. a permanent and unchanging power of a certain kind. It is a necessary condition for an individual to be ultimate that it permanently possesses constant powers, and a sufficient condition for an individual to be not ultimate that it possesses variable powers, but it is not a sufficient condition for an individual to be ultimate that it possesses constant powers. The chemical elements, though they are ultimate for chemistry, are not ultimate absolutely, though their powers, i.e. valencies, are constant.

Identity of Nature and Multiplicity of Powers

We say that the same electronic constitution confers two distinct powers upon chemical atoms: the power of entering into certain chemical combinations only, and the power to radiate a particular spectrum. Here, difference of power depends upon difference of disposition, in virtue of the same nature. So in this case identity of power seems to rest upon identity of disposition and not upon identity of nature. Similarly, if two materials exhibited the same disposition but in virtue of different natures our concept of power is such that we would say they had the same power. The power to produce a certain electromotive force may derive from a device's having a certain chemical constitution, like an electrolytic cell, or from a device's having a certain physical constitution, like an electromagnetic generator. It is a contingent, albeit *very* pervasive fact that the same power is usually found with the same nature. It is this important fact that allows the separate empirical study of dispositions and of natures, and the two stage processes of verification and falsification of power-ascriptions.

For those things which have shown their powers under our observation and lapsed back into quiescence again, we may assert the identity of the power on the assumption that the essential nature of the things or stuff has not been changed by the stimulus and reaction to it. But once a piece of paper has burned, it is no longer inflammable, so a piece of paper ash no longer has the nature of paper, and so no longer has the liability to burn. But provided we do have reason to think that the thing or stuff will respond to stimuli in a like manner again, then we may say it has the same power or powers. What would provide us with such a reason? Surely our belief that the essential nature of the stuff had not changed. So our belief in the identity of a power through time depends upon an assumption of the identity of an essential nature through time. In many cases we can determine empirically whether the essential nature in virtue of which the power is possessed has persisted, without evoking the power. There are certain signs which tell an experienced miner whether the dynamite has so changed its character that it will no longer explode or in hot conditions will explode spontaneously. But as we have seen, the fact that the essential nature may confer several different powers upon a thing or material prevents us from *identifying* the ascription of a power with the identification of an essential nature, though as we have just seen, we may ascribe identity to whatever powers are conferred by the possession of a certain nature on the basis of the identity of that nature through time.

I now come to a more complex case both for the theory of powers and for the problem of identity. So far we have seen the identification of a power to depend upon the persistence of a disposition, and our belief that a thing will continue to behave in a like fashion is backed up by identity of nature as a reason for the identity of this power through time. But there are cases where we speak of differnet manifestations of the same power, that is where what a thing or material does and can do, shows a wider diversity than the powers we ascribe to it. If this is the case, it will follow that we cannot pass in any simple-minded way from the observation of specific kinds of behaviour to the ascription of specific dispositions, to specific ascriptions of power, in all cases. Oxygen has a chemical valency of two, that is, it has a certain power to combine in certain specific ways with other chemical elements. This power is manifested in different ways on different occasions, depending upon the chemical situation. It is the power manifested in the combination of one atom of oxygen with two atoms of hydrogen, *and* in the combination of two atoms of oxygen with one atom of carbon. An electron manifests its elementary electric charge in a great variety of ways. We could hardly make sense of these cases if we identified powers with dispositions. Hydrogen is manifesting the same power in forming diatomic molecules in the gaseous state as it is in combining in the proportions two to one by volume with oxygen. Its disposition to form diatomic molecules, and its disposition to combine with oxygen are different dispositions. The power to which we refer these kinds of behaviour we call its valency. It makes better sense, I maintain, to speak in this way rather than to speak of its power of self-combination and its power to combine with oxygen, and so on, because we refer these dispositions to a common nature, that is to the structure of hydrogen as a simple one electron, one proton structure, and this is its *essential* nature, that is, it enters into the account of all its dispositions. Notice too that hydrogen and indeed all the chemical elements have something of the character of natural kinds. Their structure is supposed to be such that their differences are discontinuous. All their powers are consequences of their intrinsic natures. In these cases it is the common nature and not the identity of disposition that makes the running in the assertion of the identity of powers.

The Role of the Concept of Power in a Metaphysics for a Realistic View of Science

There are two places where a realistic point of view in the philosophy of science seemed particularly to require the concept of 'power'. These two places were (1) in explanation and causality, and (2) in observation and description, dealt with in Chapter 6. In Chapter 4, we saw that a satisfactory non-Humean and realistic account of causality, and of explanation, and so on can be achieved with the help of the idea of a causal mechanism, and this leads on to the requirement that causal knowledge, and indeed knowledge proper, must be stratified. Unless a new concept could be found, the regress of strata of explanation would either be endless, or if it were terminated (and in practice it must terminate), it would have to terminate in a species of interaction which was of the regular sequence and Humean type and we would be no further forward, having pushed the irrational back a few steps but not scotched it. Though incorporating *ad hoc* methodological recommendations to keep on trying can bring some comfort, it is obviously an unsatisfactory way to deal with the problem.

A regress of explanation is closed, in a proper way, by adverting to entities, individuals, and materials, which are characterized solely by their powers; that is, specifically by what they can do, which we do know, and only unspecifically by what they are, which for the entities which close regresses of explanation, we do not know. To ascribe a power, to characterize by powers, is to *open the question of the nature of things, without being obliged to answer it*. The ultimates can be bare powers, indeed for rational discourse, they must be bare powers, since to say that some thing or material is of an ultimate kind is to refuse, at least for the moment, to offer any account of its intrinsic nature.

But characterizing entities by their powers does not close off all possible further increments of knowledge. We would not have been wrong to ascribe powers if later the natures of such entities, in virtue of which they had the powers, were to come to light. However, were we to operate still with the old idea of trying to characterize individuals and materials by their qualities we should be in a very difficult position indeed with respect to the ultimate entities of a regress. To say what qualities things have is to say what they are and then from what they are we are supposed to be able to infer what they do. Qualitative talk draws us irresistibly towards the attempt to define individuals and materials as complexes of qualities, with all the attendant difficulties which I need not rehearse and with all the temptations to fall back into a first substance theory, even if it be disguised by such talk as of 'logical subjects', not to mention finding oneself with Humean nomological concomitance in the end. Even in the end the idea is incoherent. All this has been said once and been forgotten. This is the doctrine of Boscovich in *A Theory of Natural Philosophy* (248), and of Kant in *The Metaphysical Foundations of Natural Science* (257). We can now see, from yet another point of view, why the conceiving of iconic models is the central move in the dynamics of rational theory construction, since iconic models stand in for the real structures which are the real natures and constitutions of the things and materials of the world in virtue of which they have their powers. To discover how a thing or material behaves is to discover its powers, but since this leaves its nature and constitution undisclosed, it is not enough. Scientific knowledge also consists in the knowledge of the natures of things in virtue of which they have the powers they do.

It is incumbent upon one to offer some kind of account in terms of the traditional logic of quantifiers, variables and predicates, of the structure of sentences which might be used to ascribe a power.

We have seen that a power ascription has two components, since it is used to ascribe a specific disposition, and to ascribe some nature or other in virtue of which the thing or material has that disposition. Let us call the thing or the sample of material to which the power is being ascribed a, and the circumstances in which it exercises its power C, noting that C need not have occurred for it to be true that a has the power. The power is the power to G, so G is what does happen or might be expected to happen when a is in C, or were a to be in C.

As I understand a subjunctive conditional it is a straightforward statement about possibilities, excluding those actualized. So it is just right for part of the analysis of a power statement since to say that a thing has a power is to say what is possible for it, for that is what it is to talk of its dispositions.

To ascribe a power is also to say that the thing or material has a nature such that these possibilities are open to it. But it is precisely not to specify that nature. It is to say that there is a description of it incorporating the intrinsic, enabling conditions for the disposition to be possible, but it is not to give that description. It is to say that: There is a ϕ, such that something has ϕ, and whatever had ϕ in C, *would have to* G, i.e. if something like a did not have ϕ in C it would not, indeed *could* not G.

What about our reasons for accepting this? It is a hypothesis. It is not a law of nature because at the stage that we are ascribing powers we may or may not know what description ϕ is, or even if such a description can be found. It should be perfectly clear from this rather unsatisfactory formula, if from nothing else, that to ascribe a power is to leave the possibility of further discovery open, and in particular to leave open the discovery of which ϕ is true of a. There are further issues here, namely, the force of the modal operatior, and these raise epistemological questions of interest. This rough analysis does not, of course, allow powers to be defined out as logical functions of qualities, since by the earlier argument I offer a powers analysis of observations, so the predicates ϕ and C in this formula also stand for powers. This, I am encouraged to find, is in line with Joskes' treatment of perceptual qualities in his superb *Material Objects*.

Science, then, conceived as the study of the behaviour of things and materials, and the elucidation of their natures, and so of their powers, prompts the investigation of natures, and the analysis of natures leads to the discovery of new kinds of things and materials to which again powers are first ascribed, restarting the cycle.

The Critique of Matter

The next step in the derivation of the ultimates will be to show that they certainly cannot be corpuscles, and so to eliminate the eternal temptation to model the ultimate entities on perceived material bodies, that is, on things. And here Boscovich (248) must be followed.

The mechanical theory of corpuscles offers a very well-defined individual as the type of the ultimate entities, and endeavours to understand all interactions among sensible things as finally the product of the transfer of motion between corpuscles. In this chapter, I shall deal only with the traditional mechanical interactions. In the final chapter, the very notion of a corpuscle, of a body occupying space with its substance, and having qualities, will be disqualified as a candidate for the definitive source for the conceiving of ultimate entities i.e. for devising models of them. So even non-mechanical interactions between corpuscles, which escape the critique of this chapter, cannot provide a model for the ultimate constituents of the world. We shall conclude that neither classical mechanics nor the interactions of quantum mechanics, if they are between corpuscles, can be the ultimate interactions of the world. The transfer of motion, it was supposed, takes place only at the impact of corpuscles. All action is reduced to impulsive action by contact, and the ultimate laws of nature are those describing the redistribution of velocity among differently sized corpuscles by virtue of their mutual impact. The ultimate corpuscles were supposed to be perfectly elastic. Since all action is the transfer of motion between ultimate corpuscles, no energy can be stored within a corpuscle, or dissipated in any way other than by the transfer of motion to adjoining corpuscles. It took the genius of Boscovich to see that in terms of the corpuscular theory itself such mutual action was impossible. There can be no action by contact among ultimate corpuscles.

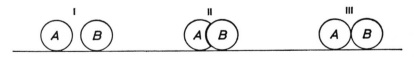

Fig. 3

Compressible bodies can interact by contact because they do not instantaneously acquire a common velocity, but by mutual compression they are enabled to slow down or accelerate in a finite time. If *A* and *B* are compressible bodies then their action by contact proceeds in a perfectly acceptable way as in the diagram. (*A* has a greater initial velocity than *B*.)

In Stage II the contacting surface of A is compressed inwards towards its centre of mass more rapidly than the contacting surface of B is compressed. And, of course, they can remain in contact while their centres of mass are differentially accelerated.

This pattern of interaction cannot be transferred to the ultimate corpuscles, since being perfectly elastic they are absolutely incompressible. In corpuscularian terms, compressability is a product of the fact that the volume of a whole body is greater than the total volume of its component corpuscles, so they can be brought into a smaller space, but an ultimate corpuscle does not itself consist of an arrangement of corpuscles. It is all body. As Boscovich sums it up (*A Theory of Natural Philosophy*, Article 20) 'Newton, and indeed many of the ancient philosophers as well, admitted the primary elements of matter to be absolutely hard and solid, possessing infinite adhesion and definite shape that it is impossible to alter.' Provided infinite forces are ruled out, or provided discontinuous action is ruled out, action by contact among ultimate corpuscles is impossible. I shall return to the proof of the provisos below, which are an important part of the whole proof. Here is the first part of the General Proof in Boscovich's own words. (*TNP*, Art. 18).

'Suppose there are two equal bodies, moving in the same straight line and in the same direction; and let the one that is in front have a degree of velocity represented by 6 and the one behind a degree represented by 12. If the latter, that is the body that was behind, should ever reach with its velocity undiminished, and come into absolute contact with the former body which was in front, then in every case it would be necessary that, at the very instant of time at which this contact happened the hindermost body should diminish its velocity and the foremost body increase its velocity, in each case by a sudden change . . . without any passage through the intermediate degrees. . . . For it cannot possibly happen that this kind of change is made by intermediate stages in some finite part, however small, of continuous time, whilst the bodies remain in contact. For if at any time the one body then had 7 degrees of velocity, the other would still retain 11 degrees, thus during the whole time that has passed since the beginning of contact when the velocities were respectively 12 and 6, until the time at which they are 11, and 7, the second body must be moved with a greater velocity than the first; hence it must traverse a greater distance in space than the other. It follows that the front surface of the second body must have passed beyond the back surface of the first body; and therefore some part of the body that follows behind must be penetrated by some part of the body that is in front. Now, on account of impenetrability, which all Physicists in all quarters recognize in matter, and which can be easily proved to be rightly attributed to it, this cannot possibly happen. There really must be, in the commencement of contact, in that indivisible instant of time which is an indivisible limit between the continuous time that preceded the contact and that subsequent to it . . . a change of velocity taking place suddenly, without any passage through intermediate stages; and this violates the Law of Continuity, which absolutely denies the possibility of a passage from one magnitude to another without passing through intermediate stages.'

This argument depends upon accepting that infinite forces are impossible or that all action is continuous. The impossibility of infinite forces will scarcely be disputed, so the argument turns on only one condition, the Law of Continuity, as Boscovich calls it. Maclaurin, Boscovich notices, accepted a similar analysis of the impact of ultimate corpuscles, but used it to deny the Law of Continuity. Boscovich's formulation, of admirable clarity, runs as follows (*TNP*, Art. 32): 'The Law of Continuity . . . consists in the idea that . . . any quantity in passing from one magnitude to another, must pass through all intermediate magnitudes of the same class.'

For the general outlines of the proof of the Law of Continuity, I follow Boscovich, supplementing his argument at some important points. For a principle of this degree of importance Boscovich argues that two different kinds of proof ought to be offered. There should be an inductive proof, to show that the Principle in fact is operative in all known actions, but there should also be a metaphysical proof. This has the aim of showing that the Principle is necessary, in the sense that it is a principle which could not be abandoned without abandoning the whole scientific enterprise. Such a proof proceeds by showing how the Principle is related to very fundamental features of space and time. In his proof Boscovich simply assumes the continuity of time. I shall try to supplement his proof by advancing what I believe are powerful reasons for that assumption.

The conditions for an inductive proof, as Boscovich outlines them in Article 40 of *A Theory of Natural Philosophy*, are

    (i) A law is followed in many cases.

    (ii) Apparent counter-instances can, without strain, be reconstrued as actually disguised instances of the law.

So, for instance, one could set about proving that light is a wave motion, by instancing interference and diffraction, and then going on to show how refraction, and finally simple reflection, which at first sight is not wavish, can be reconstrued as wave phenomena. As Boscovich puts it, 'in order that this kind of induction may be employed, it must be of such a nature that in all these cases particularly, which can be examined in a manner which is bound to lead to a definite conclusion as to whether or no the law in question is followed, in all of them the same result is arrived at; and that these cases are not merely a few. Moreover, in all other cases, if those that at first sight appeared contradictory on further and more accurate investigation, can all of them be made to agree with the law; although, whether they can be made to agree in this way better than in any other whatever, it is impossible to know directly anyhow. If such conditions obtain, then it must be considered that the induction is adapted to establishing the law.'

Finally, if we know of no reason why a law or principle should not apply in unexamined cases, then it is scientifically proper to apply it. And, Boscovich notes, if any particular class of cases are unexamined cases they must be regarded as strictly accidental, because it is a function of the kind of senses we have, and when and where our investigations are carried out. The Law of Continuity is just such a principle as meets the inductive criteria. All known cases of action take time and are continuous in that time. Those that give an impression of discontinuity are not genuinely instantaneous finite changes, but turn out on closer examination to take time, however short. And so the Law of Continuity has an Inductive Proof. In Article 43 Boscovich says, 'in the same way, then, we must deal with the Law of Continuity. The full induction that we possess should lead us to admit in general this law even in those cases in which it is impossible for us to determine directly by observation whether the same law holds good, as for instance in the collision of bodies.'

The metaphysical proof proceeds from an analysis of the conditions that must obtain for time to be continuous. But the first steps must be to lay down the most general conditions for continuity in any sequence whatever. Boscovich, anticipating Dedekind, notices that to say that in a continuous sequence any two points are contiguous will not do, since if the sequence is genuinely continuous there must be infinitely many points *between any* two points. His condition for continuity is as follows: Consider some *one* point Q, in the sequence of points. There will be a 'lower' segment with respect to Q and an 'upper' segment. A sequence of points is continuous if Q can be regarded *either* as the last point of the lower segment, in which case the upper segment has no first point, *or* as the first point of the upper segment, in which case the lower segment has no last point. But Q as the last point of the lower segment limits the upper segment, since none of its points can be on the lower side of Q. Similarly Q bounds or limits the lower segment when it is regarded as belonging to the upper segment. But the point Q cannot be considered to be both least point of the upper segment and the greatest point of the lower segment.

Suppose now that in some continuous sequence of times there should be discontinuous action. Let the action have measure a in the lower segment of the sequence of instants, and b in the upper segment, and the difference between a and b be always greater than zero. If the sequence of instants is genuinely continuous then there has to be an instant Q, the point of discontinuity of the action, which, being a point in a continuous sequence of instants, has to be able to be considered either as the upper limit of the lower segment, and thus as the first member of the upper segment, in which case the action has measure b at Q, *or* as the lower limit of the upper segment, and thus as the last member of the lower segment, in which case the action has measure a at Q. So, either at Q the action has measures of both a and b, which is a contradiction; or the action has no measure at Q which is contrary to hypothesis. In Article 49 Boscovich expresses the dilemma as follows: 'On the one hand there must be at any instant some state so that at no time can the thing be without some state of that kind, whilst on the other hand it can never have two [different] states of the same kind simultaneously,' Therefore there cannot be discontinuous action in continuous time.

But why should we suppose that time is continuous? I shall approach this question through what Kant called *Phoronomy*, the general theory of kinematics. Empirical considerations are inconclusive on such a question. Since the state of the universe can be probed empirically only at a sequence of finitely different times, the sequence of states so revealed will either be identical with each other for simple endurance, or themselves finitely different, whether or not the process probed is continuous or discontinuous. There could not be a continuous process in discontinuous time, since there would have to be infinitely many states between any two given states and in a discontinuous time there would not exist instances at which these intermediate states could occur. But it follows from the above considerations that it is impossible empirically to differentiate a genuinely continuous process. So it is impossible empirically to show that time is not discontinuous. There could not be discontinuous processes in continuous time, as Boscovich has shown, but empirical discontinuity may be but the consequence of the 'eigenstates' of processes which are actually continuous. So it is equally impossible to show empirically that time is not continuous. To advance the question further one must turn to an analysis of the concepts involved, in setting out the positions that seem at first sight to be logically possible alternatives.

Approaching this, through phoronomy, the general theory of kinematics, requires space to be considered with time and their continuity and discontinuity discussed with respect to the possibility of motion. Consider first the motion of a real particle in a continuous time $t2 - t1$, but discontinuous space $s2 - s1$. A real particle exists at all times from $t1$ to $t2$. But if space is discontinuous it occupies successively what may be an infinite number of positions, during the time $t1$ to $t2$, but there will be more instants in the continuous time interval $t1$ to $t2$ than there are points in discontinuous spatial interval $s1$ to $s2$. So the supposition of the possibility of motion in a continuous time, but discontinuous space, is subject to a destructive dilemma: either there are some times when the real particle is at no place; or it is at some place at every time. The first horn of the dilemma would require the real particle to go in and out of existence, contrary to the hypothesis of its reality. A real particle must be able, in principle, to be demonstratively referred to at all times, and so it must be at *some* place at all times. The second horn requires further analysis. Consider two sequential spatial positions, $s1$ and $s2$. If the particle is somewhere at all times, and time is continuous, then at the Boscovichean point Q, defining and distinguishing the temporal intervals at which it is at $s1$ and at which it is at $s2$, the particle must be able to be considered indifferently and arbitrarily at either $s1$ or $s2$. But since it is a real particle it must be demonstratively at $s1$ or $s2$, actually, and it cannot be arbitrarily set at either. So the second horn of the dilemma is just as unacceptable. Really both horns hinge on the same feature of a real particle, since the requirement that a real particle be always capable in principle of definite, demonstrative reference is equivalent to the condition that a real particle must exist continuously in space that is must always be at some point. Discontinuous space and continuous time define a world in which motion is impossible for a continuously existing thing. But there is motion in the real world, and so it follows that the real world cannot ultimately be one of discontinuous space and continuous time.

Now suppose space is continuous and time discontinuous. Again a destructive dilemma can be formulated, on the supposition that motion must be possible. A moving real particle in transition from place $s1$ to place $s2$ must necessarily be at every intermediate point in the course of the motion, and being a real particle must be capable, at least in principle, of being the subject of demonstrative reference at any one of those points. But if time is discontinuous there must be fewer instants in the temporal interval $t1$ to $t2$, where $t1$ is the instant the particle was at $s1$ and $t2$ the instant it is at $s2$, than there are points in the spatial interval $s1$ to $s2$. Either there is more than one spatial point at which the particle is at any given instant, or there are sequences of spatial positions, traversed by the particle in no time at all. The proposal set out in the first horn of the dilemma is objectionable because it violates the condition that no one real thing can be in more than one location at *a* time. Two spatial positions allow two simultaneous acts of demonstrative reference, which would be enough to prove that there were two real particles, and not just one at two places, contrary to hypothesis. The second horn effectively proposes that the motion of a particle be reduced to a sequence of spatial leaps at infinite velocity, since they take no time at all, though all intermediate points must be traversed. This would require infinite acceleration and deceleration, and this in turn would require infinite forces. And there are no infinite forces. It follows that motion is impossible in a world in which space is continuous and time discontinuous. Once again one can cite the existence of motion in the real world, the world scientists are studying, to refute the suggestion that the real world could ultimately be such that space was continuous and time discontinuous.

In the two arguments just set out I have been supposing only that the discontinuous partner was a sequence still of an order of infinity, say that of the rationals. Even with this relatively weak condition the possibilities of motion in either pair of continuous/discontinous space and time seems to be ruled out *a priori*. The stronger suggestion, that continuous space or time should be combined with discrete time and space, finite sequences of points and instants, seems even less plausible, and subject without much ado to stronger forms of these destructive dilemmas. I shall not therefore consider such cases separately.

The remaining possibility that would count against the Law of Continuity is that both space and time should be discontinuous. Motion in such a world would be the successive occupation, by a demonstratively identical real particle, of different successive places at different successive times. There would be no places 'between' the places the particle occupied, nor instants of time 'between' those at which it occupied the places. Neither of the destructive dilemmas, which rule out the other combinations, can be applied in this case. Discontinuous time seemed to make sense in the second proposal above because of the picture conjured up by its juxtaposition with continuous space. One thinks one could imagine a particle at several places successively while the clock stands still at noon, and then when the particle finally reaches a certain spatial position the clock jumps on to one minute past. But this picture is defective. The successive occupation of the spatial positions is itself a time, and the occupancies are its events. This suggests that perhaps the notion of discontinuous time is not really coherent, and nothing but the reflection of confusion between time and process in time. Perhaps we have a model of time as a vast cosmic process. Then picturing continuous and discontinuous processes of the increment of some quantity F, say as we imagine Discontinuous Time as a cosmic process like (B). But (B) can only be imagined against a background of continuous time. And, as Boscovich has shown, this act of the imagination is faulty, since the supposition of the acquisition of the increment $F2 - F1$ in no time at all violates the condition that T

should be continuous. We only *seem* to understand discontinuity of process with reference to continuous time. On deeper analysis, this understanding appears to be illusory, so we can hardly hypostatize an impossible process into a cosmic model. Discontinuous time is no more a

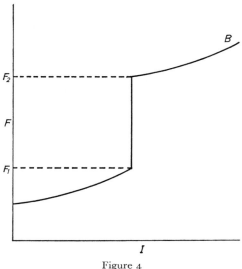

Figure 4

coherent notion, on analysis, than is discontinuous process. In short whatever may be the superficial appearances of processes in time the next instant is absolutely contiguous to that which it succeeds, and there is no time between them. The only condition that makes this possible is that time be continuous in Boscovich's sense. We are left then with the only coherent possibility, continuous space and continuous time. The Law of Continuity stands proved, provided, as has emerged, at each stage of the proof, there are no infinite forces.

Our general point of view conceives of the mention of forces as standing in for genuine explanations, as simply marking the place at which eventually a generative mechanism will be found. The action of generative mechanisms is determinate so the action of such a mechanism is adequate as an explanation of any determinate change in or modification of a state. But to suppose an infinite force acting for no time does not provide an adequate explanation since the upshot of its action is not determinate. It does not provide any reason why the new state should be any particular state. Not only is there this difficulty of indeterminacy of effect but it is hard to see what sort of generative mechanism an infinite force could stand in for.

However all these arguments notwithstanding, might it not be urged that in quantum mechanics we have an example of a system of concepts which have a central place for exactly the mode of action that Boscovich thought impossible, namely finitely discontinuous change in states of entities? What is it to talk of quanta of energy if not to acknowledge that, for example, a finite, discontinuous change in velocity can take place? If the way of thought of quantum mechanics does become permanently established, and if Boscovich's argument against discontinuous physical change in a continuous space and time is sound, then there can be only one possible way out, and that is that the space and time of quantum mechanics must also be quantized, that is that space and time must also be discontinuous. How is one to reconcile this conclusion with the grave doubts about the coherence of any concept of discontinuous time that we have just been considering? One way might be to distinguish 'empirical' and 'theoretical' time, the former of which is discontinuous and the latter continuous, the latter providing a theoretical background against which the former concept could be made intelligible. One might perhaps explain this idea by analogy with 'empirical' and 'theoretical' temperature scales, or by analogy with geometry and surveying. This way of resolving the difficulty has the further advantage that it would ensure the retention of a conceptual system in which a future advance beyond quantum mechanics would be possible.

The upshot of this whole line of argument is the conclusion that the Law of Continuity is largely exemplified in nature, and is thus deeply involved in the most fundamental conceptions by which we order our experience as to be, to all intents and puproses, immune from revision. It can be treated as a principle having Natural Necessity, as that notion has been understood in this book, but it is not a tautology. If discontinuous time as a fundamental concept is incoherent, there is no alternative within our system of thought, to our accepting the Law as true, that is the actual actions of the world are to be seen, and to be understood, as exemplifying it.

From the universal truth of the Law of Continuity follows the universal impossibility of action by contact among really solid bodies. So the hypothesis that the ultimate entities are massy, impenetrable volumes transferring motion by contact stands discredited. The Mechanical Corpuscularian Philosophy is a failure.

SUMMARY AND BIBLIOGRAPHY

The search for ultimates in science is seen to involve regresses of explanation and regresses of causes. The idea that all explanatory regresses must be micro-reductive is attacked by

239. G. SCHLESINGER, 'The Prejudice of Micro-Reduction', BJPS, 12, 215–24.

The idea that there might be independent laws for macro-structures is often discussed under the guise of the discussion of 'emergent properties'. A very clear exposition of the best argument for this view is

240. M. POLANYI, 'The Structure of Consciousness', *Anatomy of Knowledge*, edited M. Grene, Routledge and Kegan Paul, London, 1969, pp. 320–2.
See also
M. BUNGE, 'Levels' (*see* 110), Ch. 3.

Regresses of causes have been much discussed in the context of 'Final Causes' and the existence of God. The problems I am concerned with stem from Kant

241. I. KANT, *Critique of Pure Reason*, A426–7, e.g. N. Kemp-Smith, Macmillan, London, 1952, pp. 217–18.

and have been much discussed. See, for instance,

G. J. WHITROW (*see* 227), pp. 31–3.
R. SWINBURNE (*see* 228), pp. 298–301.

For general discussions of regresses of causes and knowledge of origins, see

242. P. BROWN, 'Infinite Causal Regression', *Phil. Rev.*, 75, 510–25.
M. K. MUNITZ (*see* 107), Ch. VI.
The concept of 'power' has only recently returned to fashion. For a traditional view see
J. LOCKE (*see* 11), Bk. II, Ch. 21.

The most sophisticated discussion to appear to date is

243. M. AYERS, *The Refutation of Determinism*, Methuen, London, 1968.

For the classical account of dispositionalism see

244. G. RYLE, *The Concept of Mind*, Hutchinson, London, 1949, Ch. V, 2.

For the concept of a 'nature' see

F. BACON (*see* 9).
245. T. SPRIGGE, 'Internal and External Properties', *Mind*, 71, 197–212.

For the classical world picture see

246. M. CAPEK, *The Philosophical Impact of Contemporary Physics*, Van Nostrand, New York, 1961, Part I.

For a general critique of the theory of corpuscles as ultimate entities:

247. J. B. STALLO, *The Concepts and Theories of Modern Physics*, new edition, Harvard University Press, Cambridge, Mass., 1960, Chs. IV, V, VI.
248. R. J. BOSCOVICH, *A Theory of Natural Philosophy* (Venice, 1763), M.I.T. Press, Boston, 1966, pp. 10–13, 19–68.

The attempt to substitute purely kinematic concepts for materialist concepts was made most thoroughly by

249. E. MACH, *The Science of Mechanics* (Chicago, 1893), Open Court, La Salle, 1960, Chs. II, III–IX.

My preference is for the substitution of powers. See Chapter 11. For other defences of the 'Law' of Continuity, and the continuity of space and time

250. B. ABRAMENKO, 'On Dimensionality and Continuity of Physical Space and Time', *BJPS*, 9, 89–109.
251. M. SACHS, 'The Theory of Fundamental Processes' *BJPS*, 15, 213–43.
252. E. SCHRODINGER, 'Are there quantum Jumps?', *BJPS*, 3, 109–23, 233–42.

11. The Constitution of the Ultimate

The Argument

1. *Principle.* Powers must be central to the constitution of the ultimate, not qualities.

2. *Three Conditions on the Ultimate*

    (*a*) Spatial conditions: an entity must have a place in the same system of space relations as material things.

    (*b*) Temporal conditions: some entities must endure; a power must be possessed, at least semi-permanently, though powers may wane.

    (*c*) Causal conditions: the powers of ultimate entities must be responsible for the world as manifested.

3. *What Entities meet these Conditions?*

    (*a*) They must be ontologically continuous with the world as manifested.

    Note. Phenomenalism is rejected out of hand as an account of the world as manifested.

    (*b*) Identification by reference:

        (i) Direct demonstrative reference to ultimates is impossible;

        (ii) Verbal reference is possible via the spatio-temporal framework;

        (iii) Indirect reference is possible via particular material things.

    (*c*) Not a-typical, e.g. for reidentification by direct demonstrative reference to count as proof of the existence of a material thing, a theoretical numerical identity must be assumed.

        Scepticism about use of qualitative identity as criterion for numerical identity is misplaced.

        (i) It is the best possible criterion;

        (ii) Real essences of bodies are their causal powers, manifested as qualities, therefore qualitative identity is the best possible criterion for identity of causal powers.

    (*d*) To satisfy qualitative identity, only core identification is required, and this leads to the conception of nominal essence.

4. *Could Material Things as Manifested be the Fundamental Entities?*

    No, because we frequently require to preserve identity through change from one *kind* of material to another, e.g. the atoms of an element through compounds.

5. *Distinction of Points of Space from Points in Space*

    (*a*) Points of space are defined with respect to *s-t* framework, i.e. things in general.

    (*b*) Points in space are defined with respect to particular manifested things.

6. *Distinction of Transformation from Alteration*

    (*a*) Transformation = change in real essence (causal powers), e.g. change in genotype.

L

(*b*) Alteration = change in nominal essence (manifestation of causal powers), e.g. change in phenotype (may be due to environment).

7. *Causal Powers of Manifested Things*

Can be explained by a corpuscularian theory, e.g. the motion of constituents as the cause of manifested heat.

8. *Causal Powers of Ultimate Things*

(*a*) Cannot be explained by a corpuscularian theory, because the powers of ultimate things must explain occupation of volume, solidity and other primary *qualities.*

(*b*) *Advantages of Powers as Definitive of the Ultimate*

    (i) They provide a rationale for the preservation of numerical identity through manifested changes;

    (ii) They permit a primary quality account of powers of material things as manifested without forcing a qualitative account of the ultimate.

9. *The Reduction of Primary Qualities to Powers*

(*a*) Volume and inertia;⎫
(*b*) Solidity;   ⎬ as achieved, for example, by Kant
(*c*) Mass, etc.   ⎭

The ultimate entities are point sources of mutual influence, and physical investigations must always terminate in field theories.

Ultimate entities must be both constitutive of the world and responsible for the way it manifests itself to us and to our instruments. The world, as it manifests itself, is extended in space and enduring in time, and is in a flux of change. Thus the ultimate entities must be such as to ensure both permanence and the possibility of manifested change. Permanent material atoms, in changing spatial relations, that is in continual but changing states of relative motion, have had to be rejected as possible ultimates. I now develop, in a relatively *a priori* way, the concept or notion of the only kind of ultimate entity that can be acceptable, for us.

In two recent books, P. F. Strawson's *Individuals* (95) and W. D. Joske's *Material Objects* (196), the metaphysics of the world as it is manifested to us, has, I believe, been thoroughly and authoritatively explored, and the appropriate system of concepts expounded. In what follows, the relationship between what science must offer by way of ultimate entities, and what the world as manifested to us demands of things, is elucidated so as to bring out the inevitability of the specification of ultimate entities that is developed here. The world is manifested as extended in space and enduring through changes. It could be said that space is one set of relations in which material things can stand to one another, and time another. I have argued, in effect, in earlier chapters, that relativity theories, in which both sets of relations are required for defining invariants, determine the metrics, but not the metaphysics of space and time. Connected with, but not identical to the set of relations that are time, is the causal set of relations. Each of these sets of relations determines a set of criteria for ultimate entities. Ultimate entities must satisfy three kinds of conditions:

(a) *Spatial Conditions.* As constitutive of material things ultimate entities must have places in the same system of spatial relations as material things. Just how the location of ultimate entities is achieved will be discussed in detail. It will soon emerge that they cannot be located by acts of direct, demonstrative reference. Since the world as manifested is a plurality of material things, and the ultimate entities are constitutive of them, there must be more than one ultimate entity, so there must be some way of individuating ultimate entities. It will be shown that ultimate entities cannot be individuated in themselves, but only as the constituents of material things, which can be demonstrably individuated, since they occupy different places, and always show some degree of qualitative difference. The possibility of the discovery of the location of ultimate entities by an indirect method will be shown to be compatible with the impossibility of the direct individuation by demonstrative reference.

(b) *Temporal Conditions.* The ultimate entities endure, and their endurance provides the basis of an element of permanence in the world as it manifests itself. The things in the world manifested endure for a while, but none of them is immune from decay, that is from changes that are so drastic as to exceed the permissible bounds for changes in things and substances of that kind. Somehow entities must be imagined which have no qualities, and so cannot be changed, yet are themselves responsible for change. The relation of protons, neutrons and electrons to the atoms of chemically distinguishable substances is something like the relation of the ultimate entities to the things and materials of the world as manifested. Change in the arrangements and numbers of these relatively permanent components leads to change from one chemical element to another. The atoms of chemical elements are not ultimate because they can be transmuted, one kind into another kind, or even annihilated altogether. But with respect to them protons, neutrons and electrons are ultimate.

(c) *Causal Conditions.* Since facts about the ultimate entities are the final explanations of both qualitative and relational change, in the world as manifested, the ultimate entities must be held responsible for the states of the world as manifested. The ultimate entities are causally responsible for the states of the world, without anything being causally responsible for them.

The first step in working out what kind of entities can meet these conditions, yet escape the objections raised to the most plausible previous candidate, the material atoms of the Corpuscularian Philosophy, will be the elucidation of this important concept, 'the world as manifested'.

The world is *not* manifested as a series of independent sensory fields, unified by some hazardous act of inference to an underlying material substance. The world is manifested to us as a system of qualitatively differentiable and separately locatable things and substances. Science is concerned with the system of things, the very same system of things that manifests itself to us through our instruments, and is known by the use of our senses. The electric charge which manifests itself in the reaction of an electroscope is distributed over the surface of the very pithball which *we* hung with *our* fingers on glass thread, and *we* saw was hanging evenly and which *we* heard click against the charged glass rod *we* brought up to it, and so on. There is no discontinuity between pith balls and electric charges. There *is* only one world.

I do not intend to argue the point against phenomenalism any further here. The arguments of J. L. Austin (108), W. D. Joske (196) and others go to establish the point beyond dispute. Besides I dare swear a phenomenalist will long since have thrown this book down in disgust, with its emphasis on the study of the world as the system of things, with causal powers. Science is not, as the phenomenalist would have it, a study of the flux of sensations in the sensory fields of individuals *pace* Berkeley and Mach. So I take for granted that the world as manifested is a system of material things, having causal powers, that is, being of such a nature as to be capable of bringing about changes in and among other things. I seek a general specification of the nature of the entities which must be ultimate in that world; that is which lie at the end of all possible regresses of explanation *for us*.

To refer to something is a linguistic act and depends upon a linguistic capability. It is of the essence of verbal reference that it is possible in the absence of what is being referred to. If we have the linguistic means at our disposal to make reference to an entity we can entertain the hypothesis of its existence. But as I showed in some detail in Chapter 3, the determination of questions of existence hinges upon a somewhat different capability, which makes possible what I have called 'demonstrative reference'. In an act of demonstrative reference the attention is drawn to an entity, *in its presence*. This is not a purely linguistic capability but typically involves the physical act of ostension, the pointing finger, and extensions of that. Only through demonstrative reference can existence be established.

It has been amply demonstrated by P. F. Strawson (97) that the material things of the world as manifested are reidentifiable individuals. But reidentification is a complex notion. To identify a thing is to recognize a certain complex of qualities in the same place. To reidentify something is to recognize the same or a sufficiently similar complex of qualities in the same or a different place, at a different time. Reidentification is a test of the presence of the same individual only if it is assumed that the reidentification of the appropriate set of qualities is a criterion for continuous spatial presence, either in the one place or in a continuous sequence of places, during the time that has elapsed between the first identification of the individual and its reidentification. In short, qualitative identity, for Strawsonian individuals, the ordinary things of the world, is the *empirical* criterion for a *theoretical* numerical identity, which must be supposed to exist in order that the conceptual system we have can be applied to the world as manifested. Some answer is required to a sceptic who raises inductive doubts about the viability of any alleged proof of numerical identity by the use of qualitative identity, on the grounds that what is taken to be one thing because it can be identified qualitatively on two different occasions may, for all we know, be two things very much alike.

A two-part answer seems possible. Since acts of attention are singular and there is a multiplicity of individuals, qualitative reidentification is the best *possible* criterion for numerical identity. It can also be argued that the very possibility of reference, verbal or demonstrative, to things, presupposes a spatial framework, in which there are different places to be occupied by things, and that such a framework is an intelligible construction only under the condition that there are enduring individuals. Since the only individuals for which we have any sort of endurance criterion are material things, the fact that the criterion is inductive is irrelevant, and is merely an aspect of the fact that each of us discovers the world from his own point of view, and that what we think are our discoveries may have to be revised.

Throughout this book the view has been urged that science identifies the real essence of a thing as whatever is responsible for its causal powers. So that manifestation of the same qualities under similar conditions will be a primary criterion for the identity of a thing, being the best possible empirical criterion for the fact of the continuance of the same causal powers. It has been shown further, in Chapter 9, that endurance *un*changed can rationally be supposed of those entities for which no cause of change can be found or imagined. Strawson's criterion can be strengthened, since reidentification is only one part of the satisfaction of a two-part criterion. It established a *prima facie* case for identity of causal powers and hence for the identity of the real essence of some thing. The second part of the criterion for numerical identity derives from the fact that it is changes which need to be explained while simple endurance does not call for explanation. If there is no reason why the numerical identity of the individual whose causal powers are manifested in a qualitative identity should be suspect, then it is the same individual. We are under no obligation to offer any kind of explanation of a thing remaining self-identical, if nothing happens to change it. Nothing happens spontaneously, and in particular no thing can go out of existence without being destroyed or come into existence without being generated. That requires the action of some other thing. These, of course, are facts about our conceptual scheme, which puts such demands upon our powers of experiment and of imagination as are required to satisfy the need for explanation of all change. The suggestion of spontaneous decay of radioactive atoms has to be reinterpreted to mean that the cause of decay is unknown.

It is a commonplace that the reidentification of a material thing can be accomplished even when some of its manifested qualities have changed. Some core of its totality of qualities, as manifested, serves as the *nominal essence* of a thing, that is as that set of its manifested qualities which are required to remain unchanged for it to be reidentified as the same thing. So far as I can see, the enormous variety of kinds of things that there are in the world, which can be located somewhere and which exist at various times and which can be qualitatively reidentified through minor changes, is such that no general recipe for the extraction of the nominal essence of a kind of thing from the totality of manifested qualities of individuals of the kind, is possible. For some things like cars, say, certain determinates, such as the determinate shape, form part of the nominal essence. But as convicted car thieves well know, it is no part of the nominal essence, that is it is not required for reidentification, that the determinate colour remains the same. Only the determinable must be preserved, that is it must have some colour or other. Notice that if a thief stole three or four Minis, melted them down, and made a Rolls Royce out of the metal, he would not of course have stolen the cars but the metal. This is because determinate shape is a part of the nominal essence of this or that make of car.

Perhaps the world as manifested already contains the ultimate entities I am looking for. The ordinary things of the world are Strawsonian individuals, and the next step will be to show that Strawsonian individuals are not ultimate entities. Consider the identity of a piece of silver through some chemical transformation: say into silver nitrate, and back to metallic silver again. Qualitative reidentification breaks down at the second stage, since crystals of silver nitrate are, in every respect except spatio-temporal location, unlike the silver from which they were made, and the silver which they will become. The relatively fundamental individuals in the story which makes sense of the use of the word 'silver' throughout the description of the reactions are the silver atoms. But they are never manifested to observation, in *propria persona* as atoms. They are observed only in the aggregate, as pieces of silver. Aggregates, of which atoms are the constituents, are qualitatively quite different from their constituents. Silver is a white metal, capable of conducting electricity – but silver atoms are atoms of a metal, not atoms of metal. A single atom could not be a conductor of electricity, for instance. It is the assumption of the numerical identity of the silver atoms throughout the process which binds together the qualitative differences and identities, and enables us, after the metallurgist has done his work, to exclaim with satisfaction that we have *recovered* the silver; the very same silver that was dissolved in acid and smuggled abroad in sugar packets.

Numerical identity through qualitative change can be understood as the preservation of the causal powers of the entity in question. It cannot be understood, as the above arguments show, by any extension of the notion of qualitative identity. The constitution of an entity, in virtue of which it has its causal powers, is its real essence. But such constitutions are not manifested. Powers show themselves in their effects, that is qualitatively.

The power to reflect light is possessed by a thing by virtue of its electronic constitution. The power manifests itself in shinyness or colour, and the existence of the quality is explained by reference to the electronic constitution of the stuff, but the electronic constitution is not manifested in the world.

So two conclusions of paramount importance follow:

(1) Strawsonian individuals, which include most material things, cannot be ultimate entities for science. The latter must be such as to preserve numerical identity and causal powers. They can, of course, be the permanent constituents of Strawsonian individuals, whose identity can be established inductively by qualitative reidentification.

(2) Fundamental entities, existing through qualitative change, and existing too when a thing is not manifesting its qualities, which are to be understood as the effects of its powers, cannot be qualitatively identified. But since in the perceived world the only things that instruments and observers can do is to make qualitative identifications, ultimate entities must be individuated at best via Strawsonian individuals. Such and such an atom of phosphorus is *an* atom of *this* (demonstratively identified) piece of yellow, waxy stuff. Location can be achieved by verbal reference by the use of the spatio-temporal framework, but it has been amply shown by Strawson that our capacity to construct and use that framework depends upon our capability to reidentify material things qualitatively, that is to perform acts of demonstrative reference.

In terms of the distinction used in several places in this book we can now sum up the argument, so far, in the principle that ultimate entities can be verbally referred to immediately, but can be demonstratively referred to only mediately, via demonstrative reference to the material things of which they are constituents, and for whose powers they are ultimately responsible.

I want to notice in passing, and for future reference, a distinction between points of space, and points in space. Both can be the subject of individuating reference via the spatio-temporal framework. Neither can be demonstratively referred to except via the demonstration of some material thing, but the point of space cannot be a constituent of any material thing, since different material things can occupy the same point of space. If the ultimate constituents of material things are entities having powers, even if these must be regarded as, in a certain sense, points, it is a necessary condition of the possibility of distinguishing a place from its occupants that they be regarded as points in space, identified by their effects, rather than as points of space, identified either by their occasional occupants or by reference to their relation to the spatio-temporal framework. A point in space can be at a point of space. Thus it is essential that some identifiable individuals, the points of space, be supposed to be without causal powers.

Boscovitch noticed that for the full analysis of all kinds of change possible in material things, and for properly contrasting them with their ultimate constituents, it was important to make out the distinction between a transformation; a change of real essence, and an alteration; a change in manifested qualities, which might or might not be a change in nominal essence. We have already seen this distinction operating in taxonomy, when a change in the manifested qualities of a population, the nominal essence of a species, is only denominated a change in species, as distinct from a change in variety, when it is believed that it is the product of a *change* in the genotype, rather than the same genotype differently manifested. The genotype is exactly the real essence of the species. Fundamental entities of the world will be those which, having no nominal essence of manifested qualities of any kind, cannot be altered, and being the bearers of numerical identity cannot be transformed: that is whose real essences are permanent.

Material things have causal powers. Those causal powers are manifested in various ways, but particularly as the qualities of things when they are perceived, and as the reactions of instruments when they are in some sort of relation with them. The first step in the historical development of science was the conceiving of the hypothesis at the root of the Corpuscularian Philosophy, the distinction of primary and secondary qualities. The manifestation of secondary qualities was to be explained by the actual possession by a material thing of primary qualities. In particular it was because the parts of thing were of a certain shape, in a certain arrangement and in a certain state of motion that they were capable of affecting someone who, say, saw the thing in such a way that he saw it yellow. The power of a material thing to manifest itself as coloured, or as of a certain felt warmth and so on, could be shown to be rooted, it was believed, in certain arrangements and states of motion of its parts, and its power to manifest itself as electrically charged, for instance, to an electroscope, would be similarly rooted in the 'bulk, figure, texture and motion' of its parts or perhaps in some more sophisticated set of primary qualities. This great hypothesis has been largely borne out at a first level of explanation. It can scarcely be disputed that the explanation of quite a wide range of the manifestations of material things can be given by reference to the structural arrangement and motions of their parts. Such parts are of course of a certain volume, relatively impenetrable by others, of a certain shape, and in a certain structural arrangement and in various states of motion which change by impact or other interaction among the parts. But the critique of Corpuscularianism as an account of the *ultimate* entities as constituents of things, and of their smaller parts, is unanswerable. The ultimate entities are not, and cannot be, solid volumes in motion, exchanging momentum and redistributing velocities amongst themselves by impact. Ultimate entities can have no qualities, not even primary qualities, because qualities are in the end the manifestations of powers. The ultimate entities must, therefore, be simply the bearers of powers. Since they must be constitutive of spatio-temporal things, they must be in space and time, and in principle locatable, though since they cannot themselves be manifested qualitatively they cannot be demonstrable. I shall show that within our conceptual scheme (which is perhaps better called our 'way of thinking') a kind of entity which I shall call a 'point-centre of influence' can fulfil the requirements which have been forced upon us for ultimate entities.

In short the ultimate entities will be seen to *have* to be unanalysable, locatable and permanent bearers of powers, not permanent congeries of qualities.

The preference in ultimate terms for powers over qualities, as the nature of the ultimate, seems to be based in the end upon two considerations:

(*a*) Whatever the nature of the permanent entities, that nature must be permanently possessed. Material things are manifested qualitatively in perception, but their numerical identity through time, a condition of the possibility of rational discourse about a world in which the separate objects can be the subjects only of intermittent attention by perceivers, and occasional detection by instruments, cannot be through qualitative continuity, since there is no qualitative continuity, let along perceivable qualitative identity. Manifested qualities are the product of the powers of a thing to affect other things, when their environment provides the

L 2

stimulus by which this power is occasionally evoked. So the qualities can hardly be ultimately the constituents of the things. The mode of existence of the ultimate constituents of material things cannot be the mode of existence of the qualities of things, either as manifested in perception, or as detected by instruments. But to exist as the bearer of a power is just the required mode of existence for ultimate entities, since the same power can be latent or evoked, and it is quite clear that the manifestation of material things in perception and their detection by instruments *are* evocations of powers that are latent, when the thing in question is not being perceived or detected.

(*b*) The powers of Strawsonian individuals are secondary powers, since the permanent states of the thing, in virtue of which it possesses its powers, can be described in qualitative terms. They can be described in terms of the state of its internal constitution. It has often proved possible in practice, to provide a preliminary analysis of the nature of a thing or substance by the use of a corpuscularian theory, involving entities with primary qualities, but still qualities. But (i) the ultimates cannot be corpuscularian, and (ii) the manifestation of primary qualities are as much the effects of powers as are secondary qualities.

The final step in this stage of the argument is to follow Kant and Boscovitch in the demonstration that the primary qualities are the effects of powers. It is in Kant's *Metaphysical Foundations of Natural Science* that the most thorough reduction of the corpuscularian to the dynamic philosophy is to be found. There, as will be demonstrated, the primary qualities of matter are readily shown to be the effects of forces of attraction and repulsion acting from point centres of influence. Boscovich, more profound than Kant, had already seen that the notion of force was not so fundamental as that of power. Furthermore he saw that the power to attract and repel, though truly ascribable to a single centre of influence, cannot be manifested by a point centre of influence in isolation. The manifestation of such a power must be attributed to pairs of point centres of influence, which must, then, be the minimal arrangement of ultimate entities which could be perceived, or detected, that is, could have qualities. According to Boscovitch, a centre of influence tends to move towards or away from another centre of influence, their mutual action taking place at all degrees of separation in space, but of a strength and sense which depends in a complicated way upon their distance apart. However, for the purpose of the exposition of the true nature of primary qualities, it is perhaps easier to use the metaphor of 'force', understanding it as a fictional explanatory notion for the unanalysable and inexplicable power to attract or repel.

The dynamic account of nature depends upon the single principle that all point centres of mutual influence both attract and repel all others. The degree to which the power of attraction is manifested in mutual accelerations varies with some function of distance, and is different from the degree to which the power of repulsion varies. The 'forces' of attraction and repulsion, one might say, obey different laws. A world of material things is possible only because there is a difference in the degree of dependence on mutual separation of point centres, in the two effects. From the fact of that difference alone the general lineaments of a material world will be shown to follow. Kant and Boscovich offered different laws by which this differential effect is brought about, and a thorough-going dynamic analysis of modern physics would require still other laws. Kant guessed that the power of repulsion falls off as the inverse cube, and attraction as the inverse square of the distance from the centre of influence to a test body. Boscovitch guessed that nett 'forces' between pairs of point centres of mutual influence are alternatively repelling and attracting, as the distance between the centres increases, but tend to obey an inverse square law of attraction when they are some distance apart. He conjectured that at very great distances the nett 'force' between centres of influence might again be one of repulsion, so that the grander astronomical entities would tend to move apart, and the universe seem to expand. Of course, just what laws these 'forces' obey, and whether indeed there may not be tangential accelerations as well, that is, a tendency for mutual centres of influence to move at right angles to the line joining them, are all matters for empirical study. All that a philosopher can offer is an *a priori* proof that the fundamental entities, if we discount phenomenalistic positivism as an internally consistent but absurd caricature of natural science, must be centres of mutual influence, and that the degree to which they mutually repel and attract are different functions of distance, whatever those functions are.*

Volumes of solidity and mass are the two primary qualities most inherent in and inseparable from the Corpuscularian or qualititative view of matter. I shall show that both these alleged fundamental qualities can be understood as the effect of powers of mutual attraction and repulsion between centres of influence.

Solidity is the alleged quality, the possession of which is responsible for the fact that two material things cannot occupy the same place at the same time, and is logically connected with impenetrability, the power to resist penetration, in that the possession of the former is supposed to account for the manifestation of the latter. Solidity is supposed to be the permanent state of a thing which ensures that a thing has the secondary power to resist any other body. An apparently solid

* Here we develop the simplest possible form of *field*, to explain the *manifestation* of qualities.

volume can be readily created around a centre of influence which affects other centres of influence with repulsive and attractive 'force'. The region around a centre of influence in which the repulsive 'force' is greater than the attractive will resist the penetration of other bodies. The surface around the point at which the nett 'force' is zero will seem to be the surface of a compressible solid and outside that surface there will seem to be a free space in which these solid bodies attract one another. Graphically, this could appear around one of any pair of centres of mutual influence, like this:

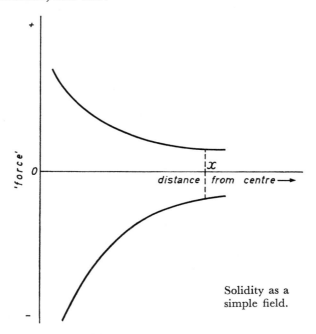

Solidity as a simple field.

The repulsive 'force' has a larger absolute value near the centre of influence than the attractive force, but falls off more rapidly, so that at x the 'forces' are equal, and there the nett 'force' is zero. A heavy body resting on the surface of the earth is subject to a nett 'force' of zero, since at the surface the repulsive 'force' of resistance to penetration exactly balances the attractive 'force' of gravity. The quality of solidity is how a nett 'force' of repulsion is manifested in the perceived world. It can no more be attributed as an ultimate quality to be constitutive of the ultimate entities or ultimate stuff, than can colour. Clearly the volume occupied by a body will depend upon the absolute magnitude of the repulsive 'force', if, as Kant was inclined to assume, the attractive 'force' is the same for all bodies. With our superior knowledge of how many interactions can be recognized between centres of influence,* that is, how many different laws of 'force' can be found to hold for the mutual action of centres of influence, both of an attractional and of a repulsional nature, we must say that volume of a body depends upon the vector sum of 'forces' of each kind.

As volume of solidity is connected with repulsion, so inertia is simply connected with the attractive powers of centres of mutual influence. Under the action of any given attractive power the rate at which the centres tend to come together, that is, their mutual accelerations, depends upon their distance apart and the ratio of their masses. In what we take to be the space between bodies, a centre of influence is under the attractive influence of all other bodies. These may be supposed to be so widely distributed that their effect upon one of a pair of mutually attracting bodies (that is, a centre of influence surrounded by a nett repulsive 'force') may be supposed to depend solely upon the absolute magnitudes of the power of attraction of each, for all the other bodies in the universe. This 'force' will oppose any change in position of a body with respect to all the other bodies in the universe. It is the power to resist change in motion called 'inertia'. It will depend upon the absolute magnitude of the attractive 'force' of a centre of influence, the ratios of which, for a pair of bodies, will appear as the ratio of their masses. So mass is no more an absolute, fundamental and primary quality of matter than is volume of solidity. It is but the effect of the mutual attraction between all centres of influence, as Mach supposed.

In a similar way differences in density between different substances can be given a dynamic explanation. Instead of the differences being due to more, or less, corpuscles being packed into a given volume, density, the ratio of mass to volume, will, in the dynamic view, depend upon the ratio of attractive to repulsive 'force', and will simply be a measure of this ratio. To take a much simplified case: suppose that attraction obeys one law in all bodies, and repulsion a different law, so that repulsion falls off more rapidly around a centre of mutual influence than attraction. The smaller the magnitude of the repulsive 'force' near the centre of influence, the smaller the 'volume' of the body. Consider a centre of repulsion with a certain absolute magnitude of repulsive 'force' near its centre, and another which repels a test body with twice that absolute magnitude. Suppose the attractive power of these centres of influence to be the same, then the 'masses' of the two bodies will be declared equal. But the former occupies a smaller 'volume' than the latter, hence it will be declared to be more dense. Of course, for bodies other than fundamental entities we know that differences in density are often to be attributed to the degree of packing of their material constituents.

*E.g. Strong and weak interactions etc.

The ultimate entities are not corpuscles constituted by their having the primary qualities of volume, solidity and mass. They are point centres of mutual influence, each of which is in continuous interaction with all other centres of mutual influence, and which tend to accelerate towards or away from each other at different rates, depending upon their distances apart. Secondary individuals which have primary qualities are groups of these primary individuals. The characteristics which individuals which are to be primary in our conceptual system must have, can be laid out *a priori* in the following six principles.

Principle One: The ultimate entities of the world, as we can understand it, must be point centres of mutual influence, that is centres of power, distributed in space. They are perpetually redistributing themselves in space, that is, they are continuously changing their spatial relations, and, consequently, their mutual influences, since these are distance dependent.

The adoption of this principle, laying down, in a general way, the specification of the concepts of ultimate entities of all possible conceptual systems, and hence the nature of the ultimate entities for all possible comprehensible worlds, is clearly forced upon us by the train of arguments that lead to the substitution of powers for qualities. The powers of mutual influence, for instance of attraction and repulsion, which are the only characteristics with which point centres of attraction need be endowed, and which are supposed to vary in their effect only with some function of distance, can be discussed via the metaphor of 'forces of attraction and repulsion'. Then following Kant and Boscovich, it can be shown that the primary qualities of materialists, supposed to be constitutive of an alleged ultimate kind of material thing or corpuscle, are not ultimate. Indeed such a conceptual scheme, depending on the notion of the redistribution of motion by contact, can, on closer analysis, be seen to be impossible. The primary qualities can be constitutive of material things functioning in causal mechanisms which are to be adverted to in less than ultimate explanations, but cannot be ultimate, ways since, like secondary qualities, they can be shown to be some of the certain balances of influence among the point centres are manifested.

Principle Two: The most fundamental material thing or atomic corpuscle, which is defined by its primary qualities, is a collection of centres of power, of mutual influence, which are jointly such that a continuous limiting surface of an infinite region of zero repulsive 'force' surrounds it. The surface is created by the mutual effect of the constituent centres of influence.

This principle unites the ultimate entities of the world with the manifestations of the world to our instruments and to ourselves. The world we perceive is made up of qualitatively distinguished things. Their qualities are divisible into groups, one group of which are to be explained in terms of mechanisms of constituent parts, made up of smaller material things, having only the qualities of the complementary group. The manifestations of the complementary or primary group are additive or some other functions of the primary qualities of the constituent *material* parts. But these, as we have seen, are not, and cannot be ultimate.

Principle Three: Perceived things are groups of corpuscles having primary qualities (though the list of these should not be thought to be confined to those mentioned by such physicists as Galileo and Newton, as the development of science has shown). Perceived things, being perceptible, necessarily have the power of manifesting secondary qualities.

The principle follows from the exegesis of Principle Two, above.

Principle Four: The mutual influences between ultimate point centres of power must be at least attraction and repulsion depending in some way or ways on the distance between the centres.

The most fundamental distinction which has to be preserved in the world as perceived and as detected by instruments is that between solid things and empty space. For this manifested distinction to be preserved there must be a differential between the variations of attraction and repulsion under changing conditions of separation of the centres of power. The only conditions that can change among ultimates are spatial relations, so attraction and repulsion must vary with some function, or functions, of distance between centres. From this, as we have seen, it follows necessarily, that solidity must be manifested in volumes, and that such solid volumes must have inertia.

Principle Five: The ultimate entities will also be the ultimate referents or logical subjects since they are qualityless points which are centres of mutual influence; they are individuated for pure numerical difference by verbal reference to their location in the spatio-temporal framework. This mode of reference depends *in general* on the existence of material things, but not on the existence of any particular material thing. The ultimate entities are individuated for any other difference, i.e., in their powers, as the constituent parts of qualitatively differentiable, secondary or Strawsonian individuals, which must be capable of being perceived or detected by instruments individually and through which mediate demonstrative reference can be made to the ultimate entities. The manifested qualities of the things we perceive arise from the mutual influence of the ultimate entities of which they are constituted. Thus, for what corresponds to qualitative identity for the ultimate entities, reference is mediated by a *particular* dependence on demonstrable material things.

Principle Six: The ultimate entities are neither perceptible nor detectable as such. This is not really an independent principle since it follows directly from Principles Two and Five. But it is of great importance for the epistemology of science. To put the principle in another way: it means that the ultimate entities can be referred to both by the mediation of the spatio-temporal system and by the mediation of particular material things. But the ultimate entities are not capable of being demonstratively referred to immediately. One can point to where they are, but one cannot point to them except by pointing to a qualitatively distinguishable and reidentifiable material thing of which they will be the ultimate constituents.

The way in which ultimate entities as the bearers of powers lie at the terminations of explanatory regresses must be stated with great care. The explanatory regress begins by the use of secondary powers to explain the manifestation of secondary qualities. These are but secondary powers because the state of nature which exists, in virtue of which the power can be exercised, is itself capable of scientific analysis in terms of the arrangements and states of constituents which can still be distinguished – at least in principle – qualitatively. These qualities are the primary qualities not necessarily confined to the mechanical list offered in the seventeenth century. But since there are no further qualities which would make a qualitative account of the state of nature responsible for the manifestation of primary qualities, that regress must terminate, if it terminates at all, in entities distinguished from empty space only by their powers.

But simply to have offered an explanation in terms of pure power does not close a regress for ever. To assert that phenobarbitone puts one to sleep because it has the power of putting one to sleep is no final termination to the regress of explanation of this phenomenon. It is always possible to ask what it is about the nature of phenobarbitone by virtue of which it has the power. And this possibility is always open whenever a regress terminates in a pure power. Whenever a power is accounted for by a state of nature, in virtue of which the power is exercised, the manner in which this state is responsible for the effect, the causal mechanism by which it acts, is either describable in terms of a more refined mechanism, which keeps the regress open, or it is closed by the adversion to powers. What can be said is that wherever our regresses stop, the fundamental material things, which having qualities might be observed, can be explanatorily underpinned by ultimate entities, which being the bearers of pure powers cannot be observed. No regress stops necessarily, but most stop contingently, and all will, in the end, be halted by insoluble technical problems of detection and observation. At such a juncture the only remaining step open to a theoretician is the postulation of ultimate entities.

Let me remind you what exactly it is to attribute a power. It is certainly to attribute more than a mere tendency, or disposition. To say of something that it has a power to produce an effect, that is, affect something in a certain way, is both to attribute to it some state or constitution in virtue of which it can produce the effect, though it is precisely *not* to say what particular state it is in; and to *specify* what it can do, that is what effect is produced. To say of A that it has the power to bring about B is to say that the causal chain of which B is an effect terminates in some state of A, and that that state, unspecified, though not unspecifiable in principle, is responsible, together with the circumstances and circumambient conditions, for the happening B. To say that the ultimate entities have powers thus leaves open the possibility of specifying the constitutive state in virtue of which they have the powers they do. Once this state is specified the entities cease to be ultimate.

It follows from this, and from the general discussion of Chapter 10, that the truth conditions for the attribution of powers are not simple. At least for what I have called 'secondary powers', a qualitative description of the state or constitution of the material in virtue of which the power is exercised is possible. The relevant characteristic of nitric acid, in virtue of which it has the power to dissolve silver, can be pretty exactly specified. So, for it to be true that nitric acid has the power of dissolving silver, it must be the case that silver is actually dissolved in the presence of nitric acid, that is, the alleged effect must occur, and it must also be the case that there is some characteristic of nitric acid, something to do with its chemical and physical nature, which can plausibly be held responsible in the relevant circumstances for the happening, i.e., the constitution of nitric acid, and not its mere presence, figures in the explanation of the happening. The truth conditions for attributions of secondary powers, being associated with characteristics of secondary individuals and with observable effects, involve both a non-inductive criterion, namely 'Does the specified state exist?', and an inductive criterion, namely 'Does the effect usually occur in similar circumstances, and can its occasional non-occurrence be explained within the same system of facts and hypotheses?'

But for the ultimate powers neither criterion is preserved. Since, *a fortiori*, no qualitative account of the states or constitutions responsible for the powers of ultimate entities can be given, the existence of these states or constitutions cannot be open to subsequent proof, since they are indemonstrable and the proof of existence depends upon the possibility of demonstrative reference. It is always open to us to say that absence of the effect is proof of the absence of the ultimate entity having power to produce the manifestation of the quality, so the truth of hypotheses as to the presence of ultimate entities is necessarily related to the truth of the particular factual statement as the occurrence of the phenomenon they explain. But this does not make their postulation trivial, since in another era the frontier of knowledge may be pushed back by the discovery of the constitution of the previously ultimate entities. This is happening right now with protons, electrons and quarks. Of course, in order to attribute an internal constitution to a point centre of mutual influence, it is often required that what was punctiform as an ultimate entity is assigned a structure of parts in space as a secondary entity, and so is seen to have a volume. This step is of course appropriate to the change in status of the entity in question. For example, the attribution of internal structure to the Clausius point-mass molecule, deprives it of its punctiform character as it deprives it of its ultimate status.

How many powers does an ultimate thing have? We have seen how powers are ordinarily individuated. Clearly in the case of ultimate individuals we can say what we like, though it is perhaps more elegant to follow Kant in attributing both a power of attraction and a power of repulsion to ultimate entities, than to follow Boscovich in attributing a power to attract or repel depending on the degree of separation of centres of power; but perhaps not. However for secondary individuals, typically material things, where the states in virtue of which the power is possessed can often be specified, a good deal more can be achieved by way of criteria of individuation of powers. Must there be a separate power for every manifested quality? It seems clear that both science and common sense uses the notion of power in such a way that powers are related to determinables, but not to determinates. A material thing has the power to be coloured, not *the* power to be red. Changing the circumstances, say by varying the colour of incident light, changes the determinate colour a thing looks, but it is the very same power which in some circumstances is manifested as red and in another as black. A rationale for this practice can surely be found in the assumption that it is virtue of the same constitutional state that, in different circumstances, the different determinate colours are manifested. Thus the difference in the manifestations can be readily explained by the manifested difference in the circumstances, supposing the state of the material thing or substance responsible for its power remains the same. And since nothing has been done to the internal constitution of the thing, it follows from the Principles of Unchange, that the internal state of the thing *is* the same. So in general, if the state of the internal constitution in virtue of which the thing has power is S, and in two sets of circumstances $C1$ and $C2$, effects $E1$ and $E2$ are produced, we can say that it is in virtue of the same power, providing the following two statements are acceptable:

(1) 'If and only if S and $C1$, then $E1$.'
(2) 'If and only if S and $C2$, then $E2$.'

These are strictly empirical hypotheses, because, since S is in principle qualitatively specifiable, being a state of a secondary individual, for instance a material thing, it is in principle observable or detectable so these statements are in principle properly empirically testable, within the general bounds of the possibilities of the confirmation of empirical hypotheses.

Lastly, the entities which are the centres of influence defined in this chapter, which are an *a priori* specification for the ultimate entities of all possible science, must be distinguished from poles, a notion proper to a specific state of the art of physics. The metaphysical theory of point sources of influence is not a polar theory. Poles are secondary individuals, since they are qualitatively distinguished. A polar theory works by distinguishing two or more kinds of poles and then treats the tendency to be attracted or to be repelled between pairs of poles as explained by their being of like or different kinds. In the familiar theory of magnetic poles, the explanation of the fact that two poles accelerate towards each other is that the poles are of different polarity, and the explanation of the fact that two other poles accelerate away from each other, is provided by the hypothesis that they are of the same polarity. Polarity then becomes a kind of quality, though not a quality that distinguishes things as perceived, but nevertheless a quality because the difference between kinds of poles can be detected. The presence of polar differences as determinates in things can be detected by suitable instruments. In principle it is possible to ask for an account of what it is about the structure and constitution of the magnetic body and its regions of polarity for one pole to manifest boreality and another pole australity. It is surely just contingent that science offers no satisfactory answer, and can proffer no entirely acceptable causal mechanism at present.

Michael Faraday, who of all scientists, has had perhaps the clearest grasp of the metaphysical point of view advocated in this book, saw that a polar theory could not be fundamental, because no qualitative theory could stand as the termination of a regress of explanation. For him each point in space was a centre of mutual influence on every other, and from this conception the true notion of the field emerged. But that is a chapter in the history of physics.

The upshot of the extended argument of this chapter is this: *Every fundamental theory must, as expressed in the language of physics, be a field theory.* But with this caveat, that the influences which are the fundamental field in one era of scientific understanding may be given an explanation in qualitative terms in another era. But then these qualities, and the mechanisms of interaction they enable to be conceived, call for a new fundamental theory, a new field structure. In contrast to this is the ever popular concept of the atom, a thing that is both ultimate and material. It is that notion that is incoherent.

Science contains two kinds of propositions, those in which specific knowledge of how things and materials behave are given, and speculations as to the natures and constitutions of things. The two groups of propositions are bound together through the concept of power, and a science which pursued either of these kinds of knowledge in the absence of the other would be no science at all. To sort out from among the indefinitely many descriptions under which a thing or material may fall those which express its essential nature for science would be impossible without relating its nature to what it can and does do.

In the last analysis, individuals and materials *are* just more or less complex centres of powers. It is logically impossible for there to be a world of wholly passive entities. The only logically possible worlds are those in which there are some active and some passive entities, for those in which all entities were wholly active are inconceivable, since for something to affect something else, the latter, be it individual or material must be acted upon. Nothing could magnetize unless something was magnetic. Entities are the centres both of powers, in the ordinary sense, and of liabilities and tendencies.

Science, conceived as the study of the behaviour of things and materials and the elucidation of their natures, will be a dialogue in which the ascription of powers prompts the investigation of natures, and the analysis of natures leads to the discovery of new kinds of things and materials to which powers are again ascribed, restarting the cycle.

For all we can know the world is infinitely old, and will endure for ever. It is made up of an enormous quantity of immensely complex things, acting and interacting, in various ways. We have no reason to think that these things are anything but ultimately infinitely complex in structure. But scientists have discovered the first level of inner structure and the mode of working of many things, and they have found other structured entities different from the ones we can perceive, both as parts of things, and as things of which our familiar world forms a part. Science is possible because human beings possess two essential capacities. Our linguistic capacity enables us to formulate general propositions, and so to grasp the flow of events. But our imaginative capacity, by which we envisage and so grasp the structures and constitutions of things, yields a deeper understanding, for it offers us, when under the discipline of truth, the capacity to understand the causal structure of the world.

SUMMARY AND BIBLIOGRAPHY

It is argued that only a 'powers' conception of ultimate entities is viable.
See P. F. Strawson (95), Ch. 1, for a general account of the connection of reference, material things and space.
For the distinction between 'points of space' and 'points in space' see

253. A. M. QUINTON, 'Matter and Space', *Mind*, **73**, 332–52.
It is argued that the first step towards the ultimate must involve primary qualities and structural conceptions. See

254. F. FRÖHLICH, 'Primary Qualities in Physical Explanation', *Mind*, **68**, 209–17.
255. L. L. WHYTE, *Aspects of Form*, Lund Humphries, London, 1951.
256. K. E. TRANNOY, *Wholes and Structures*, Theoria, Stockholm, 1956.

For the reduction of primary qualities to powers, and the dynamical philosophy generally, see

257. I. KANT, *The Metaphysical Foundations of Natural Science*, translated E. B. Bax, Bell, London, 1883.
and for an attempt at this sort of treatment of present-day physical ultimates:

258. W. A. WALLACE, 'Elementarity and Reality in Particle Physics', *Boston Studies*, III, Reidel, Dordrecht, 1967, 236–63.

Author Index

Abramenko, B., 294
Achinstein, P., 61, 62, 156, 201
Ackerman, R., 61
Aldrich, V. C., 201
Alexander, H. G., 233
Alexander, P., 128, 200, 201
Archimedes, 61
Aristotle, 133, 141, 213
Armstrong, D. M., 233, 255
Austin, J. L., 90, 91, 189, 201, 298
Ayer, A. J., 30
Ayers, M., 272, 294

Bacon, F., 26, 30, 294
Backman, C., 58, 59
Baier, K., 233
Bartley, W. W., 31
Baumer, W. M., 128
Baylis, C. A., 201
Benjamin, A. C., 200
Bennett, J., 91,177
Berkeley, G., 30, 198, 298
Beth, E. W., 31
Black, M., 61, 62
Blake, R. M., 29
Bohm, D., 127
Bohr, N., 44, 45, 48, 49
Boscovich, R. J., 85, 266, 283, 285–
    289, 293, 294, 302, 304, 305,
    308
Boyle, R., 30, 77, 265, 266
Braithwaite, R. B., 31, 61
Bridgman, P. W., 200
Brock, W. H., 30
Brodie, B., 2, 200
Brown, P., 294
Buchdahl, G., 233
Buhler, K., 32
Bunge, M., 126, 127, 255, 294
Butts, R. E., 128

Caldin, E. F., 31
Campbell, N. R., 31, 62, 97, 132

Cabanis, P. J. G., 213
Capek, M., 294
Carnap, R., 22, 30, 50, 61, 158,
    168–71, 174–5, 176, 177, 210
Chappell, V. C., 255
Chisholm, R. M., 128
Chomsky, N., 90, 180, 181
Clausius, R., 40, 312
Clavius, C., 29, 32, 126
Collins, A. W., 129
Cohen, L. J., 177
Cooper, N., 176
Coval, S. C., 96
Crawshay-Williams, R., 233
Crittenden, C., 91
Cusa, N., 243–5, 251

Darwin, C., 57, 58, 131, 170
Dedekind, R., 288
Descartes, R., 1, 8, 9, 18, 31, 96,
    266
Dietl, P., 31
Dretske, F. I., 90, 194, 196, 255
Drude, P. K. L., 60, 100, 142
Ducasse, C. J., 29, 128
Duhem, P., 61
Dumas, J. B. A., 95

Eddington, A., 256
Ellis, B., 62
Euclid, 237
Eysenck, H. J., 99

Fain, H., 32
Faraday, M., 36, 85, 98, 230, 266,
    313
Feigl, H., 233
Festinger, L., 221
Feyerabend, P. K., 201
Feynman, R. P., 171
Fine, A. I., 255
Fisk, M., 127
Fleming, M., 202

Flew, A. G. N., 31
Fodor, J. J., 233
Frege, G., 3, 141
Freud, S., 45, 219
Freudenthal, H., 62
Frölich, F., 315

Galen, 213
Galileo, G., 30, 50, 170, 253, 265, 309
Gassendi, P., 2, 23, 32, 242
Geach, P. T., 90
Gilbert, W., 10
Gillispie, C. C., 224
Goodman, N., 2, 25, 26, 32, 112, 127, 128, 274
Grene, M., 127, 294
Grunbaum, A., 31, 255

Hacking, I., 129
Hall, D. J., 202
Halle, M., 180
Hamlyn, D. W., 233
Hampshire, S., 212, 217
Hanson, N. R., 31, 195–6, 201
Hare, R., 85
Harlow, H. F., 17
Harré, R., 29, 31, 61, 62, 98
Hart, H. L. A., 102
Hawkins, D., 32
Heath, T. L., 61
Hempel, C. G., 16, 18, 21, 30, 31, 61, 71, 94, 116, 119, 120, 121, 122, 128, 134, 169, 274
Hertz, H., 31
Hesse, M. B., 31, 32, 61, 128, 200, 201
Hilbert, D., 237
Hintikka, J., 127
Hollis, M., 255
Honoré, A. M., 102
Hooker, C. A., 32
Hume, D., 3, 5, 42, 30, 99, 104, 105, 106, 109, 126, 267, 274, 275
Hutten, E. H., 61, 176

Jeffrey, R., 177

Jeffreys, H., 177
Jobe, E. J., 127
Joske, W., 201, 284, 296, 298

Kant, I., 85, 244, 247, 256, 262, 283, 289, 294, 296, 304, 305, 306, 308, 312, 315
Katz, J. J., 126
Kelvin, Lord, 39
Kemp-Smith, N., 256
Kepler, J., 29, 101, 126, 170, 195, 244
Keynes, J. M., 156
Kneale, W., 128, 177
Knight, D. M., 30
Koestler, A., 32
Körner, S., 156
Kripke, S., 128
Kuhn, T., 32, 84
Kultgen, J. H., 30
Kyberg, H. F., 177

Lamark, J. B., 170
Langer, S. K., 201
Laplace, P. S., 159
Lavoisier, A. L., 171
Lejewski, C., 90
Leplin, J., 201
Levi, I., 177
Lloyd, A. C., 200
Locke, J., 5, 10, 26, 30, 85, 201, 213, 266, 294
Lodney, D., 255
Lucas, J. R., 164

McCloskey, M., 62
Mach, E., 3, 25, 182, 200, 294, 298, 306
Mackie, J. L., 127, 128
Mackintosh, N., 50, 51
Madden, E. H., 29
Malthus, T. R., 57, 58
Manser, A. R., 20
Martin, R. M., 201
Mathur, K., 44
Maxwell, J. C., 230, 255
Maxwell, N., 32

Mayo, B., 128
Medawar, P., 32
Mellor, D. H., 156
Mendel, G., 170
Mendeleev, D. I., 18
Meyerson, E., 32
Miller, D. C., 95
Mises, R. von, 30, 177
Moore, G. E., 153–4
Morrison, P. G., 127, 256
Mullan, S., 215
Munitz, M. K., 91, 294

Nagel, E., 32, 90, 127
Nagel, T., 233
Nerlich, G. C., 128
Newton, I., 9, 10, 30, 85, 108, 109,
    144, 208, 227, 265, 266, 286,
    309
Nicod, J., 2, 24, 26, 119, 121
Nidditch, P. H., 31
North, J. D., 32

Odling, W., 55
Oppenheim, P., 18
Oresme, N., 50

Pap, A., 91
Pears, D., 128
Pearson, K., 182
Penfield, W., 215
Pitcher, G., 201
Planck, M., 170
Poincaré, H., 47, 182
Polanyi, M., 32, 294
Popper, K., 24, 127, 159, 175–6,
    177
Prior, A. N., 156
Putnam, H., 233
Putnam, R. A., 194–6, 201

Quine, W. V., 90, 156
Quinton, A. M., 255, 315

Rankin, K. W., 255
Rayliegh, Lord, 170–1
Reichenbach, H., 177, 239, 255

Rescher, N., 31
Rosen, E., 29
Rozeboom, W. W., 129
Russell, B., 3, 68
Rutherford, E., 48
Ryle, G., 217, 269, 294

Sachs, M., 294
Salmon, W. C., 129
Schachter, S., 218
Scheffler, I., 31, 90, 201
Schleiden, M. J., 173
Schlesinger, G., 61, 293
Schoenberg, J., 129
Schrödinger, E., 294
Scott-Blair, G. W., 255
Scriven, M., 31, 91
Secord, P., 58, 59
Sellars, W., 31, 127, 253
Shapere, D., 201
Singer, J. E., 218
Sklar, L., 233
Smart, J. J. C., 31, 233
Smokler, H., 129, 177
Sommers, F., 90, 233
Spector, M., 61, 201
Spinoza, B., 9, 256
Sprigge, T., 294
Stahl, G. E., 171
Stallo, J. B., 294
Stebbing, L. S., 90
Stefan, J., 170–1
Stone, D., 129
Strawson, P. F., 90, 91, 189, 212,
    217, 296, 298, 301
Suchting, W. A., 128
Sutherland, N. S., 50, 51
Swanson, J., 62
Swinburne, R., 255, 294

Tarski, A., 190
Taylor, C., 127
Teichmann, J., 233
Theobald, D. W., 91
Toulmin, S., 32, 128, 176
Trannoy, K. E., 315
Treismann, M., 50

Turner, J., 61
Tycho Brane, 195

Van der Waals, J., 144
Van Helmont, J. B., 269
Vigotsky, L. S., 32

Waismann, F., 156, 200
Wallace, W. A., 315
Watkins, J. W. N., 91, 156
Watling, J. W., 128
Weber, M., 62
Weyl, H., 256
Wheelwright, P. E., 62

Whewell, W., 26, 42, 61, 97
Whitehead, A. N., 91
Whitrow, G. J., 255, 294
Whyte, L. L., 315
Wiener, P. P., 61
Williams, L. P., 85
Wilson, F., 31
Wilson, P. R., 129
Withers, R. F. J., 201
Wittgenstein, L., 32, 37, 181, 182, 200
Woodger, J. H., 31

Yolton, J. W., 129

Subject Index

Accidents, 131, 152–3
Action, paradigms of, 257–8,
    266–8
Affirmation of the consequent, as
    a methodological principle, 23
Analysis and synthesis, 9
Art, works of, 27–8

Capability, 277–8
Causal laws, 92, 101–4
    necessity of causal laws, 93, 103–4
    universality of causal laws, 93,
    103
Causal powers, of ordinary things,
    296, 302, 304
    of ultimate things, 296
Causal reports, 92, 93, 103
Causal transforms, 52–3, 54, 59
Causation, Problem of, 6, 29
    transcategorial, 206, 330
Cause, critique of regularity theory
    of, 104–10
    meaning and criteria of, 104–10,
    113
    proximity of cause to effect, 63,
    73–4
    realist theory of, 206, 229–30
Classification, 179, 196–200
Classifying criteria, 207–8
Conditionality, 93, 111–12
Confirmation, 158
    grounds of confirmation state-
    ments, 158
    meaning of confirmation state-
    ments, 158, 165–7
    object of confirmation statements,
    158
Copernican Revolution in the
    Philosophy of Science, 15, 46–7
Corroboration, 159, 175–6

Deductive Systems, Myth of, 1, 5,
    8–10

Deductivism, critique of, 1, 2, 4,
    15, 28–9
Definition, ostensive, 17, 178,
    181–2
    taxonomic, 139–40
Disjunctive taxa, 199, 205, 216,
    217–18
    disjunctive taxa of brain-states,
    205, 218
    disjunctive taxa of mind-states,
    205, 219–21
Dualism, 204, 217

Endurance, 64, 235, 248, 249–50
Entrenchment of predicates, 93,
    112–13
Epoch Indifference, Principle of,
    235, 245–6
Equivalence Condition, and hypo-
    thetical mechanism, 120–1
Essences, 154–5
    nominal essence, 11, 131, 179, 197,
    198, 199, 203, 295, 300, 302
    real essence, 10, 11, 131, 179,
    197, 198, 199, 203, 295
Essential natures, 207, 283
Events v. durations, 235, 246–8
    Myth of, 1, 5–7, 21, 25
Evidence, 178, 191
    and fact, 178, 191
Existence,
    demonstrative criteria of, 63,
    74–5
    existence of events, 87
    existence of fictions, 72, 73
    existence and intensions, 86–8
    existence of properties, 88–9
    indirect criterion of, 72–3, 77
    non-ostensive criterion of, 63,
    70–2
    ostensive criterion of, 63, 68–70
    priority of existence, 77–8
    proof of existence, 63, 89–90

Existence – *contd.*
  recognitive criterion of, 63, 67,
    75–6
Existential hypotheses,
  falsification of, 64, 78–81
  generation of, 33, 35–6, 47–9,
    55, 64, 81–6
  subject matter of, 63, 65–6, 76
  verification of, 63, 66–76
Explanation,
  deductivist principle of, 15–21
  explanation and models, 35
  explanation and prediction, 2,
    18–21
  macroexplanation, 257, 261–2
  microexplanation, 257, 261
  of change, 7
  of difference, 11
  regress of, 257, 260–6, 310
  scientific explanation, 124–5,
    234, 261

Falsifiability, 2, 24–5, 134–5
Forces, infinite, 292
  as stand-ins for fields, 304–8
  as stand-ins for mechanisms, 292

Generality, dimensions of, 130, 131,
    135, 146–7
Generalizations,
  effects of counter instances on
    non-taxonomic, 130–1, 142–5
  effects of counter instances of
    taxonomic, 130–1, 139–41, 146
  effects of favourable instances on
    non-taxonomic, 145–6
  non-taxonomic generalizations,
    130, 136, 138, 141–2
  taxonomic generalizations, 130,
    136, 137–8, 146
Goodman's Paradox, 2, 25–6, 274

Hempel's Paradox, 94, 119–22, 274

Independence of predicates, 274
Individuation *v.* identification, 203,
    208–9

Induction, Problem of, 2, 6, 24, 29,
    134, 235, 248–9
Intensional relations, 141
Instance confirmation, Problem of,
    29, 94
Isotropy, Principle of, 235, 244–5,
    250

Language,
  and paradigms, 178, 180
  and rules, 178, 180
Law of Continuity, 259–60, 287–9
  and Quantum Mechanics, 293
Laws of Nature, 92, 99–104, 125
  acceptance conditions of, 92, 97–8
  form of, 23–25
  theory of, 92–126
Liabilities, 258, 272
Location Indifference (Cusa's
    Principle), 234, 236–7, 243

Manichean forces, 235, 249
Mathematics, as the ideal of
    knowledge, 9–10
Meaning, 178, 180
Mechanisms, 26–7
  generative, 92, 93, 107–8, 109–11,
    113, 121
  hypothetical, 34–5, 55–6, 120–1
  plausability of, 98–9
Metric of confirmation (non-taxo-
    nomic generalizations), 150–1
Mind-states,
  'free' mind-states, 204, 217
  identification of, 204, 212
  novel mind-states, 219–21
  unconscious mind-states, 204,
    221–2
Models, 2, 12, 14
  causal relation to subject, 52–3,
    54, 59, 125–6
  common notion of, 37–8
  geometrizations as, 50–3
  homoeomorphs, 39, 40–3
  iconic models, 33
  logical icons, 50
  metaphors and models, 47

Models – *contd.*
metaphysical problems of, 34
metriomorphs, 43
micro- and megamorphs, 40–1
mind-state, brain-state relation
    models, 205, 244–6
model relation to subject, 53,
    54–5, 58–9, 125–6
models of models, 60
paramorphs, 39, 43–6
    relations to source, 45
    relations to subject, 44–5
problem of single source
    identification, 45–6
protomorphs, 50
protomorphs to paramorphs, 52
real models, *v.* imagined models,
    42, 46–7
sentential models, 33, 37
source of model, *v.* subject of
    model, 38
taxonomy of models, 33–46, 50–
    52
teleiomorphs, 41–2
    abstractions, 41–2
    idealizations, 41–2, 145
ultimates, models for, 257,
    264–6, 283, 285
    conditions on models for the
        ultimate
    (i) spatial, 295, 297
    (ii) temporal, 295, 297
    (iii) causal, 295, 297–8
    critique of perceived thing as
        model for the ultimate, 259–
        260, 285–7, 295, 300–1, 303
    powers as model for the ulti-
        mate, 295–312, 303–4, 308–
        310, 311–12, 314
Monism, 210

Natural Kinds, 131, 154–5, 198–
    200, 208, 283
Necessity, natural, 93, 113–16
as 'apodeictic attitude', 203, 207,
    216, 229
potential, 137–8, 156

Necessity, natural – *contd.*
principles, 203, 207
Problem of, 29
Nicod's Criterion, 2, 24, 25–6, 119,
    121
Non-taxonomic inference, theory
    of, 131, 147–50
Numerical and Qualitative Identity
    of things, 295, 298–300, 301
Numerical Indifference, Principle
    of, 236, 252–4
causes of changes in number,
    253–4
no cause of numbers of things, 236
numbers as initial conditions, 236

Observing, as reading, 179, 192–3,
    194–5
as perceiving, 194–6
Operationism, 178, 182–4
Order, exploratory *v.* deductive, 10
logical *v.* natural, 8

Parity, 235–6, 250–2
'sense' of things and processes,
    235, 250–1
traditional conservation *v.* non-
    conservation, 236, 251–2
Parts and wholes, 257, 260, 264
Philosophy, linguistic, 4, 7–8
nature of, 3–4
Physiology and psychology, their
    relation, 205, 226–7
Powers, 2, 122–4, 257, 266, 269–84
analysis of power attributions, 178,
    179, 185–6, 189, 258, 270–1,
    284, 310–11
ascription of powers, 178, 258,
    274–7, 311
certeris paribus clauses and powers,
    123–4
dispositions *v.* powers, 258,
    269–70, 272–3
enabling conditions, 258, 271
explanation and powers, 123
identity of powers, different
    circumstances, 258, 280–2

Powers – *contd.*
  identity of powers, general, 258, 312
  identity of powers, in time, 279–80
  metaphysics of realism and powers, 259, 282–3
  natural necessity, law and powers, 258
  natures and powers, 122, 258, 277
  poles *v.* powers, 312–13
  stimulus conditions, 258, 271
  variable powers, 259, 280
  ultimate powers, 296, 313
Primary and Secondary qualities, 302–3, 308
  primary qualities as powers, 257, 264–6, 296, 304–7
Principles, 203–29
  causal principles, 203, 205, 206
  indifference principles, 234–54
  taxonomic principles, 203, 205, 206
Privileged access *v.* privileged authority, 205, 222–4
Probability, 157, 159–68
  conditions of assignment, 157, 161
  degree of confirmation and probability, 157, 158–9, 164–6
    conjunction condition, 158, 170–1
    disjunction condition, 158, 171
    equivalence condition, 158, 167–70
    evidence condition, 158, 169
    negation condition, 158, 171–2
  degrees of knowledge and probability, 157, 161–4
  existential hypotheses, probability of, 174–5
  extremal values, 158, 174
  frequency theory of probability, 157, 162–3
  grounds for probability statements, 157, 159
  meaning of probability statements, 157, 159–60

Probability – *contd.*
  numerical probability as probability of hypotheses, 167–8
  'objective' and 'subjective', 157, 164
  range theory of probability, 157, 163–4, 236
Pseudo-causal laws, 208, 227–9
Psycho-physiology,
  its possibility, 204, 213–24
  its principles, 204, 213, 214–16

Qualitative and numerical identity of things, 295, 298–300, 301

Recategorization, 206, 231–3
  and existence, 206, 232–3

Scientific description, as power attribution, 178, 179, 185–6, 189
Scientific knowledge, 125, 313
Sensationalism, 178, 182
Space, 234–5
  acausality of space, 235, 243–4
  concepts of space, 234, 237, 242–3
  continuity of space-time, 260, 289–92
  metrics of space, 234–5, 240
  minimal dynamic space, 234, 239–40
  minimal pre-dynamic space, 183, 234, 238–9
  points of space *v.* points in space, 295, 302
  uniqueness of space, 235, 240–2
Statement-picture complex, 12, 13, 34, 54, 56, 101, 119, 314
Structures, 1, 10–11, 314
  imagining structures, 12–13
Subjunctive Conditionals, 29, 94, 116–19
  general, 96, 116–18
  particular, 96, 118–19
  truth grounds of, 117
Symbol conventions, arbitrary, *v.* representational, 13–14, 33, 37–38

Taxa, *v.* classes, 197–8

Taxonomy, 179, 196–200
  of physiology and anatomy, 203, 209–10, 211
  of thoughts etc., 203, 210, 211–212

Theoretical statements, general, 92, 95, 96–9, 125
  acceptance conditions of, 92, 97–8

Theoretical terms, 2
  introduction of, 21–3
  meaning of, 49–50, 179

Theory, analysis of, 57–60
  confirmation and falsification of, 172–5
  structure of, 2, 3, 14–15, 42, 54, 56, 59–69, 172

Time, concepts of, 183–4, 235, 246–8

Time, concepts of – *contd.*
  continuity of, 260, 289–92

Transformation *v.* alteration, 295, 302

'true', common uses, 178, 189–91

Truth, 178, 188–96
  conditions for stating truly, 178, 189
  (i) location, 178, 189
  (ii) individuation, 178, 189
  evaluative analysis, 178
  traditional theories, 179, 193

Ultimate entities, nature of, 307–309
  reference to, 295, 301–2

Vehicles for thought, 12–14, 15
  Myth of, 1, 5, 7–8

An original approach to the
philosophy of science, *The
Principles of Scientific Thinking*
raises questions about the nature
of scientific thought which refute
the existing arguments in the field.
   Unlike most philosophers of
science, Rom Harré examines the
nature of scientific thought and
activity in a wide variety of
scientific disciplines, drawing as
readily on biology as on the
physical sciences. He also differs
from current schools of the
philosophy of science in a much
more central way in that he seeks
the principles of science in science
itself and not in the extra-scientific
domains of logic, language,
grammar, and syntax. In taking
this approach, he attacks as
"myths" much of the currently
popular philosophic principles and
method—most notably, the "myth
of deductivism." He offers counter
principles and opts in favor of
"realism" which resolves the
problem by preserving the
intuitions and developing a richer
system of rational principles than is
found in deductive logic. Harré
goes on to examine key concepts
of scientific thinking such as
"models, hypotheses, laws,
observation and truth, probability,
principles" in light of his schema or
perspective based on *intuitions.*
   Harré's clear, convincingly
argued case promises to provoke a
great deal of rethinking about
widely accepted notions regarding
the nature of scientific thought.